Smart Nanomaterials Technology

Series Editors

Azamal Husen, Wolaita Sodo University, Wolaita, Ethiopia

Mohammad Jawaid, Laboratory of Biocomposite Technology, Universiti Putra Malaysia, INTROP, Serdang, Selangor, Malaysia

Nanotechnology is a rapidly growing scientific field and has attracted a great interest over the last few years because of its abundant applications in different fields like biology, physics and chemistry. This science deals with the production of minute particles called nanomaterials having dimensions between 1 and 100 nm which may serve as building blocks for various physical and biological systems. On the other hand, there is the class of smart materials where the material that can stimuli by external factors and results a new kind of functional properties. The combination of these two classes forms a new class of smart nanomaterials, which produces unique functional material properties and a great opportunity to larger span of application. Smart nanomaterials have been employed by researchers to use it effectively in agricultural production, soil improvement, disease management, energy and environment, medical science, pharmaceuticals, engineering, food, animal husbandry and forestry sectors.

This book series in Smart Nanomaterials Technology aims to comprehensively cover topics in the fabrication, synthesis and application of these materials for applications in the following fields:

- Energy Systems - Renewable energy, energy storage (supercapacitors and electrochemical cells), hydrogen storage, photocatalytic water splitting for hydrogen production
- Biomedical - controlled release of drugs, treatment of various diseases, biosensors,
- Agricultural - agricultural production, soil improvement, disease management, animal feed, egg, milk and meat production/processing,
- Forestry - wood preservation, protection, disease management
- Environment – wastewater treatment, separation of hazardous contaminants from wastewater, indoor air filters

Rakesh Kumar Bachheti · Archana Bachheti ·
Azamal Husen
Editors

Nanomaterials for Environmental and Agricultural Sectors

Springer

Editors
Rakesh Kumar Bachheti
Department of Industrial Chemistry
Addis Ababa Science and Technology University
Addis Ababa, Ethiopia

Archana Bachheti
Department of Environment Science
Graphic Era University
Dehradun, Uttarakhand, India

Azamal Husen
Wolaita Sodo University
Wolaita, Ethiopia

Smart Nanomaterials Technology
ISBN 978-981-99-2873-6 ISBN 978-981-99-2874-3 (eBook)
https://doi.org/10.1007/978-981-99-2874-3

This Springer imprint is published by the registered company Springer Nature Singapore Pte Ltd.
The registered company address is: 152 Beach Road, #21-01/04 Gateway East, Singapore 189721, Singapore

Preface

Nanotechnology is one encouraging area of multidisciplinary research. Due to its extensive application in virtually every aspect of science and technology, it opens up new doors of potential. An emerging use of nanotechnology is reducing the environmental pollution. Many kinds of hazardous organic and inorganic pollutants are entering adjacent water directly or indirectly due to industrialization, endangering aquatic, and human life. So, it is necessary to innovate, improve, and upgrade existing technology to protect the environment. Environmental clean-up could benefit significantly from nanotechnology. Eliminating hazardous bacteria, heavy metals, and organic and inorganic pollutants offers enormous potential for wastewater treatment. In addition, the world's agricultural production must expand faster to meet the rising population's demand for food. In this regard, nanotechnology can significantly raise agricultural productivity and safeguard it from infections and pests.

This book consists of 13 chapters. Chapter "Role of Silver Nanoparticles on Wastewater Treatment, Environmental Implications, and Challenges" focuses on the role of silver nanoparticles in wastewater treatment, environmental implications, and challenges, while Chapter "Role of Nanoparticles in Air Quality Monitoring with Respect to Toxicity, Virus Detection and Gas Sensing" discusses the importance of nanoparticles in air quality monitoring concerning toxicity, virus detection, and gas sensing. Soil pollution has become one major problem of the world that needs remediation. Chapter "Past, Present and Possible Future Application of Nanoparticle in Contaminated Soil Remediation" covers all potential uses of NPs in rehabilitating polluted soils. Chapter "Development Strategies and Prospects of Carbon Nanotube as Heavy Metal Adsorbent" focuses on the role of carbon nanotubes as heavy metal adsorbents. Chapter "Recent Development and Importance of Nanoparticles in Disinfection and Pathogen Control" deals with recent development and the importance of nanoparticles in disinfection and pathogen control. Environmental monitoring is vital for safe human health. Chapter "Recent Advances in Nanoparticles for Environmental Monitoring and Sensing: An Overview" deals with advances in nanoparticles for environmental monitoring and sensing. Chapter "Current Trends and Future Applications of Silica Nanomaterials in Adsorption and Catalysis" discusses the current and future applications of silica nanomaterials in adsorption and catalysis. Achieving food security worldwide is a significant challenge because nearly 700 million people face hunger, and more than 2 billion are affected by mineral and vitamin deficiencies. All over the world, there is consistent and regular use of pesticides, resulting in their accumulation in food, water, and soil. The effect of nanofertilizers on plant physiology, metabolism, and associated safety issues is discussed in Chapter "Effect of Nanofertilizers on Plant Physiology, Metabolism and Associated Safety Issues". Chapter "Benefits, Future Prospective, and Problem Associated with the Use of Nanopesticides" overviews the practical benefits and pitfalls of applying nanopesticides in plants and animal sciences. Chapter "Current Applications and Future Perspectives of Nanotechnology for the Preservation and Enhancement of Grain and Seed Traits" provides an updated description of the use of nanotechnology to improve seed and grain traits. Plant growth-promoting rhizobacteria (PGPR) are free-living microbes essential in sustaining soil fertility. Chapter "Interaction Between Metal Oxide Nanoparticles and PGPR on Plant Growth and Development" focuses on the interaction between metal oxide nanoparticles and PGPR on plant growth and development. Chapter "Recent Application and Future Prospects of Nanoparticles-Based Colorimetric Sensors for Residual Pesticides Detection" focuses on nanoparticle-based colorimetric sensors for residual pesticide detection. Chapter "Applications, Opportunities and Challenges of Nanotechnology in the Food Industry" discusses the applications, opportunities, and challenges of nanotechnology in the food industries.

Overall, the recent understanding and advancements of nanomaterials and their significant application in environmental and agricultural sectors have been discussed. This book covers all potential uses of nanomaterial and nano-based hybrids in the rehabilitation of polluted water, soils, and other environment-related issues. This book also explains the nanosensors as an alternative for diagnostic applications, since they are capable of detecting viruses and bacteria. In addition to discussing the advantages and drawbacks of using nanotechnology in ecological farming, the book in hand also explains how this technology can be used in the agricultural sectors for managing soil fertility, crop production, and crop enhancement. The book in hand will be very useful to graduate students, researchers, and scientists in the fields of nanoscience, environmental science, and agricultural science.

Addis Ababa, Ethiopia — Rakesh Kumar Bachheti
Dehradun, India — Archana Bachheti
Wolaita, Ethiopia — Azamal Husen

Contents

About the Editors

Rakesh Kumar Bachheti graduated from the Hemwati Nandan Bahuguna University, Garhwal, India, in 1996. He completed his M.Sc. in Organic Chemistry from Hemwati Nandan Bahuguna University, Garhwal, India, in 1998. He had undergone a one-year Post-Graduate Diploma in Pulp and Paper Technology from Forest Research Institute, Dehradun, in 2001. He obtained his Ph.D. in Organic Chemistry from Kumaun University, Nainital, India, in 2007. He is presently working as an Associate Professor of Organic Chemistry in the Department of Industrial Chemistry at the Addis Ababa Science and Technology University (AASTU) of Ethiopia, where he teaches Ph.D., graduate, and undergraduate students. Before joining AASTU, Rakesh was working as Dean Project (Assistant) at Graphic Era University (A grade university by NACC) in Dehradun, India. Rakesh also presented papers at various international (Malaysia, Thailand, and India) and national conferences. He was also a member of various important committees, such as the Internal Quality Assurance Cell (IQAC), Anti-ragging committee.

His major research interests include natural products for Industrial application, biofuel and bioenergy, green synthesis of nanoparticles, their application and Pulp and paper technology. He retains a fundamental love for natural products, which permeates all of his research. He has also successfully advised 30 M.Sc. and 3 Ph.D. students to completion, and countless undergraduates have researched in his laboratory. Dr. Bachheti is actively involved in curriculum development for B.Sc./M.Sc./Ph.D. programs. Dr. Bachheti is an editor of 4 books and 70 publications, dealing with various aspects of natural product chemistry and nanotechnology and has twenty book chapters published by Springer, Elsevier, and Nova Publisher. Presently, he is supervising 5 Ph.D. students and 2 Master's students and works on two research projects funded by AASTU.

Archana Bachheti did B.Sc. in 1997 and M.Sc. in 1999 from HNB Garhwal University. She received her Ph.D. from Forest Research Institute, Dehradun, India, in 2006. She has carried out research projects and consultancy work in the areas of ecorestoration/ development of wasteland, physico-chemical properties of *Jatropha curcas* seed oil and their relation with altitudinal variation, and has been a consultant Ecologist to a project funded by a government agency. Dr. Joshi is currently Professor at Graphic Era University, Dehradun, India. She has also served in many capacities in academia within India and provided expertise internationally for more than 15 years. She taught Ecology and Environment, Environmental Science, Freshwater Ecology, Disaster management, and Bryophytes and Pteridophytes.

Her major research interests encompass the broad, interdisciplinary field of plant ecology, focusing on ecorestoration, green chemistry, especially the synthesis of nanomaterial, and medicinal properties of plants. The breadth of her research spans from degraded land ecological amelioration and physical and chemical properties of plant oil to plant-based nanomaterial. She guided one Ph.D. student, supervised three scholars, and guided graduate and undergraduate students with their research projects. While the fascination with forest biodiversity captured her interest, it has been her love for the exploration of values of biodiversity and social upliftment through it that has maintained that passion. Dr. Joshi is the editor of 4 books and has published over 55 research articles in international and national journals and sixteen book chapters. She organized several National seminars/conferences at Graphic Era University, India.

Azamal Husen served as Professor & Head, Department of Biology, University of Gondar, Ethiopia and is a Foreign Delegate at Wolaita Sodo University, Wolaita, Ethiopia. Earlier, he was a Visiting Faculty of the Forest Research Institute, and the Doon College of Agriculture and Forest at Dehra Dun, India. His research and teaching experience of 20 years involves studies of biogenic nanomaterial fabrication and application, plant responses to environmental stresses and nanomaterials at the physiological, biochemical and molecular levels, herbal medicine, and clonal propagation for improvement of tree species. He has conducted several research projects sponsored by various funding agencies, including the World Bank (FREEP), the National Agricultural Technology Project (NATP), the Indian Council of Agriculture Research (ICAR), the Indian Council of Forest Research Education (ICFRE); and the Japan Bank for International Cooperation (JBIC). He received four fellowships from India and a recognition award from the University of Gondar, Ethiopia, for excellent teaching, research, and community service. Husen has been on the Editorial board and the panel of reviewers of several reputed journals published by Elsevier, Frontiers Media, Taylor & Francis, Springer Nature, RSC, Oxford University Press, Sciendo, The Royal Society, CSIRO, PLOS, MDPI, John Wiley & Sons and UPM Journals. He is on the advisory board of Cambridge Scholars Publishing, UK. He is a Fellow of the Plantae group of the American Society of Plant Biologists, and a Member of the International Society of Root Research, Asian Council of Science Editors, and INPST. To his credit are over 200 publications; and he is Editor-in-Chief of the *American Journal of Plant Physiology*.

He is also working as Series Editor of *Exploring Medicinal Plants*, published by Taylor & Francis Group, USA; *Plant Biology, Sustainability, and Climate Change*, published by Elsevier, USA; and *Smart Nanomaterials Technology*, published by Springer Nature Singapore Pte Ltd. Singapore.

Role of Silver Nanoparticles on Wastewater Treatment, Environmental Implications, and Challenges

Aashna Sinha, Chetan Shrivastava, Kundan Kumar Chaubey, Shivani Tyagi, Manish Kushwah, Pranchal Rajput, Atreyi Pramanik, Sujata Hariharan, Shiv Dayal Pandey, Gaurav Pant, Deepak Kumar Verma, Maya Datt Joshi, Deen Dayal, and Anis Kumar Pal

Abstract The majority of nanoparticles (NPs) have played an important role in wastewater treatment, particularly silver nanoparticles (AgNPs) and their hybrids. AgNP hybrids with substances including cellulose, activated carbons, chitosan, alginate, graphene oxides, titanium dioxides, silicon dioxides, etc. have been widely employed for water purification. Because of their enhanced adsorption capacities and superior antibacterial qualities, AgNPs are useful options for removing contaminants from wastewater bodies. Further, the widespread use of NPs in commercial and industrial products unavoidably leads to a rise in their release into the environment, endangering both ecosystems and human health. Heavy metals, toxins, dyes, petroleum products, detergents, hospital wastes, etc. are the main components that contaminate the wastewater released by industries and households. The human immunodeficiency virus, the hepatitis B virus, the herpes simplex virus, the human parainfluenza virus, the ruminants virus, and the plant pathogenic bean yellow mosaic virus have all been successfully inhibited by AgNPs. Intriguingly, AgNPs have just recently been recognised as having significant roles to play as anti-inflammatory drugs. Inflammation is an early immune reaction by tissue against foreign particles.

A. Sinha · C. Shrivastava · K. K. Chaubey (✉) · P. Rajput · A. Pramanik · S. Hariharan · S. D. Pandey · A. K. Pal
Division of Research and Innovation, School of Applied and Life Sciences, Uttaranchal University, Arcadia Grant, P.O. Chandanwari, Prem Nagar, Dehradun, Uttarakhand 248007, India
e-mail: kundan.chaubey@uttaranchaluniversity.ac.in

S. Tyagi · M. Kushwah · M. D. Joshi · D. Dayal
Department of Biotechnology, Institute of Applied Sciences and Humanities, GLA University, Mathura, UP 281406, India

G. Pant
Department of Life Sciences, Graphic Era (Deemed to Be University), Dehradun, Uttarakhand 248002, India

D. K. Verma
Department of Biotechnology, Sanskriti University, Chhata, Distt., Mathura, UP, India

R. K. Bachheti et al. (eds.), *Nanomaterials for Environmental and Agricultural Sectors*, Smart Nanomaterials Technology,
https://doi.org/10.1007/978-981-99-2874-3_1

The main ways that NPs enter the body are through eating, inhalation, skin contact, and intraperitoneal or intravenous injections directly into the systemic circulation. The role of AgNPs in wastewater treatment, challenges, and their impact on the environment and lifestyle are covered in this chapter.

Keywords Silver nanoparticles · Wastewater Treatment · *Escherichia coli* · Paramylon · Antibacterial agent · Catalytic performance

1 Introduction

The environmental problem of water contamination has recently attracted attention on a worldwide scale. According to a recent UN assessment, the availability of clean, fresh water is a problem that will affect the entire planet in the twenty-first century since the existence of living things cannot be guaranteed in the presence of contaminated water [40, 46]. The main cause of water contamination comes from industrial waste, insufficient sewage management, sea discard problems, irradiated waste, dyeing and finishing of textiles. The textile industry finds it difficult to utilise due to the drastic changes in customer expectations, the contemporary textile processes demand significant volumes of dyes and auxiliary materials. The used water contains a variety of components, including chemicals and dyes, which are discharged directly into water sources, causing contamination and water pollution [122]. A detrimental outcome of water pollution in the environment has very threatening effects on human fitness. Researchers are continuously investigating several novel technologies to enhance low-cost water purifying processes. A promising solution to filter water with cheap cost, great operating effectiveness in eliminating contaminants, and reusable capability is now being offered by the newly developing area of nanotechnology.

As we know, nanomaterials were effectively used in a variety of fields, including catalysis, medicine, and other sciences [8, 9, 49–54, 58] (Hamid et al. 2022). Specialists discovered recently that nanomaterials are a greater option for treating wastewater due to their distinctive properties, such as their nanosize, enormous surface area, chemical sensitivity, powerful solution movement, strong mechanical property, porosity character types, hydrophilicity, dispersibility, and hydrophobicity. Due to their adaptability, ease of synthesis, low cost, and unique chemical–physical characteristics, AgNPs are frequently employed as ENMs in various sophisticated nanotechnologies. Concerns regarding their effects on people and the environment are raised by their expanding manufacturing and incorporation in environmental applications, such as water treatment. In reality, hybrids of AgNPs and cellulose have the dual benefits of being eco-safe if designed properly and being easily made utilising recycled materials, with cheap costs and potential for reuse [81]. One of the most often used artificial nanomaterials, silver nanoparticles (AgNPs), which enter wastewater during production, usage, and disposal, represent a threat to human and ecological health [128]. In wastewater treatment, AgNP hybrids with substances including cellulose, activated carbons, chitosan, alginate, graphene oxides, titanium dioxides, silicon dioxide, etc. have been widely employed [37]. The use of AgNPs is often linked to their antibacterial properties. Nanotechnology has the potential to purify water at a low cost with excellent pollution removal efficiency and reusability. Silver (Ag) is a researched metal known for its utility in industrial, household, and medical applications such as ornaments, electrical conductors, biocidal, and coins. Silver nanoparticles (AgNPs) are manufactured at the nanoscale (1–100 nm) from a known metal, silver. AgNPs are commonly employed because they are more efficient than typical silvery elements [42, 44, 103, 107]. Ag has been used to purify water for a very long time. The AgNPs-incorporated membrane has outstanding extraction efficiency and functional properties, including high antibacterial activity, photocatalytic degradation, thermal responsiveness, and electro-conductivity [141].

Two examples of how microbial communities may adapt to high levels of Ag include activating Ag^+ efflux from the cell and converting Ag^+ to a less damaging form [113]. AgNPs' resistance mechanism may be the same as that of Ag, or it may be different due to AgNPs' unique characteristics. Some of the hypothesised resistance mechanisms include the manipulation of efflux pumps, point mutations, adaptive mutations, downregulation of porins, the presence of resistance genes, pigment creation, and morphological alterations [113]. Because of their flexible, simple, and affordable arrangements, as well as their surprising chemical and physical characteristics, silver nanoparticles (AgNPs) are frequently exploited as engineered nanomaterials (ENMs) in various sophisticated nanotechnologies [123]. AgNPs have the dual advantages of being easily manufactured from reprocessed materials, with possible reuse, low costs, and being environmentally friendly [123]. Through leaching from AgNP-coated products (e.g. textiles), AgNPs enter wastewater [38]. The release of AgNPs into sewers is driven by the industrial processing of NPs [23]. During wastewater treatment, a huge percentage of AgNPs are retained in sewage sludge [59]. There are limitations to the use of AgNPs in the production and alteration of membranes. First, the effective antifouling duration of AgNPs-incorporated membranes is shortened by the excess quantity and deactivation of AgNPs on the membrane [17]. Second, the inclusion of AgNPs will reduce membrane flow by clogging the membrane pores [34]. Lastly, the production and recovery of AgNPs are no longer viable due to the excessive use of harmful and hazardous compounds. Fourth, more investigation is required into the biological toxicity and environmental impacts of AgNPs released from the composite membrane [116]. When applied in engineering systems, AgNP-containing membranes provide a number of difficulties. Although there are benefits to having AgNPs, AgNPs and Ag^+ ions impair aquatic life in freshwater. The results of in vitro studies show that AgNPs are toxic to mammalian cells taken from the skin, liver, lung, brain, vascular system, and reproductive organs. Size, shape, surface charge, coating or capping agent, dosage, oxidation state, aggregation, and the pathogens employed to assess AgNPs' toxicity are just a few of the variables that affect how dangerous they are [69]. AgNPs' effects on the environment, wastewater management, and issues are covered in this chapter.

2 Major Pollutants in Wastewater

A sizable portion of water contaminants are organic pollutants. They damage aquatic life and terrestrial life when present in water. Pesticides, organic dyes, pharmaceuticals, nitroaromatics, and mycotoxins are just a few of the several forms of organic pollutants that may be found in wastewater. In agricultural production, pesticides are used to lessen crop damage from weeds and pests [143]. Organic dyes, which are utilised in textiles, leather, paints, papers, and plastics (Table1) are made up of a sizable number of intricate aromatic compounds [142]. Due to their severe toxicity, non-biodegradability, and potential to change into agents that are carcinogenic, teratogenic, and even mutagenic, pesticides and organic dyes have garnered a lot of attention as environmental pollutants from a worldwide perspective [57]. Antibiotics are common medications used to treat bacterial infections in living things. Nonetheless, they are regarded as significant organic pollutants in water that result in a range of illnesses, including cancer, allergies, inherited genetic disorders, immune system deterioration, and allergic responses [36, 95].

Table 1 Water pollution factor and its negative effects

S. no	Factor associated with water contaminations	Harmful Impacts	References
1	Seawage contaminated water (domestic waterwaste)	Water borne disease	[104]
2	Pathogens (viruses and bacteria)	Water borne disease	[130]
3	Microscopic pollutants (agriculture chemicals)	Plastic pollution	[48]
4	Agriculture pollutants (agriculture chemicals)	Directly affect the fresh water resources	[124]
5	Nutrient pollutants (plant daberis and fertilisers)	Effect on eutrophication process	[79]
6	Inorganic pollutants (metal compound, trance elements, inorganic salts, heavy metals, minerals acids)	Aquatic flora and fauna public health problems)	[119]
7	Radioactive isotops	Bone, teeth and skin can cause	[14]
8	Organic pollutants (detergents insecticides, herbicides)	Aquatic life problems, cacogenic	[131]
9	Industrial pollutants (municiple pollutant water)	Caused water and air pollution	[75]

3 Role of Different Nanoparticles in Wastewater Treatment

The resource that is most sought-after for human survival is high-quality water. Groundwater pollution and drinking water quality have both been impacted in recent decades by a variety of anthropogenic and natural activities. One of the main causes of groundwater pollution is the discharge of hazardous effluents from the industrial sector. Adsorbents, reverse osmosis, ion exchange, and electrostatic precipitation are a few of the traditional technologies used to purify water. These techniques have a high cost, limited recyclability, and low efficiency. The application of sustainable technologies has been constrained despite advancements in their development, mostly due to the restrictions of the material's characteristics, especially their pricing. Using nanoparticles would assist in resolving this issue and alleviate the effects of pesticides and heavy metals in water. The direct bandgap, high optical absorption coefficient, layered structure, programmable band edges for improved catalysis, low cost, and low toxicity of nanoparticles are all beneficial properties [117]. The majority of applications for nanoparticles, such as the treatment of environmental pollutants such as azo dyes, chlorpyrifos, organochlorine insecticides, nitroaromatics, etc., include semiconductors, zero-valence-based metals, and some bimetallic types. Moreover, multiple studies have shown that TiO_2-based nanotubes can be utilised to remove contaminants from wastewater, including azo dyes, Congo red, phenol aromatic base pollutants, toluene, dichlorophenol trichlorobenzene, and chlorinated ethene. SiO_2, ZnO, TiO_2, Al_2O_3, and other common and significant metal oxide nano photocatalysts are available. The fact that titanium dioxide (TiO_2) is inexpensive, toxic-free, chemically stable, and readily available on Earth makes it one of the best photocatalysts among all other materials [139] (Table 2).

Table 2 A comprehensive list of some NPs used for wastewater treatment and their mode of action

Sl. no	Nanoparticles and their metal oxides	Mode of action in water treatment	References
1	Fe_3O_4, TiO_2, MnO_2, MgO, ZnO and CdO	Fe(3)O(4) nanoparticles with ascorbic acid coatings are highly selective in surface area and superparamagnetic. Fe(3)O(4) NPs with an ascorbic acid coating function as an absorbent to take out the heavy metal arsenic from wastewater	[35]
2	Chromium (Cr) (VI)	Biosorption method based on algal–bacterial aerobic granular sludge. The method is highly a pH-dependent process and the algal–bacterial aerobic sludge works great at a low pH of 2.0	[137]
3	Ferrous	Works by magnetic separation and dye removal technique	[89]
4	Silica NP	Removal of acid orange	[140]
5	Porous silica NP	Absorption and desorption of CO_2 in the bottom and fly ash stream	[112]

4 Detection of Organic Pollutants Using AgNPs

Potential nanoprobes for the detection of organic contaminants include AgNPs. Due to their large absorption band in the visible spectrum of light, which is very sensitive to the size of particles, shape, and relative permittivity of the immediate environment, they have recently gained significant attention as optical sensors for detecting various organic contaminants [39]. AgNP-based optical sensors for the detection of pharmaceuticals, nitroaromatics, mycotoxins, pesticides, and organic dyes have been described. As compared to optical sensors, electrochemical sensors, such as those based on AgNP, are thought to be more capable of detecting organic contaminants with appropriate sensitivity and selectivity [132]. They also take less time to set up and take less effort. Table 3 lists the chosen organic contaminants along with their categorization and chemical structures (Table 3).

5 Role of AgNPs in Wastewater Treatment

Impregnating other matter with silver nitrogen compounds is a real-world way to utilise the antibacterial qualities of Ag (nanotechnology). In wearing stockings, Ag nanoparticles may limit the progression of smell-producing bacteria [15]. On washing, these saturated stockings have been shown to deliver a high quantity of Ag nanoparticles. These unrestricted NPs find their way into wastewater collecting systems with ease. In bactericides, silver (Ag^{+}) is utilised in various products [87]. The highly deadly properties of ionic silvery are supposed to be due to its immersion into the bacterial cell wall which is negatively charged, cell enzymes are inactivated, and membrane permeability is disrupted, culminating in cell death and lysis [106].

5.1 Catalytic Performance of AgNPs for Monitoring Water Pollution

Because of their abundantly exposed low-coordination sites, large specific surface area, and inexpensive price, AgNPs play a part in the catalytic degradation of 4-nitrophenol [25]. However, with time, there is a propensity for decreasing catalytic activity because of the high energy of the surface. Also, it is exceedingly challenging to remove the particles from the water, which might result in secondary pollution. Immobilising AgNP in permeable oil absorbent ingredients has therefore been found to be a successful strategy for increasing AgNP's capacity to be recycled [43].

In another study, PVA and AgNP were only surface-immobilised to create multipurpose 3D filter cotton. The manufactured 3D filter cotton has a strong catalytic degradation performance that may be used to break down water-soluble

Table 3 The organic pollutants with their chemical structures

Organic Compounds	Compounds			
Dyes	Xanthene dyes		Azodyes	
	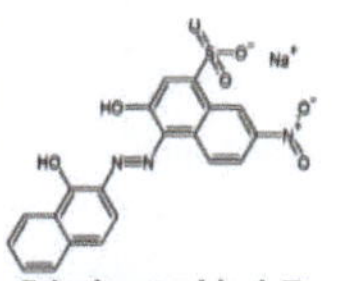 Eriochrome black T	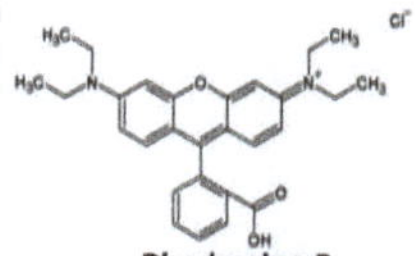 Rhodamine B	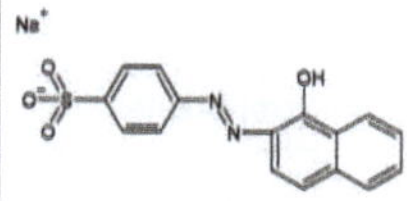 Orange II	
Drugs	Benzodiazepines	Antimetabolites	Anthracycline anticancer drug	
	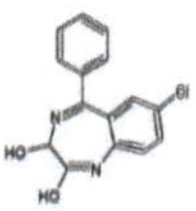 Oxazepam	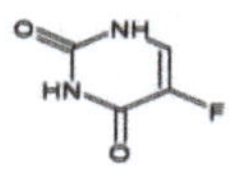 5-fluorouracil	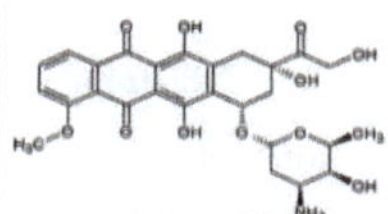 Doxorubicin	
	Antibiotics			
	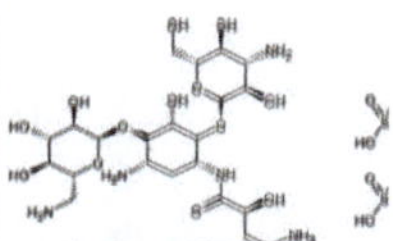 Amikacin sulphate	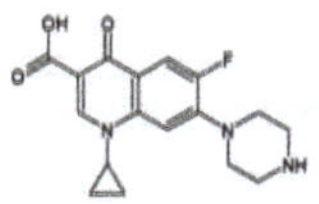 Ciprofloxacin	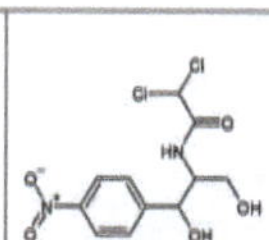 Chloramphenicol	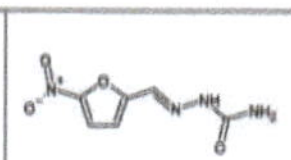 Nitrofurazone
Pesticides	Insecticides			
	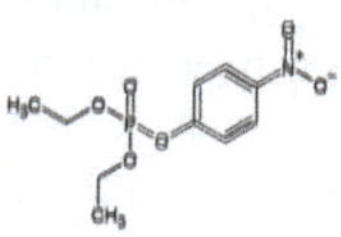 Paraoxen	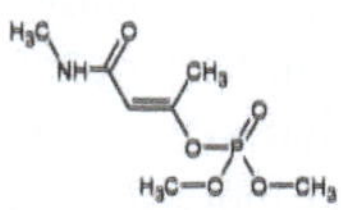 Monocrotophos	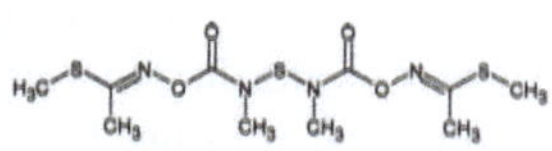 Thiodicarb	
	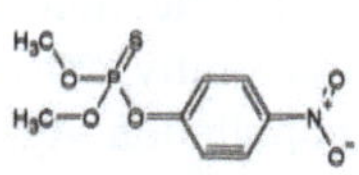 Methyl parathion	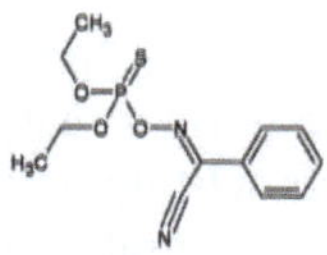 Phoxim	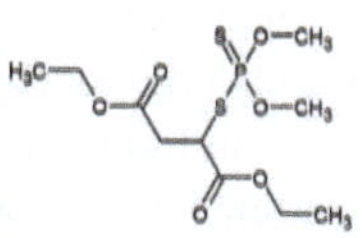 Malathion	
	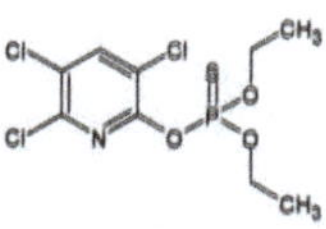 Chlorpyrifos	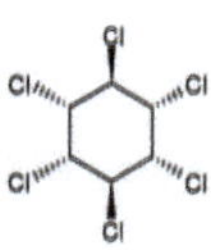 Lindane	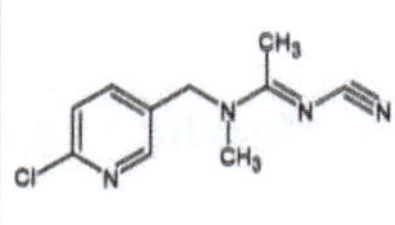 Acetamiprid	

(continued)

Table 3 (continued)

Organic Compounds	Compounds			
Nitro aromatics	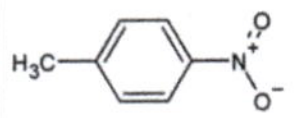 **4-Nitrotoluene**		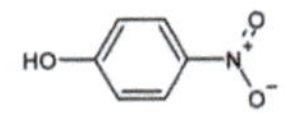 **4-Nitrophenol**	
Mycotoxins	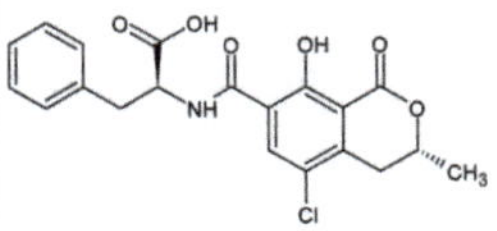 **Ochratoxin A**			
Other pollutants	Thioureas	Organohalides	Phenolics	
	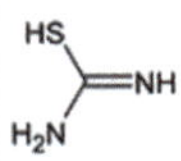 **Thiourea**	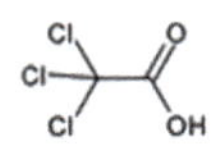 **Trichloroacetic acid**	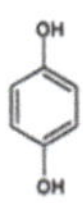 **Hydroquinone**	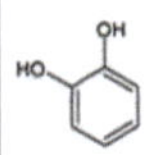 **Catechol**

4-nitrophenol. It also has outstanding oil/water separation efficiency and is recyclable [26]. The green synthesis of exceptionally stable AgNPs and their effective use as a catalyst, photocatalyst, and antibacterial agent has been described in a study to treat wastewater in a manner similar to this [105].

AgNPs are being used more and more in environmental applications, such as preventing and treating water contamination, which raises questions regarding their environmental impact. An approach to address ecological protection issues and the combination of nanoparticles is to graft AgNP onto carefully chosen polymers. When appropriately engineered to prevent and treat water pollution, AgNP-cellulosic hybrids offer the simultaneous benefits of being simply produced from recycled materials, being inexpensive and potentially reusable, and being ecologically beneficial [81].

5.2 *Photocatalytic Degradation by AgNPs*

Two components, namely photo and catalysis, can be used to link the photocatalytic mechanism. The first part deal with charge carrier generation, dynamics, and photon absorption. The second part focuses on surface radical production and surface interactions between organic molecules, O_2, and H_2O [118].

In general, Ag-based photocatalysts can be split into two groups: semiconductor photocatalysts based on Ag-containing compound bandgap excitation and plasmonic photocatalysts based on AgNPs' LSPR activity. AgNPs' plasmonic excitation in response to light irradiation generates hot electrons, whereas Ag-containing semiconductors' bandgap excitation generates electron–hole pairs. Next, before participating in the catalytic events, these electron–hole pairs undergo charge separation and transfer. Ag-based visible light-responsive photocatalysts have the potential to be extremely effective in selective organic transformation, bacterial eradication, water reduction, and oxidation, according to recent research. Free-standing metal NPs often have limited stability in heterogeneous catalysis as a result of aggregation. In order to get over this problem, plasmonic metal photocatalysts might be greatly disseminated and quickly recycled by being anchored to the support materials. Table 4 provides several instances of Ag ENMs utilised for treating dirty water, highlighting AgNP size and demonstrating that synergic effects can occur in specific circumstances (Fig. 1).

A wide range of innovative hybrid materials with synergistic or complementary behaviour were created using Ag-based photocatalysts. Bismuth vanadate ($BiVO_4$, BV) and Ag/AgO_2 hybrid nanomaterials had a significant increase in photoactivity compared to pristine BV, as determined by the photodegradation of crystal violet and Rhodamine B using commercially available low-cost blue LEDs or direct sunlight as photoexcitation sources [61]. This increase was about 28 times greater. The enhanced photoactivity of $BV/Ag/AgO_2$ photocatalysts is due to improved dye adsorption, greater visible light absorption, and enhanced charge transfer kinetics. Melinte et al. [84] created a variety of photocatalysts based on Ag, Au, or Au–Ag nanoparticles supported on photo crosslinked organic materials in order to photocatalyse the degradation of 4-nitroaniline. Researchers looked at the photocatalytic degradation of the methylene blue dye in the presence of biogenic AgNPs made from yeast (*Saccharomyces cerevisiae*) extract [109].

Table 4 Hybrid systems based on AgNP are utilised to address water contamination

AgNPs size (nm)	Treatment	Support	References
5–10	Methyl Orange	TiO_2	[65]
	Nitro derivate	Photo crosslinked matrix	[84]
10	Methylene blue	–	[109]
10–20	Crystal violet; Rhodamine B	$BiVO_4$	[61]
10–30	Rhodamine B; Methyl Orange; 4-nitrophenol	Sulfonated graphene/TiO_2	[3]
100	Rhodamine B	Cellulose nanofibrils	[28]

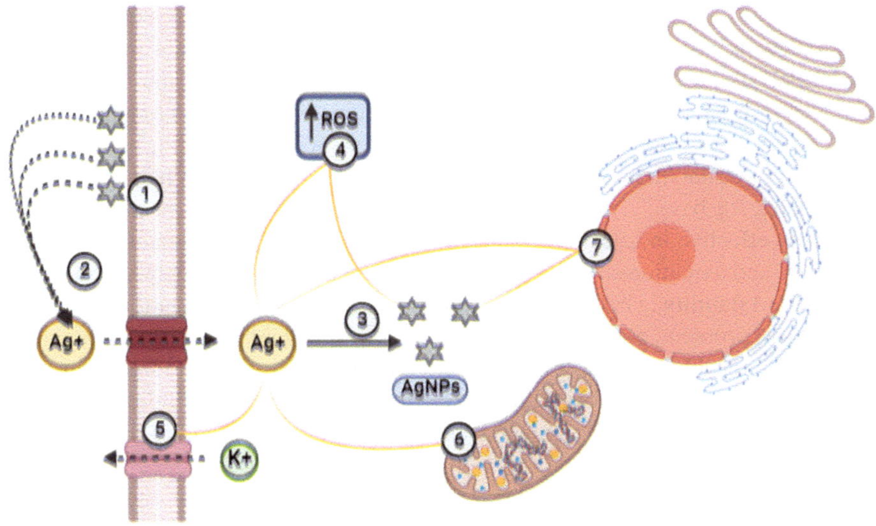

Fig. 1 (1) An electrostatic attraction exists between the fungus' surface and the AgNPs, (2) Dynamic Ag^+ release and import into the interior of the cell, (3) AgNP intracellular biosynthesis, (4) Intracellular ROS accumulation, (5) Modifications to efflux mechanisms and stimulation of K^+ release, (6) Reduced copper-dependent cytochrome c oxidase activity, (7) Modification of the metabolome, epigenome, and transcriptome

5.3 The AgNPs Antimicrobial Properties in the Treatment of Water

Due to their antibacterial activity, AgNPs are attractive candidates for use in water treatment and disinfection. The research suggests that one of these three mechanisms is what gives AgNPs their antibacterial properties: Proteins, DNA, or RNA are destroyed; membrane properties are altered; and Ag(I) is released into the cell cytoplasm. Due to AgNPs' antibacterial properties, a lot of AgNP-based products are used in water disinfection as water filters to stop hazardous bacteria and viruses [16].

In a separate study, the efficacy of equipment-covered filter modified with AgNPs to reduce fungal formation in aquacultures was examined [85]. AgNPs have been found to significantly help reduce fungal infections without having any harmful side effects when used in filters. According to [85], AgNPs' antibacterial effect is directly related to the release of Ag^+ ions into the aquatic environment.

Deshmukh, et al. developed AgNP-based biofilters for wastewater purification by eliminating ammonia and other contaminants, such as bacteria, in a manner similar to this [33]. In a separate work, carbon nanotubes were created and cyclodextrin nanotubes were created with AgNPs impregnated in them [33]. The water samples that included *E. coli* and *V. cholerae* were cleaned using these nanotubes. According to published research from the literature, AgNPs have the potential to be employed as water disinfectants for three primary reasons: (a) their extraordinarily high antibacterial activity, (b) their stability and delayed release, and (c) the feasibility of mass manufacturing using green synthesis.

5.3.1 Ag as an Efficient Antibacterial Agent

Historic understanding and usage of silver compounds and silver ions for their antibacterial properties in a range of applications [21]. Silver and its compounds have been proven to have potent bactericidal and inhibitory effects on a range of bacteria, fungi, and viruses [77, 78]. AgNPs may help in the battle against bacterial antibiotic resistance by acting as an alternative antibacterial agent to antibiotics. AgNPs have been demonstrated to have antibacterial effect against *Escherichia coli* (*E. coli*), which results in cell death as a result of AgNP building in the cell wall [66]. The size and structure of AgNPs have a substantial impact on how efficient they are against bacteria, claims the study [45, 114]. In a different study, it was shown that AgNPs synthesised from different saccharides were highly effective against both Gram-positive and Gram-negative bacteria. The significant discovery of the study was that the produced AgNPs were efficient against bacterial strains that were resistant to a variety of antibiotics, including *Staphylococcus aureus* [114] (Fig. 2).

Nevertheless, the exact method by which AgNPs kill bacteria is still not fully known. Several studies had looked at and speculated on a few potential causes for AgNPs' antibacterial function. *Staphylococcus aureus* and *E. coli* bacterial cells were used in [133] investigation of the antibacterial activity and technique of working of the silver ion. They found considerable changes in the bacterial cell membranes upon exposure to silver ions, which may be the cause or result of cell death. Mirzajani

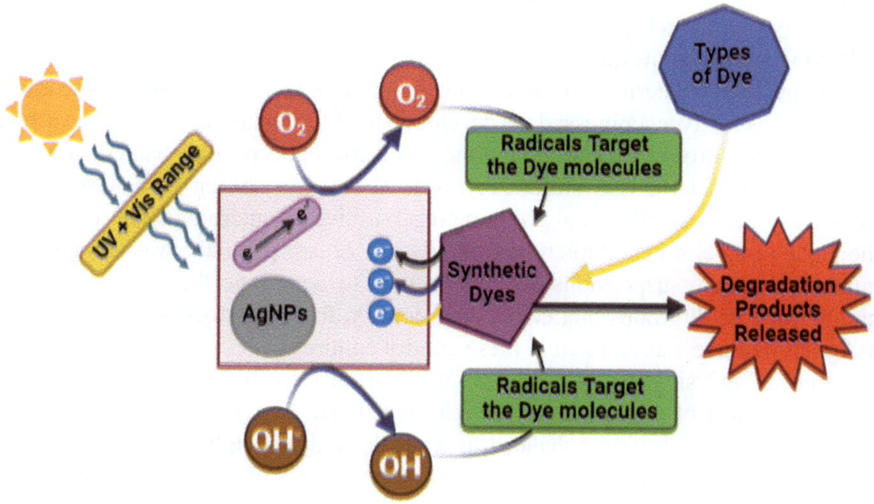

Fig. 2 Photocatalytic degradation dye

et al. [86] employed *Staphylococcus aureus* to test the antibacterial effectiveness of AgNPs (Gram-positive bacteria). They were able to comprehend the mechanism of AgNPs' antibacterial activity by looking at the growth, morphology, and molecular alterations in the cell wall. They said that 4 mg/mL of AgNPs totally suppressed bacterial growth. *E. coli* (Gram-negative bacteria) was subjected to [77] investigations into the antibacterial mechanism of AgNPs. They asserted that AgNPs impair the permeability of bacterial membranes by enabling reducing sugars and proteins to get through and inactivate the respiratory chain dehydrogenases assert that AgNPs greatly increase membrane permeability and, by creating "pits," eliminate lipopolysaccharide molecules that have accumulated inside the membrane. In addition, it has been hypothesised that silver ions interfere with DNA replication and change the cell membrane's permeability and structure [133] Studies have shown that silver ions are photoactive in the presence of UV-A and UV-C light, which speeds up the UV inactivation of bacteria and viruses [63, 73].

AgNPs Effect on *Escherichia Coli* Growth

One of the most powerful inorganic substances for fighting *E. coli* germs is silver/silver oxide nanoparticles (AgNPs). *E. coli* is moderately stress-free to achieve and works alongside additional microorganisms to collapse and digest biological supplies found in wastewater. Adverse impacts on the fitness and endurance of these advantageous microorganisms (enforced by the existence of n-Ag) will eventually disrupt wastewater management competence during the life-threatening aeration phase of treatment. Microbes are suitable experimental organisms because they produce quickly and are very economical to cultivate; they have a high surface-to-volume ratio, making them sensitive to fatal components' modest absorptions [64]. Nevertheless, it has been shown that washing these impregnated socks causes significant n-Ag particle release. These unfettered nanoparticles may get into wastewater collecting systems quite quickly. The growth of *E. coli* under nanosilver presence was studied in order to determine the microbial inhibitory effects of nanosilver. The introduction of nanosilver suppressed the development of the bacterium *E. coli*, according to the results. Aerobic wastewater treatment systems that depend on bacteria to break down organic material may suffer as a result. *E. coli* was chosen as the test subject to rule out the possibility that a competing "surviving" microorganism resides in the niche of an eliminated microbe. Initially, the *E. coli* populace was anonymous; numerous control suspensions were performed to determine the population. The number of *E. coli* groups that could fully develop from the stock solution without n-Ag interference was counted in these control cultures.

Chitosan and silver-loaded chitosan nanoparticles for bioactive polyester have been the subject of preliminary investigation [5]. The activity of the silver-loaded chitosan nanoparticle was enhanced as a result of the complimentary impacts of silver and chitosan nanoparticles. Several studies have coupled chitosan with AgNPs to increase antibacterial efficiency at lower silver concentrations and decrease the toxicity of silver.

Zhao et al. 146] used AgNPs/Polyvinyl alcohol (PVA)/chitosan nanofibres with AgNO3 concentrations of 5 and 10 mol L^1 to create an electrospun web. In-situ formation of the AgNPs was placed in a PVA/chitosan mix solution. AgNPs, PVA, and chitosan nanofibres produced as a consequence had flat surfaces and homogeneous diameters. Also, it was noted that when the silver level dropped, the average fibre diameter somewhat shrank. On the surface of the nanofibers, circular AgNPs with a diameter ranging from 4 to 14 nm were visible. The AgNPs/PVA/chitosan nanofibres showed a 98% inhibitory ratio against *E. coli* when used in a 50 mL bacterial feeding solution with an initial concentration of around 104 CFU mL^{-1}. The AgNPs/PVA/ chitosan nanofibres showed a 98% inhibition ratio against *E. coli* up to a contact period of 12 h when added to a 50 mL bacterial feeding solution with an initial concentration of roughly 104 CFU mL. These findings point to the potential of using silver-coated electrospun chitosan nanofibrous webs as water filters for bacterial contamination.

5.3.2 Ag as an Efficient Antifungal Agent

Nanoparticles have a propensity to interfere with a wide range of biological functions, including microorganism cell membrane structure and function [92]. Similar to this, AgNPs also block the expression of proteins involved in ATP synthesis [136]. These characteristics of AgNPs have drawn several professionals from all over the world to employ them in a variety of applications, including water purification [98]. AgNP concentration was raised while nanosize was lowered to maximise antifungal activity. At 25 ppm, 37.5 ppm, and 50 ppm concentrations of AgNPs, the ability of the mycelium to grow was reduced by roughly 50%, 75%, and 90%, respectively. It was shown that the productivity of fungal biomass in the liquid growth media was too low at the 25–37.5 ppm of AgNPs concentrations for all sizes. In addition, it was discovered that the application of silver nanoparticles gradually reduced the size and quantity of septations on both micro- and macroconidia [6]. The antifungal effectiveness of biogenic AgNPs is significantly influenced by their size, shape, and coating materials. The wide diversity of biogenic AgNPs makes it particularly difficult to identify a single mechanism of action. So, the majority of study focus has been on figuring out the mechanism of action of chemically synthesised AgNPs, which is connected to the attachment of AgNPs to the surface of the fungus as a result of electrostatic attraction (Fig. 3). Studies [29] (Le et al. 2012; Vazquez-Muoz et al. 2014) [68, 88] have shown that the extracellular accumulation of AgNPs results in a dynamic release of Ag + , which actively enters the cell and raises the intracellular concentration as well as the intracellular biosynthesis of AgNPs. There are currently no known membrane channels or cell receptors for the uptake of silver. Nevertheless, it has been found that the Ag^+ importer high-affinity copper transporter (Ctr1) [47, 110].

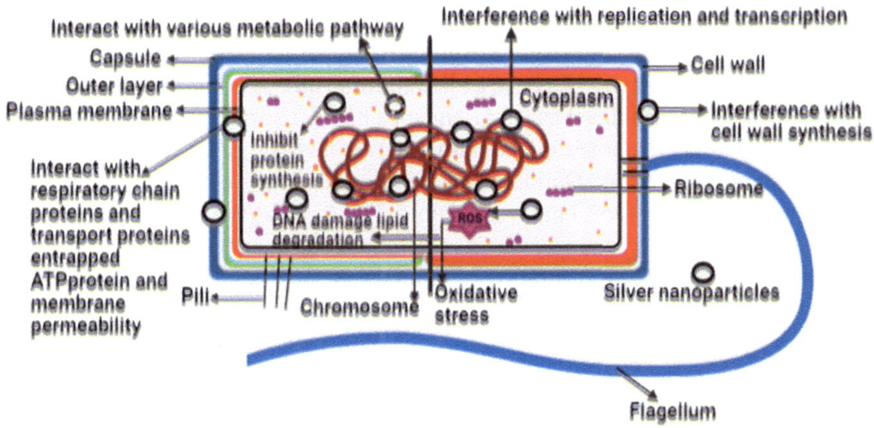

Fig. 3 Antibacterial mechanism of AgNPs

6 Use of Paramylon in Water Purification

The heterogeneous glucose polymers known as -glucans, which are present in a number of species, can stimulate the innate immune system. Purity, molecular structure, degree of polymerization, and source of the material are all factors that affect how effectively an activator works. The paramylon isolated from the WZSL non-chloroplastic mutant of Euglena gracilis is one of the clearest varieties of crystalline-(1–3)-glucan to be discovered in nature [12]. Since it is free of any cellular component contamination, paramylon from the non-chloroplast WZSL mutant of *Euglena gracilis* may be regarded as being unusually pure (membranes, proteins, pigments). Optical, scanning electron, and NMR spectroscopy analysis prove the presence of 100% glucose and indicate the absence of any general immune-modulating effects. Paramylon has been used to investigate the connection between structure and function and the mechanism of action in animal and plant models. It has also been utilised to determine whether the polysaccharide, whether in granular form or as linear (1,3)-linked glucan nanofibers, has the potential to be beneficial [13]. The storage polysaccharide paramylon is produced by both Euglena gracilis wild type and WZSL mutant cells, and it is then deposited as granules in the cytoplasm. Due to the helical deposition of higher order aggregates of elemental 4–10 nm nanofibers made of unbranched triple helices of -(1–3)-glucan chains, these granules, which range in size from 1 to 2 m, exhibit a very high degree of crystallinity. The wide range of cellular compositions found in Paramylon varies according to the carbon supply and growth environment (in the light or the dark). The maximum concentration of the polysaccharide is attained for both cell types after 24 h in batch culture settings, under both growth circumstances, with greater values in the dark compared to the light, and with glucose as the best carbon source. There is more paramylon in the mutant than in the wild type: 2.5 ng/cell, or 85% of the dry biomass of the mutant, as opposed to 1.4 ng/cell, or 48% of the dry biomass of the wild type [13]. The WZSL mutant can be grown aseptically using the fed-batch method, a growth strategy frequently used in the fermentation industry, to achieve high cell densities and improve the quality of the finished product [27, 135]. This is crucial for algal biomass utilised in the extraction of important pure compounds. A fed-batch culture with 10 g/L glucose and Cramer and Myers media [31] yields 8 g L^{-1} of biomass after 20 days of growth, with paramylon production frequently exceeding 80% of the dry biomass [31]. Global food production is negatively impacted by low water availability, a critical environmental element that prevents photosynthesis and causes the world's arable land to become more salinized [71].

One of the most crucial crops in terms of nutrition and economics is the tomato, which is particularly vulnerable to water shortages that lower fruit yield and quality. Research mostly focuses on choosing genotypes that are drought-tolerant and have lower water requirements in order to mitigate its effects. Paramylon nanofibers have been shown to alter hormone levels, photosynthesis, and water use efficiency when applied to the roots of tomato plants [115]. Plants of the *Solanum lycopersicum* cultivar Micro-Tom, chosen for its short lifespan (3 months) and small size, were subjected to crop failure stress either through or without root treatment with paramylon nanofibers in order to test the viability of using paramylon as a novel root diagnosis to help tomato plants with low water availability [11]. Aeroponics, a soil-free air–water system ideal for studying plant roots since the nutrient solution is sprayed directly into the roots via atomizers, was used to grow plants in controlled environments.

7 AgNPs Potentially Harm the Environment and Life

Despite the fact that AgNPs and their composites offer a variety of uses, they do have significant drawbacks. AgNPs are used in the food sector, toys, cleaning products, digital gadgets, and scientific equipment [39]. The extensive use of AgNPs as antibacterial agents and disinfectants may increase bacterial resistance, and they may prove to be hazardous to aquatic creatures [138]. Water is the wildest way to spread pollution over huge areas, and once silver (Ag) nanoparticles have accumulated in reservoirs, there is a risk that they will be spread across a large area, posing a warning not only to marine animals, but also to creatures whose natural home is coastal areas. Because water is home to numerous species that make up the first stage of food chains, Ag nanoparticles might simply permeate higher levels of the trophic chain hierarchy [101]. The highest allowed concentration of AgNPs additives in drinking water is 100 g/l, although it is anticipated that the release of nano-sized silver from AgNPs will cause the quantities of dissolved silver and nanosilver to reach dangerous levels. Aquatic species exhibit more AgNP-induced toxicity than terrestrial animals and people [91]. As the outcome of their washing, silver nanoparticles rooted in fabrics find their way into wastewater. This raises the risk of it escaping and causing harm to living things [20]. Silver nanoparticles were shown to be attached to sludge and sewage, which might lead to their dispersion in agricultural areas in the future, posing a toxicological and bioaccumulation danger [20].

Consecutive releases of nano-sized silver are linked to environmental toxicity, and the products' dispersion and emissions have an impact on marine life [99]. According to a literature study, the surface processes of aggregation, sulfidation, and phase transition in aquatic species are what cause the AgNPs' toxicity. Also, it shows that several terrestrial and aquatic plants, fungi, algae, human cells, pores, and skin, as well as other vertebrates, are poisonous to silver [32]. According to a different research, AgNPs are hazardous because of the buildup of particles in the cell medium, cellular uptakes, localisation inside the cell, and their discharge and release in the lung cells [108].

Fish gills constantly interact with the available water, creating a budding target for silver nanoparticles [19]. When perch were exposed to silver nanoparticles, the researchers assessed how much oxygen they utilised. The researchers discovered that exposure to silver nanoparticles diminishes tolerance to oxygen or hypoxia scarcity [19]. Silver nanoparticles in soil may permeate plant tissue, which can subsequently feed people and animals [100]. The tissue of creatures like annelids, which are eaten by species higher up the food chain, may also contain silver nanoparticles. Nonetheless, compared to Ag^+ ions, AgNPs have a higher potential for toxicity [24].

Because of their biocidal characteristics, which make them extremely effective antimicrobial agents, AgNPs are the most frequently and widely utilised ENMs in consumer items. According to a research done on two German wastewater treatment facilities, AgNPs can be removed up to 96.4% of the time, with residual concentrations in effluents ranging from 0.7 to 11.1 ng/L. The authors projected a total discharge in the environment of 33 kg AgNPs/year for the entire nation based on such data, notwithstanding the great clearance efficiency [72]. AgNPs have drawn a lot of interest because of the possible environmental dangers associated with their manufacturing, use, disposal, and application.

The release of Ag^+ ions, which are known as one of the most hazardous metal ions in the aquatic environment [33], is strongly related to the antibacterial action of AgNPs [94]. Due to the constant and extended release of Ag^+ ions near the target organism using AgNPs as opposed to $AgNO_3$ results in a considerably more effective and long-lasting antibacterial action [106, 144]. As their biocidal effects are not just limited to microorganisms, this trait raises the risk of AgNPs being released into the environment. In fact, the harmful effects of AgNPs on aquatic life, including freshwater [60, 70, 90] and marine species, have been extensively reported [97, 103].

Regarding microorganisms, dissolution is primarily responsible for AgNPs' ecotoxicity to non-target species, even though other studies claim that this explanation may not be complete and that nano-specific toxicity may also be at play [76, 126]. Although there is no agreement on the relative contributions of each element [55], additional impacts can be seen as particle size decreases [4, 41]. Smaller particles may have an impact on toxicity by increasing the surface area available for the particle to dissolve [43] or by allowing the particle to enter the cell [102], which may then interact with molecular systems or dissolve intracellularly [96]. The behaviour of the particles and their dissolution in the media, which is a crucial component for comprehending the mechanism of toxicity and, consequently, their environmental safety, are not well understood in many research looking into the hazardous effect of AgNPs on non-target animals. Chlorine or sulphur, which are oxidising substances, may promote the release of Ag^+ ions and, as a result, considerably alter the harmful effect of Ag^+, not just against bacteria, but also generally towards all potentially exposed organisms [83].

8 Challenges in Water Purification

Each wastewater treatment process has its own advantages and disadvantages, and the effectiveness is influenced by the elements stated in Table 5. Due to their huge surface area, high adsorption capacity, low resistance to diffusion, and quick rates of equilibrium, nanoparticles have become more important in the last ten years for the removal of dyes from wastewater [2]. More precisely, various research teams from across the world have looked at the function of silver nanoparticles in dye removal. Because of their exceptional physical, chemical, and biological characteristics, silver nanoparticles may be used in a variety of sectors. In addition to the qualities listed above, silver nanoparticles have antibacterial and antifungal capabilities that make it easier for them to be used in wastewater treatment [127].

9 Conclusion

Consideration has been given heavily to AgNPs and their hybrids in the field of wastewater treatment methods. In wastewater treatment, AgNP hybrids with components including cellulose, activated carbon, chitosan, alginate, graphene oxides, titanium dioxides, silicon dioxide, etc. have been frequently used. The main mechanisms of action of silver nanoparticles on the dye effluent treatment, including adsorption and photocatalytic degradation, are highlighted with an emphasis on the elimination of dyes that are often used in the textile sector. It has been proven that preparing silver nanoparticles using natural or environmentally friendly reducing chemicals effectively removes a variety of colours from wastewater. The poisonous effects of heavy metal is a severe issue for water effluents. The existence of destructive microbes and their poisons also poses a severe threat to human health as well as the aquatic environment. Because of their enhanced adsorption capacities and superior antibacterial qualities, AgNPs are useful options for eliminating impurities from wastewater bodies. AgNPs' having advantages in cleaning up water however, its reckless utilisation might exacerbate environmental deterioration. Yet, current study findings demonstrate the danger of nanoparticle discharge into the environment and

Table 5 An overview of wastewater treatment techniques

S. no	Mechanism	Type	Advantages	Factors influencing treatment	Limitations	Key references
1	Biological treatment method	Aerobic	Specificity, effective and the possibility of immobilised	Specificity towards substrate	Cost-intensive, and more extended detention periods	[62, 125]
		Combined aerobic and anaerobic				[56, 121]
		Enzyme				[18, 93]
		Anaerobic				[22, 129],
2	Radiation treatment	Radiation	Affordable can be renewed for the upcoming use	Reactivity, photosensitivity, and the presence of dissolved oxygen in the dye	More amounts and a longer contact time are needed	[67, 80, 120]
3	Physical treatment method	Adsorption	Affordable can be renewed for the upcoming use	Interaction between the dye and the sorbent, as well as sorbent surface area, particle size, temperature, pH, and contact time	Larger quantities required, and longer contact time required	[1, 82, 145]
4	Chemical treatment methods	Oxidation	Efficient for all types of dyes, has a brief detention period, and requires less capital	Agents that oxidise, oxidising environments, and pH	Process pH maintenance issues water removal and sludge handling	[7, 10]
		Electrochemical oxidation		Electrolyte type, concentration, and time of electrolysis		[111, 134]
		Precipitation		pH and precipitation		[30, 74]

their toxicity to aquatic and terrestrial biological systems at low quantities. Thus, cautious manipulation of special characteristics of AgNPs is essential for the actual implementation of such nanoparticles in the treatment of wastewater.

References

1. Aboua KN, Yobouet YA, Yao KB, Goné DL, Trokourey A (2015) Investigation of dye adsorption onto activated carbon from the shells of Macoré fruit. J Environ Manage 156:10–14. https://doi.org/10.1016/J.JENVMAN.2015.03.006
2. Ahmad A, Mohd-Setapar SH, Chuong CS, Khatoon A, Wani WA, Kumar R, Rafatullah M (2015) Recent advances in new generation dye removal technologies: novel search for approaches to reprocess wastewater. RSC Adv 5(39):30801–30818. https://doi.org/10.1039/C4RA16959J
3. Alamelu K, Jaffar Ali BM (2020) Ag nanoparticle-impregnated sulfonated graphene/TiO_2 composite for the photocatalytic removal of organic pollutants. Appl Surf Sci 512:145629. https://doi.org/10.1016/J.APSUSC.2020.145629
4. Ale A, Liberatori G, Vannuccini ML, Bergami E, Ancora S, Mariotti G, Bianchi N, Galdopórpora JM, Desimone MF, Cazenave J, Corsi I (2019) Exposure to a nanosilver-enabled consumer product results in similar accumulation and toxicity of silver nanoparticles in the marine mussel *Mytilus galloprovincialis*. Aquatic Toxicol (Amsterdam, Netherlands) 211:46–56. https://doi.org/10.1016/J.AQUATOX.2019.03.018
5. Ali SW, Rajendran S, Joshi M (2011) Synthesis and characterization of chitosan and silver loaded chitosan nanoparticles for bioactive polyester. Carbohyd Polym 83(2):438–446. https://doi.org/10.1016/J.CARBPOL.2010.08.004
6. Akpinar I, Unal M, Sar T (2021) Potential antifungal effects of silver nanoparticles (AgNPs) of different sizes against phytopathogenic *Fusarium oxysporum* f. sp. radicis-lycopersici (FORL) strains. SN Appl Sci 3(4):1–9. https://doi.org/10.1007/S42452-021-04524-5/TABLES/2
7. Azbar N, Yonar T, Kestioglu K (2004) Comparison of various advanced oxidation processes and chemical treatment methods for COD and color removal from a polyester and acetate fiber dyeing effluent. Chemosphere 55(1):35–43. https://doi.org/10.1016/J.CHEMOSPHERE.2003.10.046
8. Bachheti RK, Abate L, Bachheti A, Madhusudhan A, Husen A (2021) Algae-, fungi-, and yeast-mediated biological synthesis of nanoparticles and their various biomedical applications. In: Handbook of greener synthesis of nanomaterials and compounds. Elsevier, Amsterdam, pp 701–734. https://doi.org/10.1016/B978-0-12-821938-6.00022-0
9. Bachheti RK, Fikadu A, Bachheti A, Husen A (2020) Biogenic fabrication of nanomaterials from flower-based chemical compounds, characterization and their various applications: a review. Saudi J Biol Sci 27(10):2551–2562
10. Balcioglu IA, Arslan I, Sacan MT (2011) Homogenous and heterogenous advanced oxidation of two commercial reactive dyes. Environ Technol 22(7):813–822. https://doi.org/10.1080/09593332208618323
11. Barsanti L, Coltelli P, Gualtieri P (2019) Paramylon treatment improves quality profile and drought resistance in *Solanum lycopersicum* L. cv. Micro-Tom. Agronomy 9(7):394. https://doi.org/10.3390/AGRONOMY9070394
12. Barsanti L, Gualtieri P (2019) Paramylon, a potent immunomodulator from WZSL mutant of *Euglena gracilis*. Molecules, 24(17). https://doi.org/10.3390/MOLECULES24173114
13. Barsanti L, Passarelli V, Evangelista V, Frassanito AM, Gualtieri P (2011) Chemistry, physico-chemistry and applications linked to biological activities of β-glucans. Nat Prod Rep 28(3):457–466. https://doi.org/10.1039/C0NP00018C
14. Bayoumi TA, Saleh HM (2018) Characterization of biological waste stabilized by cement during immersion in aqueous media to develop disposal strategies for phytomediated radioactive waste. Prog Nucl Energy 107:83–89. https://doi.org/10.1016/J.PNUCENE.2018.04.021
15. Benn TM, Westerhoff P (2008) Nanoparticle silver released into water from commercially available sock fabrics. Environ Sci Technol 42(11):4133–4139. https://doi.org/10.1021/ES7032718/SUPPL_FILE/ES7032718_SI.PDF

16. Bhardwaj AK, Sundaram S, Yadav KK, Srivastav AL (2021) An overview of silver nanoparticles as promising materials for water disinfection. Environ Technol Innov. https://doi.org/10.1016/J.ETI.2021.101721
17. Bi Y, Han B, Zimmerman S, Perreault F, Sinha S, Westerhoff P (2018) Four release tests exhibit variable silver stability from nanoparticle-modified reverse osmosis membranes. Water Res 143:77–86. https://doi.org/10.1016/J.WATRES.2018.06.036
18. Bilal M, Rasheed T, Iqbal HMN, Hu H, Wang W, Zhang X (2018) Horseradish peroxidase immobilization by copolymerization into cross-linked polyacrylamide gel and its dye degradation and detoxification potential. Int J Biol Macromol 113:983–990. https://doi.org/10.1016/J.IJBIOMAC.2018.02.062
19. Bilberg K, Malte H, Wang T, Baatrup E (2010) Silver nanoparticles and silver nitrate cause respiratory stress in Eurasian perch (Perca fluviatilis). Aquatic Toxicol (Amsterdam, Netherlands) 96(2):159–165. https://doi.org/10.1016/J.AQUATOX.2009.10.019
20. Blaser SA, Scheringer M, MacLeod M, Hungerbühler K (2008) Estimation of cumulative aquatic exposure and risk due to silver: contribution of nano-functionalized plastics and textiles. Sci Total Environ 390(2–3):396–409. https://doi.org/10.1016/J.SCITOTENV.2007.10.010
21. Bosetti M, Massè A, Tobin E, Cannas M (2002) Silver coated materials for external fixation devices: in vitro biocompatibility and genotoxicity. Biomaterials 23(3):887–892. https://doi.org/10.1016/S0142-9612(01)00198-3
22. Bras M, Queenan B, Susin SA (2005) Programmed cell death via mitochondria: different modes of dying. Biochemistry (Moscow) 70:231–239
23. Brar SK, Verma M, Tyagi RD, Surampalli RY (2010) Engineered nanoparticles in wastewater and wastewater sludge–evidence and impacts. Waste Manage (New York, N.Y.) 30(3):504–520. https://doi.org/10.1016/J.WASMAN.2009.10.012
24. Budama-Kilinc Y, Cakir-Koc R, Zorlu T, Ozdemir B, Karavelioglu Z, Egil AC, Kecel-Gunduz S (2018) Assessment of nano-toxicity and safety profiles of silver nanoparticles. In: Khan M (ed) Silver nanoparticles—fabrication, characterization and applications. IntechOpen, London, UK, p 75645
25. Cao E, Duan W, Wang F, Wang A, Zheng Y (2017) Natural cellulose fiber derived hollow-tubular-oriented polydopamine: in-situ formation of Ag nanoparticles for reduction of 4-nitrophenol. Carbohyd Polym 158:44–50. https://doi.org/10.1016/J.CARBPOL.2016.12.004
26. Cheng D, Zhang Y, Bai X, Liu Y, Deng Z, Wu J, Bi S, Ran J, Cai G, Wang X (2020) Mussel-inspired fabrication of superhydrophobic cotton fabric for oil/water separation and visible light photocatalytic. Cellulose 27(9):5421–5433. https://doi.org/10.1007/S10570-020-03149-Y
27. Coelho RS, Vidotti ADS, Reis ÉM, Franco TT (2014) High cell density cultures of microalgae under fed-batch and continuous growth. Chem Eng Trans 38:313–318. https://doi.org/10.3303/CET1438053
28. Chook SW, Chia CH, Chan CH, Chin SX, Zakaria S, Sajab MS, Huang NM (2015) A porous aerogel nanocomposite of silver nanoparticles-functionalized cellulose nanofibrils for SERS detection and catalytic degradation of rhodamine B. RSC Adv 5(108):88915–88920. https://doi.org/10.1039/C5RA18806G
29. Chwalibog A, Sawosz E, Hotowy A, Szeliga J, Mitura S, Mitura K et al (2010) Visualization of interaction between inorganic nanoparticles and bacteria or fungi. Int J Nanomed 5:1085–1094. https://doi.org/10.2147/IJN.S13532
30. Coro E, Laha S (2001) Color removal in groundwater through the enhanced softening process. Water Res 35(7):1851–1854. https://doi.org/10.1016/S0043-1354(00)00440-1
31. Cramer M, Myers J (1952) Growth and photosynthetic characteristics of *Euglena gracilis*. Arch Mikrobiol 17(1–4):384–402. https://doi.org/10.1007/BF00410835/METRICS
32. DeAlba-Montero I, Ruiz-Torres CA, Portales-Pérez DP, Martínez-Gutierrez F, Echeverría F, Compeán-Jasso ME, Cata-ño-Cañizales YG, Ruiz F (2020) Atmospheric corrosion, antibacterial properties, and toxicity of silver nanoparticles synthesized by two different routes. Bioinorg Chem Appl 2020:8891069

33. Dong F, Mohd Zaidi NF, Valsami-Jones E, Kreft JU (2017) Time-resolved toxicity study reveals the dynamic interactions between uncoated silver nanoparticles and bacteria. Nanotoxicology 11(5):637–646. https://doi.org/10.1080/17435390.2017.1342010
34. Fang X, Li J, Ren B, Huang Y, Wang D, Liao Z, Li Q, Wang L, Dionysiou DD (2019) Polymeric ultrafiltration membrane with in situ formed nano-silver within the inner pores for simultaneous separation and catalysis. J Membr Sci 579:190–198. https://doi.org/10.1016/J.MEMSCI.2019.02.073
35. Feng L, Cao M, Ma X, Zhu Y, Hu C (2012) Superparamagnetic high-surface-area Fe_3O_4 nanoparticles as adsorbents for arsenic removal. J Hazard Mater 217–218:439–446. https://doi.org/10.1016/J.JHAZMAT.2012.03.073
36. Gai S, Zhang J, Fan R, Xing K, Chen W, Zhu K, Zheng X, Wang P, Fang X, Yang Y (2020) Highly stable zinc-based metal-organic frameworks and corresponding flexible composites for removal and detection of antibiotics in water. ACS Appl Mater Interfaces 12(7):8650–8662. https://doi.org/10.1021/ACSAMI.9B19583
37. Ganguly K, Dutta SD, Patel DK, Lim KT (2020) Silver nanoparticles for wastewater treatment. Aquananotechnol Appl Nanomater Water Purif 385–401. https://doi.org/10.1016/B978-0-12-821141-0.00016-1
38. Geranio L, Heuberger M, Nowack B (2009) The behavior of silver nanotextiles during washing. Environ Sci Technol 43(21):8113–8118. https://doi.org/10.1021/ES9018332/SUPPL_FILE/ES9018332_SI_001.PDF
39. Ghosh S, Mondal A (2020) Aggregation chemistry of green silver nanoparticles for sensing of Hg^{2+} and Cd^{2+} ions. Colloids Surf, A. https://doi.org/10.1016/J.COLSURFA.2020.125335
40. Gitis V, Hankins N (2018) Water treatment chemicals: trends and challenges. J Water Process Eng 25:34–38. https://doi.org/10.1016/J.JWPE.2018.06.003
41. Gliga AR, Skoglund S, Odnevall Wallinder I, Fadeel B, Karlsson HL (2014) Size-dependent cytotoxicity of silver nanoparticles in human lung cells: the role of cellular uptake, agglomeration and Ag release. Part Fibre Toxicol 11(1):1–17. https://doi.org/10.1186/1743-8977-11-11/FIGURES/8
42. Gonfa YH, Tessema FB, Bachheti A, Tadesse MG, Eid EM, Abou Fayssal S, Adelodun B, Choi KS, Širić I, Kumar P, Bachheti RK (2022) Essential oil composition of aerial part of *Pluchea ovalis* (Pers.) DC., silver nanoparticles synthesis, and larvicidal activities against fall armyworm. Sustainability 14(23):15785
43. Guo M, Zhang Y, Du F, Wu Y, Zhang Q, Jiang C (2019) Silver nanoparticles/polydopamine coated polyvinyl alcohol sponge as an effective and recyclable catalyst for reduction of 4-nitrophenol. Mater Chem Phys
44. Guzman M, Dille J, Godet S (2012) Synthesis and antibacterial activity of silver nanoparticles against gram-positive and gram-negative bacteria. Nanomed Nanotechnol Biol Med 8(1):37–45. https://doi.org/10.1016/J.NANO.2011.05.007
45. Haque MA, Imamura R, Brown GA, Krishnamurthi VR, Niyonshuti II, Marcelle T, Mathurin LE, Chen J, Wang Y (2017) An experiment-based model quantifying antimicrobial activity of silver nanoparticles on Escherichia coli. RSC Adv 7(89):56173–56182. https://doi.org/10.1039/C7RA10495B
46. Hodges BC, Cates EL, Kim JH (2018) Challenges and prospects of advanced oxidation water treatment processes using catalytic nanomaterials. Nat Nanotechnol 13(8):642–650. https://doi.org/10.1038/s41565-018-0216-x
47. Horstmann C, Campbell C, Kim DS, Kim K (2019) Transcriptome profile with 20 nm silver nanoparticles in yeast. FEMS Yeast Res 19:1–15. https://doi.org/10.1093/femsyr/foz003
48. Hlongwane GN, Sekoai PT, Meyyappan M, Moothi K (2019) Simultaneous removal of pollutants from water using nanoparticles: a shift from single pollutant control to multiple pollutant control. Sci Total Environ 656:808–833. https://doi.org/10.1016/J.SCITOTENV.2018.11.257
49. Husen A, Rahman QI, Iqbal M, Yassin MO, Bachheti RK (2019) Plant-mediated fabrication of gold nanoparticles and their applications. In: Nanomaterials and plant potential. Springer, Cham, pp 71–110. https://doi.org/10.1007/978-3-030-05569-1_3

50. Husen A (2023a) Smart nanomaterials from agricultural and horticultural products. Springer Nature Singapore Pte Ltd., 152 Beach Road, #21-01/04 Gateway East, Singapore 189721, Singapore
51. Husen A (2023b) Secondary metabolites based green synthesis of nanomaterials and their applications. Springer Nature Singapore Pte Ltd., 152 Beach Road, #21-01/04 Gateway East, Singapore 189721, Singapore
52. Husen A, Siddiqi KS (2023) Advances in smart nanomaterials and their applications. Elsevier Inc., 50 Hampshire St., 5th Floor, Cambridge, MA 02139, USA
53. Husen A (2022) Engineered nanomaterials for sustainable agricultural production, soil improvement and stress management. Elsevier Inc., 50 Hampshire St., 5th Floor, Cambridge, MA 02139, USA
54. Husen, A, Iqbal, M (2019). Nanomaterials and Plant Potential. Springer International Publishing AG, Gewerbestrasse 11, 6330 Cham, Switzerland. https://doi.org/10.1007/978-3-030-05569-1
55. Ivask A, Kurvet I, Kasemets K, Blinova I, Aruoja V, Suppi S, Vija H, Kakinen A, Titma T, Heinlaan M, Visnapuu M, Koller D, Kisand V, Kahru A (2014) Size-dependent toxicity of silver nanoparticles to bacteria, yeast, algae, crustaceans and mammalian cells in vitro. PloS One 9(7). https://doi.org/10.1371/JOURNAL.PONE.0102108
56. Jayapal M, Jagadeesan H, Shanmugam M, Danisha JP, Murugesan S (2018) Sequential anaerobic-aerobic treatment using plant microbe integrated system for degradation of azo dyes and their aromatic amines by-products. J Hazard Mater 354:231–243. https://doi.org/10.1016/J.JHAZMAT.2018.04.050
57. Jiang XH, Wang LC, Yu F, Nie YC, Xing QJ, Liu X, Pei Y, Zou JP, Dai WL (2018) Photodegradation of organic pollutants coupled with simultaneous photocatalytic evolution of hydrogen using quantum-dot-modified g-C_3N_4 catalysts under visible-light irradiation. ACS Sustain Chem Eng 6(10):12695–12705. https://doi.org/10.1021/ACSSUSCHEMENG.8B01695
58. Jin-Chul K, Madhusudhan A, Husen A (2021) Smart nanomaterials in biomedical applications. Springer Nature Switzerland AG, Gewerbestrasse 11, 6330 Cham, Switzerland
59. Kaegi R, Voegelin A, Sinnet B, Zuleeg S, Hagendorfer H, Burkhardt M, Siegrist H (2011) Behavior of metallic silver nanoparticles in a pilot wastewater treatment plant. Environ Sci Technol 45(9):3902–3908. https://doi.org/10.1021/ES1041892/SUPPL_FILE/ES1041892_SI_001.PDF
60. Kalantzi I, Mylona K, Toncelli C, Bucheli TD, Knauer K, Pergantis SA, Pitta P, Tsiola A, Tsapakis M (2019) Ecotoxicity of silver nanoparticles on plankton organisms: a review. J Nanoparticle Res 21(3). https://doi.org/10.1007/S11051-019-4504-7
61. Khan AA, Marchiori L, Ferreira-Neto EP, Wender H, Parveen R, Muneeb M, Mattos BO, Rodrigues-Filho UP, Ribeiro SJL, Ullah S (2022) Enhanced photoredox activity of $BiVO_4$/Prussian blue nanocomposites for efficient pollutant removal from aqueous media under low-cost LEDs illumination. Catalysts 12(12):1–18. https://doi.org/10.3390/catal12121612
62. Kiayi Z, Lotfabad TB, Heidarinasab A, Shahcheraghi F (2019) Microbial degradation of azo dye carmoisine in aqueous medium using *Saccharomyces cerevisiae* ATCC 9763. J Hazard Mater 373:608–619
63. Kim JY, Lee C, Cho M, Yoon J (2008). Enhanced inactivation of E. coli and MS-2 phage by silver ions combined with UV-A and visible light irradiation. Water Res 42(1–2):356–362. https://doi.org/10.1016/J.WATRES.2007.07.024
64. Klaine SJ, Alvarez PJJ, Batley GE, Fernandes TF, Handy RD, Lyon DY, Mahendra S, McLaughlin MJ, Lead JR (2008) Nanomaterials in the environment: behavior, fate, bioavailability, and effects. Environ Toxicol Chem 27(9):1825–1851. https://doi.org/10.1897/08-090.1
65. Kodom T, Rusen E, Cəlinescu I, Mocanu A, Şomoghi R, Dinescu A, Diacon A, Boscornea C (2015) Silver nanoparticles influence on photocatalytic activity of hybrid materials based on TiO_2 P25. J Nanomater. https://doi.org/10.1155/2015/210734
66. Krce L, Šprung M, Maravić A, Umek P, Salamon K, Krstulović N, Aviani I (2020) Bacteria exposed to silver nanoparticles synthesized by laser ablation in water: modelling *E. coli* growth and inactivation. Materials 13(3). https://doi.org/10.3390/MA13030653

67. Kumar M, Mehta A, Mishra A, Singh J, Rawat M, Basu S (2018) Biosynthesis of tin oxide nanoparticles using Psidium Guajava leave extract for photocatalytic dye degradation under sunlight. Mater Lett 215:121–124. https://doi.org/10.1016/J.MATLET.2017.12.074
68. Lara HH, Romero-Urbina DG, Pierce C, Lopez-Ribot JL, Arellano-Jiménez MJ, Jose-Yacaman M (2015) Effect of silver nanoparticles on *Candida albicans* biofilms: an ultrastructural study. J Nanobiotechnol 13:91. https://doi.org/10.1186/s12951-015-0147-8
69. Lekamge S, Miranda AF, Abraham A, Li V, Shukla R, Bansal V, Nugegoda D (2018) The toxicity of silver nanoparticles (AgNPs) to three freshwater invertebrates with different life strategies: *Hydra vulgaris*, *Daphnia carinata*, and *Paratya australiensis*. Front Environ Sci 6:152. https://doi.org/10.3389/FENVS.2018.00152/BIBTEX
70. Lekamge S, Miranda AF, Pham B, Ball AS, Shukla R, Nugegoda D (2019) The toxicity of non-aged and aged coated silver nanoparticles to the freshwater shrimp *Paratya australiensis*. J Toxicol Environ Health A 82(23–24):1207–1222. https://doi.org/10.1080/15287394.2019.1710887
71. Leng G, Hall J (2019) Crop yield sensitivity of global major agricultural countries to droughts and the projected changes in the future. Sci Total Environ 654:811–821. https://doi.org/10.1016/J.SCITOTENV.2018.10.434
72. Li L, Stoiber M, Wimmer A, Xu Z, Lindenblatt C, Helmreich B, Schuster M (2016) To what extent can full-scale wastewater treatment plant effluent influence the occurrence of silver-based nanoparticles in surface waters? Environ Sci Technol 50(12):6327–6333. https://doi.org/10.1021/ACS.EST.6B00694
73. Li Q, Mahendra S, Lyon DY, Brunet L, Liga MV, Li D, Alvarez PJJ (2008) Antimicrobial nanomaterials for water disinfection and microbial control: potential applications and implications. Water Res 42(18):4591–4602. https://doi.org/10.1016/J.WATRES.2008.08.015
74. Lin SH, Chen ML (1997) Treatment of textile wastewater by chemical methods for reuse. Water Res 31(4):868–876
75. Liu C, Hong T, Li H, Wang L (2018) From club convergence of per capita industrial pollutant emissions to industrial transfer effects: an empirical study across 285 cities in China. Energy Policy 121:300–313. https://doi.org/10.1016/J.ENPOL.2018.06.039
76. Liu H, Wang X, Wu Y, Hou J, Zhang S, Zhou N, Wang X (2019) Toxicity responses of different organs of zebrafish (*Danio rerio*) to silver nanoparticles with different particle sizes and surface coatings. Environ Pollut (Barking, Essex: 1987) 246:414–422. https://doi.org/10.1016/J.ENVPOL.2018.12.034
77. Li WR, Xie XB, Shi QS, Zeng HY, Ou-Yang YS, Chen YB (2010) Antibacterial activity and mechanism of silver nanoparticles on *Escherichia coli*. Appl Microbiol Biotechnol 85(4):1115–1122. https://doi.org/10.1007/S00253-009-2159-5
78. Lok CN, Ho CM, Chen R, He QY, Yu WY, Sun H, Tam PKH, Chiu JF, Che CM (2006) Proteomic analysis of the mode of antibacterial action of silver nanoparticles. J Proteome Res 5(4):916–924. https://doi.org/10.1021/PR0504079
79. Ma H, Guo Y, Qin Y, Li YY (2018) Nutrient recovery technologies integrated with energy recovery by waste biomass anaerobic digestion. Biores Technol 269:520–531. https://doi.org/10.1016/J.BIORTECH.2018.08.114
80. Mahanthappa M, Kottam N, Yellappa S (2019) Enhanced photocatalytic degradation of methylene blue dye using CuSCdS nanocomposite under visible light irradiation. Appl Surf Sci 475:828–838. https://doi.org/10.1016/J.APSUSC.2018.12.178
81. Mantanis GI, Lykidis C, Papadopoulos AN (2020) Silver nanoparticles for water pollution monitoring and treatments: ecosafety challenge and cellulose-based hybrids solution. Polymers 12(8):1635. https://doi.org/10.3390/POLYM12081635
82. Mall ID, Upadhyay SN, Sharma YC (2007) A review on economical treatment of wastewaters and effluents by adsorption. Int J Environ Stud 51(2):77–124. https://doi.org/10.1080/00207239608711074
83. McGillicuddy E, Murray I, Kavanagh S, Morrison L, Fogarty A, Cormican M, Dockery P, Prendergast M, Rowan N, Morris D (2017) Silver nanoparticles in the environment: sources, detection and ecotoxicology. Sci Total Environ 575:231–246. https://doi.org/10.1016/J.SCITOTENV.2016.10.041

84. Melinte V, Stroea L, Buruiana T, Chibac AL (2019) Photocrosslinked hybrid composites with Ag, Au or Au-Ag NPs as visible-light triggered photocatalysts for degradation/reduction of aromatic nitroderivatives. Eur Polym J. https://doi.org/10.1016/J.EURPOLYMJ.2019.109289
85. Meneses MJC, Partida AH, Mejía JC, Amadeo J, Martínez B, Del Carmen M, Dosta M, Bustos Martínez JA (2018) Silver nanoparticles applications (AgNPS) in aquaculture. IJFAS 6(2):5–11. www.fisheriesjournal.com
86. Mirzajani F, Ghassempour A, Aliahmadi A, Esmaeili MA (2011) Antibacterial effect of silver nanoparticles on *Staphylococcus aureus*. Res Microbiol 162(5):542–549. https://doi.org/10.1016/J.RESMIC.2011.04.009
87. Morones JR, Elechiguerra JL, Camacho A, Holt K, Kouri JB, Ramírez JT, Yacaman MJ (2005) The bactericidal effect of silver nanoparticles. Nanotechnology 16(10):2346–2353. https://doi.org/10.1088/0957-4484/16/10/059
88. Mussin JE, Roldán MV, Rojas F, Sosa M, de LÁ, Pellegri N, Giusiano G (2019) Antifungal activity of silver nanoparticles in combination with ketoconazole against *Malassezia furfur*. Amb Express 9:131–139. https://doi.org/10.1186/s13568-019-0857-7
89. Nakum J, Bhattacharya D (2022) Various green nanomaterials used for wastewater and soil treatment: a mini-review. Front Environ Sci 9:420. https://doi.org/10.3389/FENVS.2021.724814/BIBTEX
90. Navarro E, Wagner B, Odzak N, Sigg L, Behra R (2015) Effects of differently coated silver nanoparticles on the photosynthesis of *Chlamydomonas reinhardtii*. Environ Sci Technol 49(13):8041–8047. https://doi.org/10.1021/ACS.EST.5B01089/SUPPL_FILE/ES5B01089_SI_001.PDF
91. Paladini F, Pollini M (2019) Antimicrobial silver nanoparticles for wound healing application: progress and future trends. Materials 12:2540
92. Pal S, Tak YK, Song JM (2007) Does the antibacterial activity of silver nanoparticles depend on the shape of the nanoparticle? A study of the Gram-negative bacterium *Escherichia coli*. Appl Environ Microbiol 73(6):1712–1720. https://doi.org/10.1128/AEM.02218-06
93. Pandi A, Marichetti Kuppuswami G, Numbi Ramudu K, Palanivel S (2019) A sustainable approach for degradation of leather dyes by a new fungal laccase. J Clean Prod 211:590–597. https://doi.org/10.1016/J.JCLEPRO.2018.11.048
94. Pang C, Brunelli A, Zhu C, Hristozov D, Liu Y, Semenzin E, Wang W, Tao W, Liang J, Marcomini A, Chen C, Zhao B (2016) Demonstrating approaches to chemically modify the surface of Ag nanoparticles in order to influence their cytotoxicity and biodistribution after single dose acute intravenous administration. Nanotoxicology 10(2):129–139. https://doi.org/10.3109/17435390.2015.1024295
95. Peng Y, Gautam L, Hall SW (2019) The detection of drugs of abuse and pharmaceuticals in drinking water using solid-phase extraction and liquid chromatography-mass spectrometry. Chemosphere 223:438–447. https://doi.org/10.1016/J.CHEMOSPHERE.2019.02.040
96. Pem B, Pongrac IM, Ulm L, Pavičić I, Vrček V, Jurašin DD, Ljubojević M, Krivohlavek A, Vrček IV (2019) Toxicity and safety study of silver and gold nanoparticles functionalized with cysteine and glutathione. Beilstein J Nanotechnol 10(1):1802–1817. https://doi.org/10.3762/BJNANO.10.175
97. Pham TL (2019) Effect of silver nanoparticles on tropical freshwater and marine microalgae. J Chem. https://doi.org/10.1155/2019/9658386
98. Pradeep T, Anshup (2009) Noble metal nanoparticles for water purification: a critical review. Thin Solid Films 517(24):6441–6478. https://doi.org/10.1016/J.TSF.2009.03.195
99. Prasath S, Palaniappan K (2019) Is using nanosilver mattresses/pillows safe? A review of potential health implications of silver nanoparticles on human health. Environ Geochem Health 41:2295–2313
100. Pulit-Prociak J, Banach M (2016) Silver nanoparticles: a material of the future...? Open Chem 14(1):76–91. https://doi.org/10.1515/CHEM-2016-0005/ASSET/GRAPHIC/J_CHEM-2016-0005_FIG_004.JPG
101. Pulit-Prociak J, Stokłosa K, Banach M (2015) Nanosilver products and toxicity. Environ Chem Lett 13(1):59–68. https://doi.org/10.1007/S10311-014-0490-2

102. Radwan IM, Gitipour A, Potter PM, Dionysiou DD, Al-Abed SR (2019) Dissolution of silver nanoparticles in colloidal consumer products: effects of particle size and capping agent. J Nanopart Res Interdisc Forum Nanoscale Sci Technol 21(7). https://doi.org/10.1007/S11051-019-4597-Z
103. Rai M, Yadav A, Gade A (2009) Silver nanoparticles as a new generation of antimicrobials. Biotechnol Adv 27(1):76–83. https://doi.org/10.1016/J.BIOTECHADV.2008.09.002
104. Rajasulochana P, Preethy V (2016) Comparison on efficiency of various techniques in treatment of waste and sewage water: a comprehensive review. Resour-Efficient Technol 2(4):175–184. https://doi.org/10.1016/J.REFFIT.2016.09.004
105. Rani P, Kumar V, Singh PP, Matharu AS, Zhang W, Kim KH, Singh J, Rawat M (2020) Highly stable AgNPs prepared via a novel green approach for catalytic and photocatalytic removal of biological and non-biological pollutants. Environ Int 143:105924. https://doi.org/10.1016/J.ENVINT.2020.105924
106. Ratte HT (1999) Bioaccumulation and toxicity of silver compounds: a review. Environ Toxicol Chem 18(1):89–108. https://doi.org/10.1002/ETC.5620180112
107. Ray PC, Yu H, Fu PP (2009) Toxicity and environmental risks of nanomaterials: challenges and future needs. J Environ Sci Health. Part C Environ Carcinog Ecotoxicol Rev 27(1):1–35. https://doi.org/10.1080/10590500802708267
108. Rodriguez-Garraus A, Azqueta A, Vettorazzi A, De Cerain AL (2020) Genotoxicity of silver nanoparticles. Nanomaterials 10:251
109. Roy K, Sarkar CK, Ghosh CK (2015) Rapid colorimetric detection of Hg^{2+} ion by green silver nanoparticles synthesized using *Dahlia pinnata* leaf extract. Green Process Synth 4(6):455–461. https://doi.org/10.1515/GPS-2015-0052/PDF
110. Ruta LL, Banu MA, Neagoe AD, Kissen R, Bones AM, Farcasanu IC (2018) Accumulation of Ag(I) by *Saccharomyces cerevisiae* cells expressing plant metallothioneins. Cells 7:266. https://doi.org/10.3390/cells7120266
111. Sadeghi S, Raki G, Amini A, Mengelizadeh N, Amin MM, Hashemi M (2018) Study of the effectiveness of the third generation polyamideamine and polypropylene imine dendrimers in removal of reactive blue 19 dye from aqueous solutions. Environ Health Eng Manage J 5(4):197–203. https://doi.org/10.15171/EHEM.2018.27
112. Sai Bhargava Reddy M, Ponnamma D, Sadasivuni KK, Kumar B, Abdullah AM (2021) Carbon dioxide adsorption based on porous materials. RSC Adv 11(21):12658–12681. https://doi.org/10.1039/D0RA10902A
113. Salas-Orozco M, Niño-Martínez N, Martínez-Castañón GA, Méndez FT, Jasso MEC, Ruiz F (2019) Mechanisms of resistance to silver nanoparticles in endodontic bacteria: a literature review. J Nanomater. https://doi.org/10.1155/2019/7630316
114. Sathiyaseelan A, Saravanakumar K, Mariadoss AVA, Wang MH (2020) Biocompatible fungal chitosan encapsulated phytogenic silver nanoparticles enhanced antidiabetic, antioxidant and antibacterial activity. Int J Biol Macromol 153:63–71. https://doi.org/10.1016/J.IJBIOMAC.2020.02.291
115. Scartazza A, Picciarelli P, Mariotti L, Curadi M, Barsanti L, Gualtieri P (2017) The role of *Euglena gracilis* paramylon in modulating xylem hormone levels, photosynthesis and water-use efficiency in *Solanum lycopersicum* L. Physiol Plant 161(4):486–501. https://doi.org/10.1111/PPL.12611
116. Shao J, Yu N, Kolwijck E, Wang B, Tan KW, Jansen JA, Walboomers XF, Yang F (2017) Biological evaluation of silver nanoparticles incorporated into chitosan-based membranes. Nanomedicine (Lond) 12(22):2771–2785. https://doi.org/10.2217/NNM-2017-0172
117. Singh S, Kumar V, Romero R, Sharma K, Singh J (2019) Applications of nanoparticles in wastewater treatment. Nanotechnol Life Sci. https://doi.org/10.1007/978-3-030-17061-5_17/COVER
118. Sinha T, Ahmaruzzaman M (2015) High-value utilization of egg shell to synthesize silver and gold-silver core shell nanoparticles and their application for the degradation of hazardous dyes from aqueous phase: a green approach. J Colloid Interface Sci 453:115–131. https://doi.org/10.1016/J.JCIS.2015.04.053

119. Sizmur T, Fresno T, Akgül G, Frost H, Moreno-Jiménez E (2017) Biochar modification to enhance sorption of inorganics from water. Biores Technol 246:34–47. https://doi.org/10.1016/J.BIORTECH.2017.07.082
120. Sreedhar Ready S, Kotaiah B (2005) Decolorization of simulated spent reactive dye bath using solar/TiO_2/H_2O_2. Int J Environ Sci Technol 2(3):245–251. https://doi.org/10.1007/BF03325883/METRICS
121. Supaka N, Juntongjin K, Damronglerd S, Delia ML, Strehaiano P (2004) Microbial decolorization of reactive azo dyes in a sequential anaerobic–aerobic system. Chem Eng J 99(2):169–176. https://doi.org/10.1016/J.CEJ.2003.09.010
122. Susan PM, Mathew B (2018) Treatment of water effluents using silver nanoparticles. Mater Sci Eng Int J 2(5). https://doi.org/10.15406/MSEIJ.2018.02.00050
123. Syafiuddin A, Salmiati S, Hadibarata T, Kueh ABH, Salim MR, Zaini MAA (2018) Silver Nanoparticles in the water environment in Malaysia: inspection, characterization, removal, modeling, and future perspective. Sci Rep 8(1):1–15. https://doi.org/10.1038/s41598-018-19375-1
124. Tang K, Gong C, Wang D (2016) Reduction potential, shadow prices, and pollution costs of agricultural pollutants in China. Sci Total Environ 541:42–50. https://doi.org/10.1016/J.SCITOTENV.2015.09.013
125. Tan L, He M, Song L, Fu X, Shi S (2016) Aerobic decolorization, degradation and detoxification of azo dyes by a newly isolated salt-tolerant yeast *Scheffersomyces spartinae* TLHS-SF1. Biores Technol 203:287–294. https://doi.org/10.1016/J.BIORTECH.2015.12.058
126. Tortella GR, Rubilar O, Durán N, Diez MC, Martínez M, Parada J, Seabra AB (2020) Silver nanoparticles: toxicity in model organisms as an overview of its hazard for human health and the environment. J Hazard Mater. https://doi.org/10.1016/J.JHAZMAT.2019.121974
127. Tran QH, Nguyen VQ, Le AT (2013) Silver nanoparticles: synthesis, properties, toxicology, applications and perspectives. Adv Nat Sci Nanosci Nanotechnol 4(3):033001. https://doi.org/10.1088/2043-6262/4/3/033001
128. Truu M, Ligi T, Nõlvak H, Peeb A, Tiirik K, Devarajan AK, Oopkaup K, Kasemets K, Kõiv-Vainik M, Kasak K, Truu J (2022) Impact of synthetic silver nanoparticles on the biofilm microbial communities and wastewater treatment efficiency in experimental hybrid filter system treating municipal wastewater. J Hazard Mater 440:129721. https://doi.org/10.1016/J.JHAZMAT.2022.129721
129. Türgay O, Ersöz G, Atalay S, Forss J, Welander U (2011) The treatment of azo dyes found in textile industry wastewater by anaerobic biological method and chemical oxidation. Sep Purif Technol 79(1):26–33. https://doi.org/10.1016/J.SEPPUR.2011.03.007
130. Umar K, Parveen T, Khan MA, Ibrahim MNM, Ahmad A, Rafatullah M (2019) Degradation of organic pollutants using metal-doped TiO_2 photocatalysts under visible light: a comparative study. Desalin Water Treat 161:275–282. https://doi.org/10.5004/DWT.2019.24298
131. Wang J, Wang Z, Vieira CLZ, Wolfson JM, Pingtian G, Huang S (2019) Review on the treatment of organic pollutants in water by ultrasonic technology. Ultrason Sonochem 55:273–278. https://doi.org/10.1016/J.ULTSONCH.2019.01.017
132. Wei T, Dong T, Wang Z, Bao J, Tu W, Dai Z (2015) Aggregation of individual sensing units for signal accumulation: conversion of liquid-phase colorimetric assay into enhanced surface-tethered electrochemical analysis. J Am Chem Soc 137(28):8880–8883. https://doi.org/10.1021/JACS.5B04348
133. Woo KJ, Hye CK, Ki WK, Shin S, So HK, Yong HP (2008) Antibacterial activity and mechanism of action of the silver ion in *Staphylococcus aureus* and *Escherichia coli*. Appl Environ Microbiol 74(7):2171–2178. https://doi.org/10.1128/AEM.02001-07
134. Xiao W, Yan B, Zeng H, Liu Q (2016) Dendrimer functionalized graphene oxide for selenium removal. Carbon 105:655–664. https://doi.org/10.1016/J.CARBON.2016.04.057
135. Xie Y, Jin Y, Zeng X, Chen J, Lu Y, Jing K (2015) Fed-batch strategy for enhancing cell growth and C-phycocyanin production of Arthrospira (Spirulina) platensis under phototrophic cultivation. Biores Technol 180:281–287. https://doi.org/10.1016/J.BIORTECH.2014.12.073

136. Yamanaka M, Hara K, Kudo J (2005) Bactericidal actions of a silver ion solution on *Escherichia coli*, studied by energy-filtering transmission electron microscopy and proteomic analysis. Appl Environ Microbiol 71(11):7589–7593. https://doi.org/10.1128/AEM.71.11.7589-7593.2005
137. Yang X, Zhao Z, Yu Y, Shimizu K, Zhang Z, Lei Z, Lee DJ (2020) Enhanced biosorption of Cr(VI) from synthetic wastewater using algal-bacterial aerobic granular sludge: batch experiments, kinetics and mechanisms. Sep Purif Technol 251:117323. https://doi.org/10.1016/J.SEPPUR.2020.117323
138. Yang Y, Zheng S, Li R, Chen X, Wang K, Sun B, Zhang Y, Zhu L (2021) New insights into the facilitated dissolution and sulfidation of silver nanoparticles under simulated sunlight irradiation in aquatic environments by extracellular polymeric substances. Environ Sci Nano 8:748–757
139. Yaqoob AA, Parveen T, Umar K, Ibrahim MNM (2020) Role of nanomaterials in the treatment of wastewater: a review. Water 12(2):495. https://doi.org/10.3390/W12020495
140. Yoon S, Calvo JJ, So MC (2019). Removal of acid orange 7 from aqueous solution by metal-organic frameworks. Crystals 9(1). https://doi.org/10.3390/CRYST9010017
141. Yu Y, Zhou Z, Huang G, Cheng H, Han L, Zhao S, Chen Y, Meng F (2022) Purifying water with silver nanoparticles (AgNPs)-incorporated membranes: Recent advancements and critical challenges. Water Res 222:118901. https://doi.org/10.1016/J.WATRES.2022.118901
142. Zahran M, Khalifa Z, Zahran MAH, Abdel Azzem M (2021) Recent advances in silver nanoparticle-based electrochemical sensors for determining organic pollutants in water: a review. Mater Adv 2(22):7350–7365. https://doi.org/10.1039/D1MA00769F
143. Zhang B, Li B, Wang Z (2020) Creation of carbazole-based fluorescent porous polymers for recognition and detection of various pesticides in water. ACS Sens 5(1):162–170. https://pubmed.ncbi.nlm.nih.gov/31927991/
144. Zhang C, Hu Z, Deng B (2016) Silver nanoparticles in aquatic environments: physiochemical behavior and antimicrobial mechanisms. Water Res 88:403–427. https://doi.org/10.1016/J.WATRES.2015.10.025
145. Zhao B, He M, Chen B, Hu B (2018) Ligand-assisted magnetic solid phase extraction for fast speciation of silver nanoparticles and silver ions in environmental water. Talanta 183:268–275. https://doi.org/10.1016/J.TALANTA.2018.02.081
146. Zhao Y, Zhou Y, Wu X, Wang L, Xu L, Wei S (2012) A facile method for electrospinning of Ag nanoparticles/poly (vinyl alcohol)/carboxymethyl-chitosan nanofibers. ApSS 258(22):8867–8873. https://doi.org/10.1016/J.APSUSC.2012.05.106

Role of Nanoparticles in Air Quality Monitoring with Respect to Toxicity, Virus Detection and Gas Sensing

Pranchal Rajput, Aashna Sinha, Kundan Kumar Chaubey, Chetan Shrivastava, Manish Kushwah, Atreyi Pramanik, Anis Kumar Pal, Sujata Hariharan, Shiv Dayal Pandey, Deen Dayal, Maya Datt Joshi, Mansi Singh, and Sanjesh Kumar

Abstract Nanoparticles (NPs) are particles having less than 100 nm in size with at least one dimension. They may have some physical qualities which are desirable in biological processes and other studies to check the homogeneity, conductivity, or unique optical qualities. The unique physical and chemical characteristics of NPs, including their mechanical, thermal, magnetic, electronic, optical, and catalytic features, make them appropriate for a variety of applications. Literature on silver nanoparticles supports that it may effectively remove bacterial bioaerosols from the air during the air filtering process. Nano-sensors are the best alternative for diagnostic applications since they can detect viruses and bacteria at low concentrations. Three separate mechanisms—surface contact with the filter structure, the inertial force from flowing gas and electric charge produced by particle and filter construction—allow the membrane of the filter to extract toxic materials from the environment. Nanomaterial-enabled sensors are also used for the monitoring of hazardous gases, such as hydrogen sulphide, sulphur dioxide and nitrogen dioxide. In this chapter, we are focussing on various forms of nanoparticles playing a vital role in monitoring and maintenance of air quality as well as detection and cleaning of various toxins, viruses and harmful gases.

P. Rajput · A. Sinha · K. K. Chaubey (✉) · C. Shrivastava · A. Pramanik · A. K. Pal · S. Hariharan
Division of Research and Innovation, Uttaranchal University, Arcadia Grant, P.O. Chandanwari, Premnagar, Dehradun, Uttarakhand 248007, India
e-mail: kundan.chaubey@uttaranchaluniversity.ac.in

M. Kushwah · D. Dayal · M. D. Joshi
Department of Biotechnology, GLA University, Chaumuhan, Mathura, Uttar Pradesh 281406, India

S. D. Pandey
Uttaranchal Institute of Technology, Uttaranchal University, Arcadia Grant, P.O. Chandanwari, Premnagar, Dehradun, Uttarakhand 248007, India

M. Singh · S. Kumar
Institute of Pharmaceutical Research, GLA University, Chaumuhan, Mathura, Uttar Pradesh 281406, India

R. K. Bachheti et al. (eds.), *Nanomaterials for Environmental and Agricultural Sectors*, Smart Nanomaterials Technology,
https://doi.org/10.1007/978-981-99-2874-3_2

Keywords Nanoparticles · Toxicity · Nanotechnology · Air pollution · Gas sensing · Virus detection

1 Introduction

Particles with at least one dimension and a size of less than 100 nm are included in the large category of materials known as nanoparticles (NPs). Depending on the overall form, these materials exist in 0D, 1D, 2D and 3D varieties (Fig. 1). It has been noticed that a substance's size may alter its physiochemical properties, such as its optical capabilities, the relevance of these materials became apparent. The wine red, yellowish grey, black, and deep black colours of the 20-nm NPs of gold (Au), platinum (Pt), silver (Ag), and palladium (Pd) are unique to each metal [1]. NPs are often categorized into three groups based on their composition: organic, carbon-based and inorganic [2]. The distinct physical and chemical features of NPs like mechanical, thermal, magnetic, electronic, optical and catalytic properties make them suitable for a wide range of applications [3–11]. One of the most pressing issues on the planet is air pollution, which is described as a change in the natural structure of the atmosphere brought about by the entry of multiple contaminants (physical, biological and chemical) resulting from industrial activities or human activity [12–15]. Bioaerosols or aerosols are the categories of indoor air pollutants

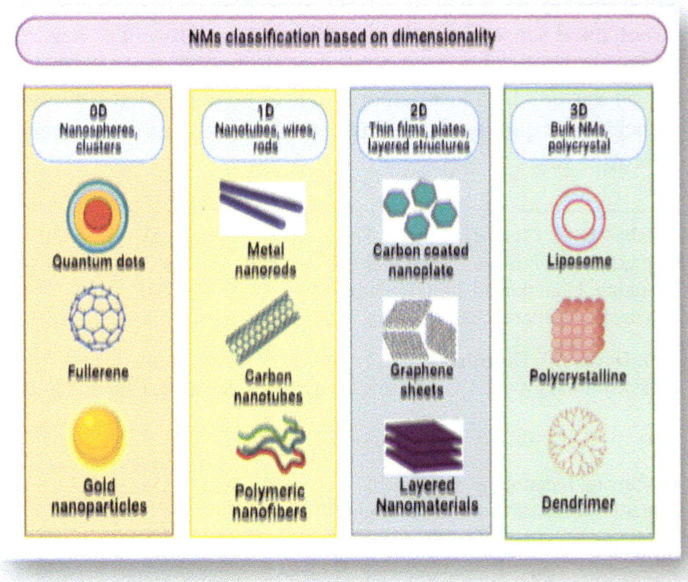

Fig. 1 Diagram showing the relative sizes of nanoparticles with illustrations for each group

having biological origins like viruses, bacteria and fungi which are more attracted to scientific and societal attention. These aerosols can cause a number of ailments, including infections and allergies, and may spread swiftly with airflow [16].

Three separate mechanisms—surface contact with the filter structure, the inertial force from moving gas and electric charge produced by particle and filter construction—allow the membrane of the filter to extract poisons from the environment [17]. The extraordinary photoelectrochemical and conductivity properties of nanomaterials, as well as their portability and simplicity of use, are exploited to improve the efficacy and sensitivity of nano-sensors for the detection of respiratory viruses. Nano-sensors are the best alternative for diagnostic applications since they can detect viruses and bacteria at low concentrations [18]. It's possible that the virus will initially become adsorbed on the metal surface, where it can interact with the metal particles via its glycoproteins and become inactive [19, 20]. Nanomaterial-enabled sensors are also used for the monitoring of hazardous gases, such as hydrogen sulphide, sulphur dioxide and nitrogen dioxide. On the other hand, because of uncertainties and aberrations in size, structure and chemical composition, the existence of some nanoparticles could be harmful to the environment and people's health. Concerns about the transportation and modification of nanoparticles emitted into the environment were consequently raised [21]. Here, we summarized the significant role of the nanoparticles in the monitoring of air quality. So, using NPs we will monitor the air quality and eliminate these air toxins with little harm to the environment.

2 Air Pollutions Present in Environment

Air pollution, which is described as a change in the atmosphere's natural structure brought on by an assortment of pollutants (physical, biological and chemical) coming from industrial processes or human activities, is one of the world's most urgent problems. One of the main environmental issues today is global warming, and air pollutants like carbon dioxide, fuel soot particles, methane, halocarbons, ozone, CO, CFC, heavy metals (As, Cr, Pb, Cd, Hg), hydrocarbons, 23q' nitrogen oxides, organic compounds (VOCs and dioxins), SO_2 and chlorofluorocarbons contribute to this problem [12–15]. When certain pollutants, such as sulphur and nitrogenous oxides from burning, interact with the atmosphere, they leave behind acidic precipitation in the form of rain, fog or snow [22]. A range of ailments, including infections and allergies, can be brought on by bioaerosols or aerosols with biological origins from viruses, bacteria and fungi [16].

3 Nanoparticles Involves in Air Monitoring

Using nanotechnology to minimize air pollution has the potential to have a number of positive environmental effects. The three main aspects can include detection and sensing, remediation and treatment, and pollution prevention [23, 24]. Adsorption on nanomaterials has emerged as a more effective and cost-effective technique due to the enormous surface area of nanoparticles, which may significantly boost the adsorption capacity, as well as the availability of nanomaterials and their capacity for regeneration [25]. Many substances, such as zinc oxide (ZnO), titanium dioxide (TiO_2) and tungsten trioxide (WO_3), are used for air purification and self-cleaning, as well as composites like silver (Ag)-TiO_2-graphene and nanomaterials based on bismuth (Bi) [26]. Apart from carbon- or inorganic-based sources, organic matter is also a source of organic nanomaterials, which can be in the form of micelles, liposomes or polymeric particles. One example of how composite nanoparticles, which are multiphasic in nature, may mix one of their phases with other similar or dissimilar nanoparticles in bulk, complex structures, is a gold nanoparticle in composition with ceramic or a polymer. Carbon-based nanoparticles including fullerenes, nanotubes and nanofibers are frequently seen in two-dimensional morphological shapes [22]. Several kinds of nano-structured materials and their uses are demonstrated in Table 1.

Air filtration technology, which employs antimicrobial components including silver nanoparticles, copper nanoparticles, CNTs and natural objects, is the most popular and effective way to remove bioaerosols through ventilation methods. Many researchers have suggested that the air filtering mechanism may remove bacterial bioaerosols from the air with silver nanoparticles. Several factors that affect the

Table 1 Types of nano-structured materials and functions

Sr. no.	Nano-structural materials	Function	References
1	Silver nanoparticle array membranes	Water quantity monitoring	[27]
2	Carbon nanotubes (CNTs)	Electrochemical sensors	[28]
3	CNTs as a building block	The electrical resistance of CNTs varies significantly when exposed to gases like NO_2, NH_3 or O, either as a result of physical adsorption	[29]
4	CNTs with enzymes	Create a rapid electron transfer from an electrode to the enzyme's active site using a CNT, often boosting the electrochemical activity of the biomolecule	[30]
5	CNT sensors	Developed for the investigation of the sequence-specific DNA, ethanol, glucose and sulphur	[31]
6	Magnetic nanoparticles	Useful for quickly detecting microorganisms in complicated matrix	[32]

Table 2 Various types of nanoparticles are used for the detection and cleaning of different gases

Sr. no.	Nanoparticle	Feature	References
1	Carbon-based nano-adsorbents	Utilized to collect carbon dioxide at a high temperature based on the reversible carbonation reaction of calcium oxides (CaO)	[35]
2	Titanium dioxide (TiO_2)-coated nano-calcium carbonate	Effectively absorb carbon dioxide while preventing the sintering of nano-calcium carbonate	[36]
3	Alkali-metal-based nanoparticles	Demonstrated the capacity to absorb carbon dioxide at high temperatures	[37]
4	Lithium ortho-silicate based	Showed a sintering-resistant quality and absorbed carbon dioxide (CO_2) at 700 °C	[37]
5	Carbon nanotubes (CNTs)	Utilized to increase adsorption effectiveness when there is moisture and to absorb CO_2	[38]
6	Metallic nickel nanoparticles	Used as catalysts in the thermal breakdown of methane to create hydrogen	[39]
7	Titanate nanotubes	The photocatalytic oxidation of NO_x has been extensively documented	[40]
8	Carbon nanostructures like graphene nanosheets	Used to boost TiO_2's photocatalytic efficiency	[41]
9	Activated carbon-coated iron nanoparticles	The SO_2 adsorption process includes formation of weakly acidic groups on the adsorbent surface, increasing the surface acidity	[42]
10	Polyacrylonitrile (PAN)-based carbon nanofiber (CNF)	At a low concentration, a sizable amount of formaldehyde was adsorbed	[33]

antibacterial effect of silver nanoparticles include the kind of bacteria, concentration, relative humidity (RH), size distribution and exposure time [33, 34]. The functionalities of many types of nano-structured materials are listed in Table 2.

3.1 Nanoparticles Used in Monitoring of Toxins in Air

Because of their toxin, mutagenic and carcinogenic potential, poly-aromatic hydrocarbons are a type of persistent and widespread environmental pollutants that are also harmful to people. Both the demand for and the emissions from petroleum products have increased. Burning organic resources like coal and firewood when there is a lack of oxygen is one of the contributing factors [22]. Filters are helpful in the fight against air pollution. The porous structure of a filter is constrained, allowing gas to pass through while trapping particles inside where they belong. The membrane of the filter removes toxins by three different mechanisms: surface contact with the filter structure, inertial force provided by flowing gas and electric charge generated by particle and filter construction [17]. A nanosensor is a cutting-edge device that can detect and react to a physical change that can be felt on a nano-scale. Little 3D circular molecules known as nanoparticles are used as sensors to find poisons in the environment [43]. The production of semiconductors and optoelectronic devices frequently uses isopropyl alcohol (IPA). Due to a lack of air pollution controls, IPA vapour is untreated and discharged into the atmosphere. Because it is unpleasant and carcinogenic, the discharge of IPA vapour can be harmful to human health. When utilized as an adsorbent to absorb IPA vapour, HNO_3 and NaClO solution oxidized SWNTs [24].

3.2 Nanoparticles Used in Virus Detection in Air

Several sensor technologies are now being investigated to detect viruses, and nanomaterial-based approaches have emerged as viable options. They have demonstrated a considerable improvement in detection devices termed nano-sensors. The sensors work by combining interacting recognition elements with the sensing system, which enables very accurate and sensitive target identification. Nanomaterials' remarkable photoelectrochemical and conductivity capabilities, portability and ease of use are taken advantage of to increase the effectiveness and sensitivity of nano-sensors for the detection of respiratory viruses. The greatest option for diagnostic purposes is the use of nano-sensors since they can identify viruses and bacteria at low concentrations (Fig. 2) [18]. It's possible that the virus will initially become adsorbed on the metal surface, where it can interact with the metal particles via its glycoproteins and become inactive. Another possibility is that metal nanoparticles could enter cells and interact with viral nucleic acids to activate their antiviral capabilities. Nano-aggregates have been shown to be among the most frequently used antibacterial materials due to their enormous specific surface area and unique chemical and physical features (Fig. 3) [44].

The sharp edges of GO may harm and inactivate virus protein casings during the virus disinfection process [19, 20]. It has been shown that GO has broad-spectrum antiviral activity against the DNA virus that causes pseudorabies and the swine epidemic diarrhoea virus (PEDV, RNA virus). A study showed that GO can considerably reduce PRV and PEDV infection at doses below cytotoxicity. The non-ionic polymer PVP and GO together show antiviral action (Fig. 4) [44].

3.3 Nanoparticles Used in Gas Sensing in Air

The monitoring of dangerous gases, including hydrogen sulphide, sulphur dioxide and nitrogen dioxide, also uses nanomaterial-enabled sensors. Conversely, the existence of certain nanoparticles might harm the environment and people's health

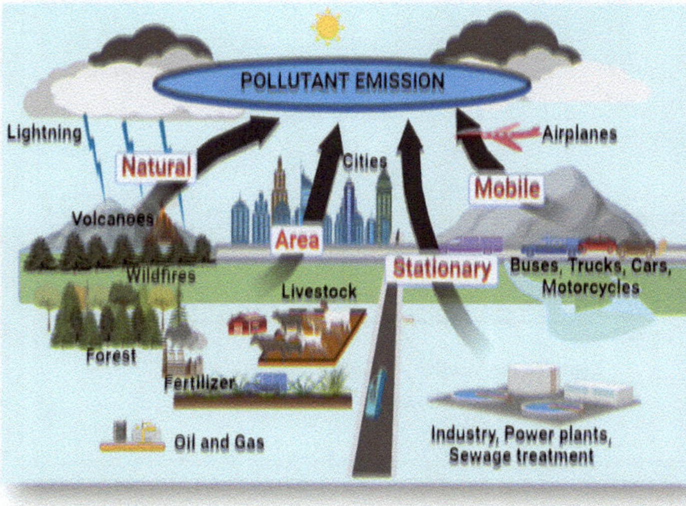

Fig. 2 Different types of air contaminants

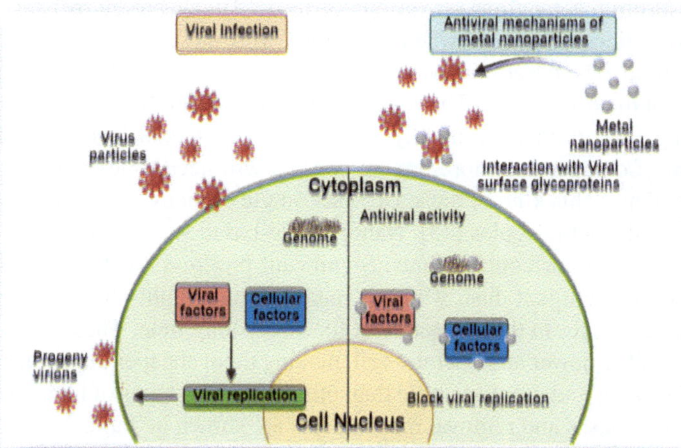

Fig. 3 Eukaryotic cell attack by a virus and the antiviral action of metal nanoparticles are shown schematically

due to uncertainties and aberrations in size, structure and chemical composition. As a result, worries concerning the movement and transformation of nanoparticles released into the environment were raised [21]. The production of nano-sensors suitable for continuous detection will be made possible by advances in nanoelectronics.

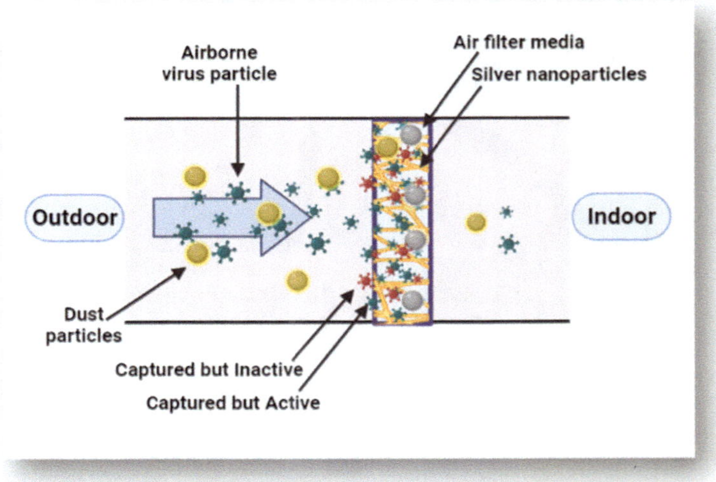

Fig. 4 A nano-Ag-loaded filter's schematic illustration for removing viruses from the air

Variable photocatalytic conversion pathways might result from air contaminants with different adsorption activations on various surfaces. The electrical resistance of the nanotubes is noticeably altered when in contact with these gases, and this change is then measured. The possibility of an unpleasant and dangerous leak is one of the risks in the commercial sector. These gases have lately been found in trace concentrations using sensors that are mostly utilized at commercial levels. For this, the development of quicker and more precise sensors will need to leverage technological developments. This kind of biosensor uses single or many membrane nanotubes to catch harmful gas molecules [43]. Nanoparticle-based sensors are useful tool for promptly identifying air contaminants. Significant progress has been made in this field because of intelligent dust, which is made of many light-computerized nanosensors and may linger in the environment for a very long time. These nano-sensors use extremely little power and operate well, making them cheaper than other types of sensors and smaller and more accurate than others. Various types of nano-adsorptive materials, target gases and removal mechanisms are shown in Table 3.

4 Future Prospects

Nanoparticles have altered the legal strategy for future environmental and technological cases. Together with human progress in nature, industrialization and resource consumption have increased. Although the negative effects of older methods, such as adsorption, are not regretted, it is urgently important to eliminate these toxins with minimal harm to the environment [22]. It can contribute in the creation of innovative

Table 3 Types of nano-adsorptive materials, target gases and removal mechanisms

Nano-adsorptive material	Types of nanoparticles	Target pollutant gases	Removal mechanism	Key references
Carbon nanotubes (CNTs)	(SWNTs and MWNTs)	NO_x (mixture of NO and NO_2)	CNTs allow the passage of NO and O_2, with NO being converted to NO_2 and subsequently absorbing on the surface of nitrite species	[45]
	Modified CNTs utilizing 3-aminopropyltriethoxysilane are known as (CNTs-APTS) (APTS)	CO_2	The many amine groups on the surface of CNTs create a large number of chemical site for CO_2 adsorption, allowing CNTs to absorb more CO_2 genes at low temperature (20–100 °C)	[38]
	NaClO/SWNTs	Isopropyl vapour	Chemical adsorption onto surface functional groups of the absorbent and physical adsorption via van der Waals forces	Hsu and Lu [46]
	CNTs deposited on quartz filters	VOSs	It is carried out by π–π interactions	Amade et al. (2014)
	Boron-doped SWCNTs and Si-doped	CO and CH_3OH gases	The electronic characteristics of SWCNT considerably enhance the gas adsorption, whether it be through physisorption or chemisorption	[47]

(continued)

Table 3 (continued)

Nano-adsorptive material	Types of nanoparticles	Target pollutant gases	Removal mechanism	Key references
Fullerene	Fullerene B40	CO_2	Strong chemisorptions with high CO_2 adsorption capability	Dong et al. (2015)
	Fullerene-like boron nitride nanocage	N_2O	Adsorption and decomposition of N_2O	Esrafili (2017)
Graphene	Nanocomposites/graphene oxide (GO)	NH_3, SO_2, H_2S, CO_2 and N_2	Adsorption of gases is caused by functional groups on GO, which is aided by the interaction of metal centres with GO surface	Petit and Bandosz [48], Mabayoje et al. [49], Wu [50]

cleanup methods and improve the sensing and detection of toxins. Nanomaterials make good adsorbents, catalysts and sensors due to their large specific surface areas and high reactivates. Most nanocatalysts for long-term air filtration are still in development or are only now becoming usable. The results of further developments in nanocatalysis and the variety of applications in the environment show the value of this area for future investments in catalysis research. It will be important to develop improved processes for catalyst scale-up and large-scale manufacture in order to broadly allow the entry of nano-engineered materials into new commercial catalyst applications. Its development ought to be encouraged by scientifically established processes for creating and managing composite particle formations. As a result, there are currently several nanomaterials and applications on the market, and many more are projected in the upcoming years. The ultimate goal of future nanotechnology is to ensure continued optimal utilization of gaseous waste by having the outputs of one manufacturing process becomes the inputs for another one in the industrial process chain [14].

5 Conclusion

Nanoparticles' various forms are available for monitoring of air quality and are used for the detection and cleaning of various toxins present in the air, it is also used for the detection of various types of biological particles like viruses, bacteria and fungi and also used for gas sensing and cleaning of various harmful gases like CO_2,

NO_2, SO_2, HNO_3, etc. Nanoparticle applications in cutting-edge sensing technologies enable the identification of pollutants in smaller concentrations. It is feasible to create highly focussed, sensitive and selective sensors by coating nanoparticles with different ligands and altering their shape to vary the surface area-to-volume ratio and with the development of an ecologically benign chemical or material, the replacement of commonly used harmful compounds and decreased environmental effect, nanotechnology is also utilized to avoid the production of pollutants or contaminants. The advancement of quick and precise environmental processes to reduce or avoid emissions or transform toxins into beneficial by-products is another area in which nanotechnology holds great promise.

References

1. Khan I, Saeed K, Khan I (2019) Nanoparticles: properties, applications and toxicities. Arab J Chem 12(7):908–931
2. Ealias AM, Saravanakumar MP (2017) A review on the classification, characterisation, synthesis of nanoparticles and their application. MS&E 263(3):032019
3. Husen A, Rahman QI, Iqbal M, Yassin MO, Bachheti RK (2019) Plant-mediated fabrication of gold nanoparticles and their applications. In: Nanomaterials and plant potential. Springer, Cham, pp 71–110. https://doi.org/10.1007/978-3-030-05569-1_3
4. Husen A (2023b) Smart nanomaterials from agricultural and horticultural products. Springer Nature Singapore Pte Ltd., 152 Beach Road, #21–01/04 Gateway East, Singapore 189721, Singapore
5. Husen A (2023a) Secondary metabolites based green synthesis of nanomaterials and their applications. Springer Nature Singapore Pte Ltd., 152 Beach Road, #21–01/04 Gateway East, Singapore 189721, Singapore
6. Husen A, Siddiqi KS (2023) Advances in smart nanomaterials and their applications. Elsevier Inc., 50 Hampshire St., 5th Floor, Cambridge, MA 02139, USA
7. Husen A (2022) Engineered nanomaterials for sustainable agricultural production, soil improvement and stress management. Elsevier Inc., 50 Hampshire St., 5th Floor, Cambridge, MA 02139, USA
8. Husen, A, Iqbal, M (2019) Nanomaterials and plant potential. Springer International Publishing AG, Gewerbestrasse 11, 6330 Cham, Switzerland. https://doi.org/10.1007/978-3-030-05569-1
9. Jin-Chul K, Madhusudhan A, Husen A (2021) Smart nanomaterials in biomedical applications. Springer Nature Switzerland AG, Gewerbestrasse 11, 6330 Cham, Switzerland
10. Hamid R. Taghiyari, Jeffrey J Morrell, Husen A (2022) Emerging nanomaterials (Opportunities and challenges in forestry sectors). Springer Nature Switzerland AG, Gewerbestrasse 11, 6330 Cham, Switzerland. https://doi.org/10.1007/978-3-031-17378-3
11. Joudeh N, Linke D (2022) Nanoparticle classification, physicochemical properties, characterization, and applications: a comprehensive review for biologists. J Nanobiotechnol 20(1):1–29
12. Araújo IPS, Costa DB, de Moraes RJB (2014) Identification and characterization of particulate matter concentrations at construction jobsites. Sustainability 6(11):7666–7688
13. Daly A, Zannetti P (2007) An introduction to air pollution—definitions, classifications, and history. In: Zannetti P, Al-Ajmi D, Al-Rashied S (eds) Chapter 1 of Ambient air pollution. The Arab School for Science and Technology (ASST) and The Enviro Comp Institute
14. Mohamed EF (2017) Nanotechnology: future of environmental air pollution control. Environ Manag Sustain Dev 6(2):429

15. Yu BF, Hu ZB, Liu M, Yang HL, Kong QX, Liu YH (2009) Review of research on air-conditioning systems and indoor air quality control for human health. Int J Refrig 32(1):3–20
16. Stark PC, Burge HA, Ryan LM, Milton DK, Gold DR (2012) Fungal levels in the home and lower respiratory tract illnesses in the first year of life. Ats J 168(2):232–237
17. Zhu M, Han J, Wang F, Shao W, Xiong R, Zhang Q et al (2017) Electrospun nanofibers membranes for effective air filtration. Macromol Mater Eng 302(1):1600353
18. Naikoo GA, Awan T, Hassan IU, Salim H, Arshad F, Ahmed W et al (2021) Nanomaterials-based sensors for respiratory viral detection: a review. IEEE Sens J 21(16):17643
19. Ye S, Shao K, Li Z, Guo N, Zuo Y, Li Q et al (2015) Antiviral activity of graphene oxide: how sharp edged structure and charge matter. ACS Appl Mater Interfaces 7(38):21578–21579
20. Ye Y, Guo Y, Yue Y, Zhang Y (2015) Facile colorimetric detection of nitrite based on anti-aggregation of gold nanoparticles. Anal Methods 7(10):4090–4096
21. Saleem H, Zaidi SJ, Ismail AF, Goh PS (2022) Advances of nanomaterials for air pollution remediation and their impacts on the environment. Chemosphere 287:132083
22. Roy A, Sharma A, Yadav S, Jule LT, Krishnaraj R (2021) Nanomaterials for remediation of environmental pollutants. Bioinorg Chem 2021
23. Yadav K, Singh J, Gupta N, Kumar V (2017) A review of nanobioremediation technologies for environmental cleanup: a novel biological approach
24. Yunus IS, Harwin, Kurniawan A, Adityawarman D, Indarto A (2013) Nanotechnologies in water and air pollution treatment. 1(1):136–148
25. Ibrahim RK, Hayyan M, AlSaadi MA, Hayyan A, Ibrahim S (2016) Environmental application of nanotechnology: air, soil, and water. Environ Sci Pollut Res 23(14):13754–13788
26. Rafique MS, Tahir MB, Rafique M, Shakil M (2020) Photocatalytic nanomaterials for air purification and self-cleaning. Nanotechnol Photocatal Environ Appl 203–219
27. Mantanis GI, Lykidis C, Papadopoulos AN (2020) Silver nanoparticles for water pollution monitoring and treatments: ecosafety challenge and cellulose-based hybrids solution. Polymers 12(8):1635
28. Zaporotskova IV, Boroznina NP, Parkhomenko YN, Kozhitov LV (2016) Carbon nanotubes: sensor properties. A review. Mod Electron Mater 2(4):95–105
29. Norizan MN, Moklis MH, Ngah Demon SZ, Halim NA, Samsuri A, Mohamad IS et al (2020) Carbon nanotubes: functionalisation and their application in chemical sensors. RSC Adv 10(71):43704
30. Yang C, Denno ME, Pyakurel P, Venton BJ (2015) Recent trends in carbon nanomaterial-based electrochemical sensors for biomolecules: a review. Anal Chim Acta 887:17
31. Schroeder V, Savagatrup S, He M, Lin S, Swager TM (2019) Carbon nanotube chemical sensors. Chem Rev 119(1):599
32. Han H, Sohn B, Choi J, Jeon S (2021) Recent advances in magnetic nanoparticle-based microfluidic devices for the pretreatment of pathogenic bacteria. Biomed Eng Lett 11(4):297
33. Lee BU, Yun SH, Jung JH, Bae GN (2010) Effect of relative humidity and variation of particle number size distribution on the inactivation effectiveness of airborne silver nanoparticles against bacteria bioaerosols deposited on a filter. J Aerosol Sci 41(5):447–456
34. Lee KJ, Shiratori N, Lee GH, Miyawaki J, Mochida I, Yoon SH et al (2010) Activated carbon nanofiber produced from electrospun polyacrylonitrile nanofiber as a highly efficient formaldehyde adsorbent. Carbon N Y 48(15):4248–4255
35. Abanades JC, Alvarez D (2003) Conversion limits in the reaction of CO_2 with lime. Energy Fuels 17(2):308–315
36. Wang Y, Zhu Y, Wu S (2013) A new nano CaO-based CO2 adsorbent prepared using an adsorption phase technique. Chem Eng J F 218:39–45
37. Wang K, Zhao P, Guo X, Li Y, Han D, Chao Y (2014) Enhancement of reactivity in Li4SiO4-based sorbents from the nano-sized rice husk ash for high-temperature CO2 capture. Energy Convers Manag 81:447–454
38. Su F, Lu C, Cnen W, Bai H, Hwang JF (2009) Capture of CO2 from flue gas via multiwalled carbon nanotubes. Sci Total Environ 407(8):3017–3023

39. Wang HY, Lua AC (2012) Development of metallic nickel nanoparticle catalyst for the decomposition of methane into hydrogen and carbon nanofibers. J Phys Chem C 116(51):26765–26775
40. Nguyen NH, Bai H (2015) Effect of washing pH on the properties of titanate nanotubes and its activity for photocatalytic oxidation of NO and NO2. Appl Surf Sci 355:672–680
41. Low J, Cheng B, Yu J (2017) Surface modification and enhanced photocatalytic CO2 reduction performance of TiO2: a review. Appl Surf Sci 392:658–686
42. Arcibar-Orozco JA, Rangel-Mendez JR, Bandosz TJ (2013) Reactive adsorption of SO2 on activated carbons with deposited iron nanoparticles. J Hazard Mater 246–247:300–309
43. Peng F, Su Y, Zhong Y, Fan C, Lee ST, He Y (2014) Silicon nanomaterials platform for bioimaging, biosensing, and cancer therapy. Acc Chem Res 47(2):612–623
44. Li R, Cui L, Chen M, Huang Y (2021) Nanomaterials for airborne virus inactivation: a short review. Aerosol Sci Eng 5(1):1
45. Zhang Y, Yuan S, Feng X, Li H, Zhou J, Wang B (2016) Preparation of nanofibrous metal-organic framework filters for efficient air pollution control. J Am Chem Soc 138(18):5785–5788
46. Hsu CL, Lu HP (2007) Consumer behavior in online game communities: a motivational factor perspective. Comput Hum Behav 23(3):1642–1659
47. Azam MA, Alias FM, Tack LW, Seman RNAR, Taib MFM (2017) Electronic properties and gas adsorption behaviour of pristine, silicon-, and boron-doped (8, 0) single-walled carbon nanotube: a first principles study. J Mol Graph Model 75:85–93
48. Petit C, Bandosz TJ (2009) MOF–graphite oxide composites: combining the uniqueness of graphene layers and metal–organic frameworks. Adv Mater 21(46):4753–4757
49. Mabayoje O, Seredych M, Bandosz TJ (2012) Enhanced reactive adsorption of hydrogen sulfide on the composites of graphene/graphite oxide with copper (hydr) oxychlorides. ACS Appl Mater Interfaces 4(6):3316–3324
50. Wu J (2013) Landscape sustainability science: ecosystem services and human well-being in changing landscapes. Landscape Ecol 28:999–1023

Past, Present and Possible Future Application of Nanoparticle in Contaminated Soil Remediation

Sapna Yadav, Aashna Sinha, Atreyi Pramanik, Shivani Tyagi, Chetan Shrivastava, Pranchal Rajput, Anis Kumar Pal, Kundan Kumar Chaubey, Sujata Jayaraman, Manish Kushwah, Deen Dayal, Deepak Kumar Verma, Rajesh Bahuguna, Shalini Sharma, and Maya Datt Joshi

Abstract Many types of pollution have long threatened soil, the base of ecology. The distinctive physicochemical and biological properties of nanotechnology (NT), which differ from those in a large-scale model for the same substance, have drawn the attention of numerous scientists. Nanoparticles (NPs) have many uses in industries, including health, medication delivery, electronics, fuel cells, solar cells, food, space exploration and more. NPs have demonstrated several advantages for treating various soil contaminants in these applications. The development of NPs might pave the way for a more environmentally friendly and sustainable method of removing hazardous substances from contaminated soils more quickly. Numerous initiatives have been taken to boost the effectiveness of phytoremediation, including the use of rhizobacteria, genetic engineering and chemical additions. Herein, a combination of NT and bioremediation has opened up new possibilities for modernizing the

S. Yadav · A. Sinha · A. Pramanik · C. Shrivastava · P. Rajput · A. K. Pal · K. K. Chaubey (✉) · S. Jayaraman
Division of Research and Innovation, School of Applied and Life Sciences, Uttaranchal University, P.O. Chandanwari, Prem Nagar, Arcadia Grant, Dehradun, Uttarakhand 248007, India
e-mail: kundan2006chaubey@gmail.com

S. Tyagi · M. Kushwah · D. Dayal · M. D. Joshi
Department of Biotechnology, Institute of Applied Sciences and Humanities, GLA University, Mathura, UP 281406, India

D. K. Verma
School of Basic and Applied Sciences, Sanskriti University, Uttar Pradesh, Chhata, District Mathura 281401, India

R. Bahuguna
Law College, Uttaranchal University, Prem Nagar, P.O. Chandanwari, Arcadia Grant, Dehradun, Uttarakhand 248007, India

S. Sharma
Department of Veterinary Physiology and Biochemistry, Lala Lajpat Rai University of Veterinary and Animal Sciences, Hisar, Haryana 125004, India

R. K. Bachheti et al. (eds.), *Nanomaterials for Environmental and Agricultural Sectors*, Smart Nanomaterials Technology,
https://doi.org/10.1007/978-981-99-2874-3_3

remedy processes. Since the nanoscale process enhances the adsorption and degradation of contaminants, innovative remediation systems integrate nano-technological and biological remediation strategies. Due to their special surface qualities, NPs may absorb/adsorb a wide range of pollutants and accelerate processes by reducing the energy needed to break them down. Consequently, this remediation procedure prevents contaminants from building up and spreading from one medium to another. This chapter covers all potential uses of NPs in the rehabilitation of polluted soils and any related environmental issues.

Keywords Nanoparticles · Nanotechnology · Soil remediation · Heavy metals · Zero-valent iron (nZVI) · Phytoremediation

1 Introduction

Land contamination challenges sustainable development, quality of life and livelihoods [48]. Many trace elements, rare earth elements, and other raw minerals are mined from specific areas and used to make commonplace things consumed and eventually thrown away worldwide, such as paints, construction materials, and electrical and electronic equipment. As a result, these pollutants are distributed differently across the ecosystem, changing their normal distribution. Particularly in soils near the primary sources of pollutants, such as mining, industrial, and urban regions, close to transportation networks, landfills; and unofficial recycling zones, higher concentrations than the natural values can be discovered [26, 34, 48, 62]. Due to the demands of an expanding population and the reduction of arable land due to technological activity, soil conservation is thus of the highest importance for the contemporary period.

NT creates a global opportunity for the efficient remediation or restoration of damaged soil [52, 60]. The use of nanotechnologies in repairing contaminated soils is now a growing field with enormous potential to enhance the effectiveness of conventional remediation technologies. However, when NTs are introduced to ecosystems, environmental worries have also surfaced addressing both human and environmental health. Additionally, soils are one of the many environmental matrices where NT has been identified as a viable approach for the remediation of contaminants [60]. One of the key areas where nano-technological methods have been applied extensively in this perspective is soil remediation. As a result of their huge specific surface area, NPs have recently been used to eliminate pollutants in several methods, including adsorption, redox reactions, precipitation, and co-precipitation [7, 39, 59]. NT employed a variety of small particles with properties such as high specific surface area, high reactivity, and flexibility, ranging in size from 1 to 100 nano-meters. In Fig. 2, several kinds of nanomaterials are displayed. Nanomaterials can remove toxins from soil, water, and air due to their diverse characteristics. The large specific surface area of NPs greatly improves the efficiency of the decontamination process. The nanometre size of the particles increased their capacity to penetrate polluted soil. It was explored if using NT in clean-up would be very effective, affordable, flexible, and environmentally friendly [21]. This chapter discusses the potential applications of NPs in the remediation of polluted soils and any related environmental problems.

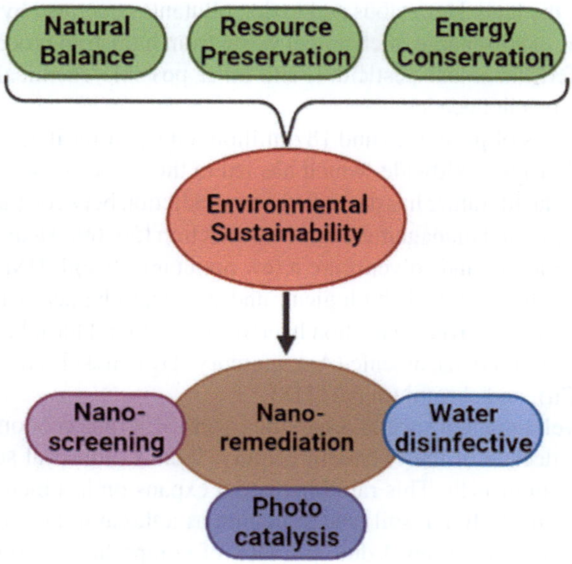

Fig. 1 Cycle of nanoparticles in environment

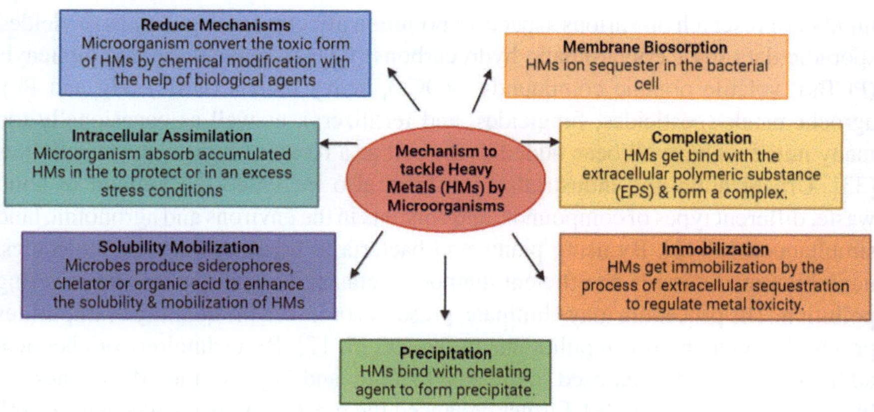

Fig. 2 Mechanisms to tackle HMs

2 Sources of Soil Pollution

One of the essential components of the environment that supports life's ecosystem is soil. However, due to urbanization, industrialization has expanded quickly, negatively affecting the land. Hazardous and toxic pollutants generated by anthropogenic sources, such as the discharge of chemical waste from industrial processes, the abandoned use of fertilizers and pesticides, and other possible chemical sources, have polluted soil in recent decades.

4 million tonnes of pesticides and 187 million tonnes of fertilizer are used yearly for agricultural crops worldwide, which has led to the causes of soil pollution [75]. For a long time, the literature has described a contradiction between the advantages of viable agricultural land management and its protection [56, 69]. Heavy metals (HM), pesticides, mineral oil, and solvents are a few pollutants in soil. HM concentrations adversely affect the chemical, biological, and physical characteristics of soil and plant development [9]. Over 20 million hectares (ha) of land have been poisoned by HM such as cadmium (Cd), arsenic (As), mercury (Hg), nickel (Ni), zinc (Zn), lead (Pb), copper (Cu), and chromium (Cr) [18].

With the development of special economic zones (SEZ) and economic law reforms since the latter decade of the twentieth century, India's industrial sector has experienced significant growth. This rapid industrial expansion has increased the threat to the environment. In India, soil contamination as a threat to human life is mainly overlooked at the national level due to a lack of comprehensive understanding of the subject, despite the fact that remediating it is far more difficult than remediating polluted air and water.

Despite the need for a coordinated national effort to analyse soil contamination, a number of research on various aspects of pollution affecting soil quality have yielded sporadic data [64]. Polyaromatic hydrocarbons (PAHs), polychlorinated biphenyls (PCBs), volatile organic compounds (VOCs), heavy metals (HMs) (Hg and Pb), agrochemicals (pesticides, fungicides, and fertilizers), as well as occasionally too many nutrients have all been added to the soil as a result of human-made activities [33]. Urbanization and industrialization have also increased the amount of solid waste, different types of compounds, and solvents in the environs and agronomic land simultaneously [42]. By using plants and bacteria to break down toxic molecules, nano-bioremediation is an efficient method of enhancing soil quality and lowering pollution. The procedure may eliminate, preserve, or lessen the quantity of impurities present by breaking down pollutants in the soil [5, 17]. Biotechnology or chemical additives have both been used in the past to study and improve the effectiveness of bioremediation [28], but NT further enhanced the practice with a newer aspect [32].

3 Heavy Metals in Soil

Increased human activity is primarily to blame for the presence of HMs in the environment. However, HMs are released into the environment as contaminants due to disrupted biogeochemical cycles. There are no biological tasks for elements like As, Cd, Cr, Hg, and Pb to carry out in the biological system. Since heavy metals have significant harmful effects on biota even at low concentrations, they make up a significant portion of inorganic contaminants [3, 40]. The bioavailability and absorption of HMs also affect how harmful they are [61]. Acidic conditions cause HM toxicity, mainly if the soil organization is weak and there are few nutrients present (e.g. mining areas) [54].

Heavy metals predominantly impact plants and lower soil creatures by causing the production of reactive oxygen species (ROS), which further leads to the degradation of macromolecules like proteins and nucleic acids [40]. Crops, vegetation, their nutritional value, and the ecological elements related to them are all impacted by the presence of HMs in the soil. HMs have different effects on crops based on the kind of crop, the soil's physicochemical properties, and the HM [47]. The production of ROS, which has an impact on the cell organelles, macromolecules like proteins and nucleic acids, and other elements of the plant's structure and function, is the general mechanism by which HMs are harmful to agricultural plants [37, 47]. Additionally, it has been shown to have an impact on respiration and photosynthesis,decrease enzyme activities; increase oxidative stress; decrease biomass; lower crop production; and change the variety, activity, and genetic make-up of beneficial soil micro-flora [4, 23].

Among the most hazardous elements in the environment are As, Cd, Cr, Hg, and Pb, according to the Environmental Protection Agency (EPA) [79]. Furthermore, only four heavy metals are of particular worry about human health, according to the US, Agency for Toxic Substances and Disease Registry (ATSDR): The four most frequent causes of acute heavy metal poisoning are As, Pb, Cd, and Hg. As is the most prevalent cause of acute heavy metal poisoning and hence ranks first on the ATSDR's "top 20 list", Pb is second, and Cd is third [24, 25]. Excessive Cr levels in soil, for example, prevent critical elements like phosphorus (K), calcium (Ca), iron (Fe), and magnesium (Mg), by forming insoluble compounds and obscuring the absorption site from being absorbed [1, 80]. The presence of humic material in soils has a significant impact on chemical sorption [68]. Because humic and fulvic acids can exist in a dissociated state, they are more significant than the interlayer gaps in montmorillonite clay. As a result, the external surface features of clays are far more significant, for example, clay's inner particle crystal structure is better at binding organic stuff than its outer particle crystal structure [41]. Much recent research has examined the effectiveness of metal NPs for soil remediation. Jalali et.al investigated that using SiO_2 NPs, two polluted non-calcareous and calcareous soils were tested for zinc (Zn), nickel (Ni), and cadmium (Cd) [55]. The reduction of Cd was greatest with 3 percent SiO_2 (56.1%) and 1 percent Al_2O_3 (38.3%) in both calcareous and non-calcareous soils, according to the results [58].

4 Soil Remediation

Contaminated soil is the consequence of human-made pollution contaminating the ecosystem of natural soil. Compromise soil may have disastrous environmental repercussions, whether it is caused by industrial activities, agricultural chemicals, or other carelessly handled trash. Heavy metals, which may persist in soil for decades, represent the ultimate pollution sink in soil. When contaminants enter the soil system, they transform into more stable solid phases and remain in the organic phase of the soil [45]. Removing heavy metals (HMs) from soils is necessary to protect the environment and living things [30]. HMs must be left in the soil for a long time since physiochemical processes cannot break them down because they are non-biodegradable. Previous studies have discovered that plants' uptake of more nutrients and water from the soil may make it possible for HMs to enter food chains [2]. Adsorption, Fenton, immobilization, Fenton-like oxidation, reduction reactions, and various combinations of NT and bioremediation are the principal soil remediation strategies utilized today [72]. The mechanism of NT and bioremediation has arisen as a great concern recently. Table 1 presents a summary of nanomaterials and NT applied for in situ removal of contaminants from soils, including heavy metals, organic compounds, and metalloids. The removal of heavy metals and organic compounds from contaminated soil and water and groundwater pollution has a substantial potential with the application of nano-zero-valent iron NPs. The properties of the contaminants and the condition of the soils heavily influenced the nanomaterials used for immobilization treatment. The immobilization of organic and inorganic contaminants in soil settings has been widely accomplished using several nanoparticle additives [57]. The use of nanoscale iron oxides in water treatment and environmental remediation has attracted interest recently due to its potent ability to adsorb several significant water contaminants, such as As. The study of native soil bacteria demands a different approach than the majority of toxicological studies. In a recent experiment, 1 mg/L Ag NPs (20–30 nm in K-media) were administered to the worm *Caenorhabditis elegans* [15, 38].

On the other hand, soil-dwelling microbes are probably well adapted to certain soil conditions, such as concentrations of [51] metals. Metals and metalloids in contaminated soil can move to other ecological compartments through various solubilization, mobilization, and transport mechanisms. The components must be in a free condition for microbial or fungal access. The pH of the soil, the presence of clay minerals, the amount of organic matter, and other pertinent elements all play a role in heavy metal sedimentation and microbial degradation of the amount of contaminated organic soil will be proportional to the affinity of the organic component for the soil [83].

Table 1 A list of the many NT applications' mechanisms for in situ soil remediation

Mechanism	Nano-particle	Pollutants conc	Pollutants	Nano-particle conc	Results	References
Immobilization	nZVI	70.2 mg/g	As	5 wt%	70% removal of As	[29]
	Carbon nanotubes (CNTs)	4000 µg/L	Polycyclic aromatic hydrocarbons (PAHs)	0.25 wt%	66.1% removal within 24 h	[82]
	Nano-sized hydroxyapatite nHAP	547 µg/g	Copper (Cu)	5 wt%	Fully removal of Cu	[70]
	Iron (III) oxide	185 mg/kg	Cobalt (Co)	0.5 wt%	31% removal within 60 days	[11]
	Titanium dioxide TiO_2	20 µg/g	Diphenylarsinic acid (DPAA)	5 wt%	82% removal within 3 h	[77]
	nHAP	70.4 µg/g	Zinc (Zn)	5 wt%	Completely reduction of Zn	[70]
Fenton-like oxidation	Fe_3O_4	100 µg/g	PAHs	18 mM	85.2% of removal at pH 3.5	[8]
Reduction reaction	CMC/HA @nZVI	110 µg/g	Chromium (VI)	3 wt%	74.5% reduction within 3 months	[36]
	PVP-nZVI	727 µg/kg	Tri-chloroethylene (TCE)	0.4 g	84.73% dechlorination of TCE	[73]
	nZVI	24 mg/kg	Dichloro-diphenyl-trichloroethane (DDT)	20% aqueous	25% of DDT removal within 3 days	[22]

(continued)

Table 1 (continued)

Mechanism	Nano-particle	Pollutants conc	Pollutants	Nano-particle conc	Results	References
	CMC-nZVI	800 µg/g	Chromium (VI)	1.1 wt%	95% reduction within a week	[81]
	nZVI	50 mg/L	Chromium (VI)	45.7 m2/g	62% removal within 60 min	[13]
	nZVI-activated persulphate	17 µg/g	Polycyclic aromatic hydrocarbons (PAHs)	3 wt% of persulphate and 0.35 wt% of nZVI	82.2% removal within 104 days	[66]
	nZVI	0.15 g	Dichloro-diphenyl-trichloroethane (DDT)	5 wt%	2.5% of DDT removal per nZVI	[13]

5 Nano-bioremediation in Soil Pollution

A combination of NT with bioremediation carried out at the nanoscale is referred to as nano-bioremediation (NB). Due to the distinct physicochemical characteristics of the NPs, which also serve as catalysts and aid in lowering the activation energy needed to break down the compounds, the target pollutants are adsorbed, degraded, or changed [53]. The most popular NPs employed in NB are formed of carbon and metal, according to research and examination of the procedure [35].

6 Microbial Nano-bioremediation

In context to NB, microbial NB is a similar process where microorganisms are involved in the remedy process. Metals and microbial enzymes combine to generate effective nanoparticles for NB. Abiotic and biotic processes are the two phases of the microbial NB process. In phase one, NPs initially enter the system, contaminants are subjected to a variety of processes, including adsorption, absorption, dissolution, and photocatalysis. These particles are removed from the system in the second phase by a number of biotic mechanisms like biostimulation and biotransformation. The second step, the biotic phase, is crucial to the bioremediation of contaminants. The application of microbial NB for both inorganic and organic contaminants is widespread [19, 65]. However, because HMs are not biodegradable and are extremely likely to penetrate biological systems and food chains, they provide another difficulty [14].

Traditional techniques for removing HMs from contaminated soils include biosorption and bioaccumulation, which involve plants and bacteria (Fig. 2). However, new data indicate that the use of NPs in HM repair has shown outstanding results [71]. According to reports, particular microorganisms and NPs have been used in conjunction simultaneously or sequentially, and the outcomes have been persuasive [43, 78]. By serving as microbial bio-sorbents or nano-carriers of microorganisms, they could hasten the removal of HMs [6].

The three processes of membrane complication, hydrophobic separation, and ion exchange are the basic processes in soil that take part in the contamination transfer enabled by colloid [27]. The possibility of aided transfer is also a result of mobilization mechanisms mediated by colloids. Nanoscale particles may significantly impact trace metal transport, either by slowing it down while they are caught in the matrix or speeding it up while it is travelling [74]. The presence of NPs has been reported in water [12, 31], in soil [76], and also in the air [44, 46]. Continuing NP exposures would very certainly lead to their ubiquity, the possibility that they would enter the food chain at different trophic levels, or the chance that they would have cytotoxic effects on numerous aquatic and terrestrial creatures, as well as on people's health [16, 63]. Because of their distinct physicochemical features, NPs may interact with biological systems in the environment, which has a detrimental impact on the environment [84].

NPs behave as contaminants once they reach water bodies and are consumed by lower aquatic biota. NPs therefore affect biochemical pathways at the molecular and tissue-organ levels after entering the biological system [10, 67]. According to the results of the study, acute toxicity and filter-feeding were produced by metal NPs in three distinct animal models that represented various trophic levels, including the *Danio rerio* (zebra fish), and *Pseudokirchneriella subcapitata* (microalgae). *Daphnia pulex* (daphnia), *Daphnids* are particle filter feeders and more sensitive to exposure to NPs. Therefore it makes sense that the NPs were more dangerous to daphnia and microalgae than to zebrafish [20]. Several investigations have demonstrated that SiO_2 and TiO_2 NPs caused immunosuppressive and immunomodulatory effects in biological systems [49, 50]. Therefore, these environmental issues should be given the utmost priority prior to the deployment of NPs for nano-bioremediation, and NPs should be created in ways that support sustainable usage.

7 Conclusion

The fast release of toxins into the environment in the modern era poses a serious risk to human and ecological health. As a result, limiting their detrimental effects will depend on developing efficient ways to eliminate them from different environmental media. Novel ways are thus needed to completely eradicate pollutants because many existing processes cannot effectively eliminate various categories of contaminants. The reduction reaction and immobilization mechanisms are used in the use of technology for soil remediation. Carbon NPs, nZVI, and metal oxide nanomaterials are the most effective methods for reducing soil pollutants. According to the findings of a plethora of studies in this area, NT is a highly likely way to create solutions for dirty media. Closing the real-world remediation data gap, made possible by increased business and research collaboration, is a positive outcome. Consequently, it will be helpful for guiding future soil clean-up initiatives, which are occasionally required and barely wait for the outcomes of the experiment to be acquired, using ways to be achieved in the meantime. Thus, continued research into the use of NT in soil remediation is still necessary to protect the environment worldwide.

References

1. Ali B, Gill RA, Yang S, Gill MB, Farooq MA, Liu D, Daud MK, Ali S, Zhou W (2015) Regulation of cadmium-induced proteomic and metabolic changes by 5-aminolevulinic acid in leaves of brassica napus L. PLoS ONE 10(4):123328. https://doi.org/10.1371/JOURNAL.PONE.0123328
2. Ali H, Khan E (2018) Trophic transfer, bioaccumulation, and biomagnification of non-essential hazardous heavy metals and metalloids in food chains/webs—Concepts and implications for wildlife and human health 25(6):1353–1376. https://doi.org/10.1080/10807039.2018.1469398

3. Alissa EM, Ferns GA (2011) Heavy metal poisoning and cardiovascular disease. J Toxicol. https://doi.org/10.1155/2011/870125
4. Arif N, Yadav V, Singh S, Singh S, Ahmad P, Mishra RK, Sharma S, Tripathi DK, Dubey NK, Chauhan DK (2016) Influence of high and low levels of plant-beneficial heavy metal ions on plant growth and development. Front Environ Sci 4(NOV):69. https://doi.org/10.3389/FENVS.2016.00069/BIBTEX
5. Ashraf S, Ali Q, Zahir ZA, Ashraf S, Asghar HN (2019) Phytoremediation: environmentally sustainable way for reclamation of heavy metal polluted soils. Ecotoxicol Environ Saf 174:714–727. https://doi.org/10.1016/J.ECOENV.2019.02.068
6. Ayangbenro AS, Babalola OO (2017) A new strategy for heavy metal polluted environments: a review of microbial biosorbents. Int J Environ Res Pub Health 14(1). https://doi.org/10.3390/IJERPH14010094
7. Bachheti R, Fikadu A, Bachheti A, Husen A (2020) Biogenic fabrication of nanomaterials from flower-based chemical compounds, characterization and their various applications: a review. Saudi J Biol Sci 27(10):2551–2562. https://doi.org/10.1016/J.SJBS.2020.05.012
8. Barzegar G, Jorfi S, Soltani RDC, Ahmadi M, Saeedi R, Abtahi M, Ramavandi B, Baboli Z (2017) Enhanced Sono-Fenton-like oxidation of PAH-contaminated soil using nano-sized magnetite as catalyst: optimization with response surface methodology 26(5):538–557. https://doi.org/10.1080/15320383.2017.1363157
9. Basheer AA (2018) Chemical chiral pollution: impact on the society and science and need of the regulations in the 21st century. Chirality 30(4):402–406. https://doi.org/10.1002/CHIR.22808
10. Behzadi S, Serpooshan V, Tao W, Hamaly MA, Alkawareek MY, Dreaden EC, Brown D, Alkilany AM, Farokhzad OC, Mahmoudi M (2017) Cellular uptake of nanoparticles: journey inside the cell. Chem Soc Rev 46(14):4218. https://doi.org/10.1039/C6CS00636A
11. Bidast S, Golchin A, Baybordi A, Zamani A, Naidu R (2020) The effects of non-stabilised and Na-carboxymethylcellulose-stabilised iron oxide nanoparticles on remediation of Co-contaminated soils. Chemosphere 261. https://doi.org/10.1016/J.CHEMOSPHERE.2020.128123
12. Blinova I, Ivask A, Heinlaan M, Mortimer M, Kahru A (2010) Ecotoxicity of nanoparticles of CuO and ZnO in natural water. Environ Pollut (Barking, Essex : 1987) 158(1):41–47. https://doi.org/10.1016/J.ENVPOL.2009.08.017
13. Blundell SP, Owens G (2021) Evaluation of enhancement techniques for the dechlorination of DDT by nanoscale zero-valent iron. Chemosphere 264(Pt 1). https://doi.org/10.1016/J.CHEMOSPHERE.2020.128324
14. Briffa J, Sinagra E, Blundell R (2020) Heavy metal pollution in the environment and their toxicological effects on humans. Heliyon 6(9). https://doi.org/10.1016/J.HELIYON.2020.E04691
15. Cecchin I, Reddy KR, Thomé A, Tessaro EF, Schnaid F (2017) Nanobioremediation: integration of nanoparticles and bioremediation for sustainable remediation of chlorinated organic contaminants in soils. Int Biodeterior Biodegradation 119:419–428. https://doi.org/10.1016/J.IBIOD.2016.09.027
16. Cedervall T, Hansson LA, Lard M, Frohm B, Linse S (2012) Food chain transport of nanoparticles affects behaviour and fat metabolism in fish. PLoS ONE 7(2):e32254. https://doi.org/10.1371/JOURNAL.PONE.0032254
17. Chandra RNK, Dubey VK (2017) Phytoremediation of industrial pollutants and life cycle assessment 441–470. https://doi.org/10.1201/9781315161549-18
18. Dehghani MH, Sanaei D, Ali I, Bhatnagar A (2016) Removal of chromium(VI) from aqueous solution using treated waste newspaper as a low-cost adsorbent: Kinetic modeling and isotherm studies. J Mol Liq 215:671–679. https://doi.org/10.1016/J.MOLLIQ.2015.12.057
19. Desiante WL, Minas NS, Fenner K (2021) Micropollutant biotransformation and bioaccumulation in natural stream biofilms. Water Res 193. https://doi.org/10.1016/J.WATRES.2021.116846

20. Dietz KJ, Herth S (2011) Plant nanotoxicology. Trends Plant Sci 16(11):582–589. https://doi.org/10.1016/J.TPLANTS.2011.08.003
21. Ekins P, Zenghelis D (2021) The costs and benefits of environmental sustainability. Sustain Sci 16(3):949–965. https://doi.org/10.1007/S11625-021-00910-5/FIGURES/5
22. El-Temsah YS, Sevcu A, Bobcikova K, Cernik M, Joner EJ (2016) DDT degradation efficiency and ecotoxicological effects of two types of nano-sized zero-valent iron (nZVI) in water and soil. Chemosphere 144:2221–2228. https://doi.org/10.1016/J.CHEMOSPHERE.2015.10.122
23. Enez A, Hudek L, Bräu L (2018) Reduction in trace element mediated oxidative stress towards cropped plants via beneficial microbes in irrigated cropping systems: a review. Appl Sci 8(10):1953. https://doi.org/10.3390/APP8101953
24. Fay RM, Mumtaz MM (1996) Development of a priority list of chemical mixtures occurring at 1188 hazardous waste sites, using the HazDat database. Food Chem Toxicol Int J Publ Br Ind Biol Res Assoc 34(11–12):1163–1165. https://doi.org/10.1016/S0278-6915(97)00090-2
25. Flora SJS, Pachauri V, Saxena G (2009) Arsenic, cadmium and lead. Reprod Dev Toxicol 415–438. https://doi.org/10.1016/B978-0-12-382032-7.10033-5
26. Fujimori T, Takigami H (2014) Pollution distribution of heavy metals in surface soil at an informal electronic-waste recycling site. Environ Geochem Health 36(1):159–168. https://doi.org/10.1007/S10653-013-9526-Y/METRICS
27. Gao J, Wang Y, Du Y, Zhou L, He Y, Ma L, Yin L, Kong W, Jiang Y (2017) Construction of biocatalytic colloidosome using lipase-containing dendritic mesoporous silica nanospheres for enhanced enzyme catalysis. Chem Eng J C(317):175–186. https://doi.org/10.1016/J.CEJ.2017.02.012
28. Gerhardt KE, Gerwing PD, Greenberg BM (2017) Opinion: taking phytoremediation from proven technology to accepted practice. Plant Sci An Int J Experiment Plant Biol 256:170–185. https://doi.org/10.1016/J.PLANTSCI.2016.11.016
29. Gil-Díaz M, Alonso J, Rodríguez-Valdés E, Gallego JR, Lobo MC (2017) Comparing different commercial zero valent iron nanoparticles to immobilize As and Hg in brownfield soil. Sci Total Environ 584–585:1324–1332. https://doi.org/10.1016/J.SCITOTENV.2017.02.011
30. Glick BR (2010) Using soil bacteria to facilitate phytoremediation. Biotechnol Adv 28(3):367–374. https://doi.org/10.1016/J.BIOTECHADV.2010.02.001
31. Gondikas AP, Von Der Kammer F, Reed RB, Wagner S, Ranville JF, Hofmann T (2014) Release of TiO2 nanoparticles from sunscreens into surface waters: a one-year survey at the old danube recreational lake. Environ Sci Technol 48(10):5415–5422. https://doi.org/10.1021/ES405596Y/SUPPL_FILE/ES405596Y_SI_001.PDF
32. Gong X, Huang D, Liu Y, Peng Z, Zeng G, Xu P, Cheng M, Wang R, Wan J (2018) Remediation of contaminated soils by biotechnology with nanomaterials: bio-behavior, applications, and perspectives. Crit Rev Biotechnol 38(3):455–468. https://doi.org/10.1080/07388551.2017.1368446
33. Gorovtsov A, Demin K, Sushkova S, Minkina T, Grigoryeva T, Dudnikova T, Barbashev A, Semenkov I, Romanova V, Laikov A, Rajput V, Kocharovskaya Y (2022) The effect of combined pollution by PAHs and heavy metals on the topsoil microbial communities of Spolic Technosols of the lake Atamanskoe Southern Russia. Environ Geochem Health 44(4):1299–1315. https://doi.org/10.1007/S10653-021-01059-X
34. Guagliardi I, Cicchella D, De Rosa R (2012) A geostatistical approach to assess concentration and spatial distribution of heavy metals in urban soils. Water Air Soil Pollut 223(9):5983–5998. https://doi.org/10.1007/S11270-012-1333-Z/METRICS
35. Hidangmayum A, Debnath A, Guru A, Singh BN, Upadhyay SK, Dwivedi P (2022) Mechanistic and recent updates in nano-bioremediation for developing green technology to alleviate agricultural contaminants. Int J Environ Sci Technol 2022:1–26. https://doi.org/10.1007/S13762-022-04560-7
36. Hou S, Wu B, Peng D, Wang Z, Wang Y, Xu H (2019) Remediation performance and mechanism of hexavalent chromium in alkaline soil using multi-layer loaded nano-zero-valent iron. Environ Pollut 252:553–561. https://doi.org/10.1016/J.ENVPOL.2019.05.083

37. Huang H, Ullah F, Zhou DX, Yi M, Zhao Y (2019) Mechanisms of ROS regulation of plant development and stress responses. Front Plant Sci 10:800. https://doi.org/10.3389/FPLS.2019.00800/BIBTEX
38. Huber DL (2005) Synthesis, properties, and applications of iron nanoparticles. Small 1(5):482–501. https://doi.org/10.1002/SMLL.200500006
39. Husen A, Siddiqi KS (2023) Advances in smart nanomaterials and their applications. Elsevier Inc., 50 Hampshire St., 5th Floor, Cambridge, MA 02139, USA
40. Jaishankar M, Tseten T, Anbalagan N, Mathew BB, Beeregowda KN (2014) Toxicity, mechanism and health effects of some heavy metals. Interdiscip Toxicol 7(2):60. https://doi.org/10.2478/INTOX-2014-0009
41. Kononova, MM (2013) Soil Organic Matter: Its nature, its role in soil formation and in soil fertility 545
42. Krishna D, Sachan HK, Jatav HS (2022) Management of sewage sludge for environmental sustainability. Sustain Manage Util Sew Sludge 353–381. https://doi.org/10.1007/978-3-030-85226-9_17
43. Lahiri D, Nag M, Sheikh HI, Sarkar T, Edinur HA, Pati S, Ray RR (2021) Microbiologically-synthesized nanoparticles and their role in silencing the biofilm signaling cascade. Front Microbiol 12:180. https://doi.org/10.3389/FMICB.2021.636588/BIBTEX
44. Liu NM, Miyashita L, Maher BA, McPhail G, Jones CJP, Barratt B, Thangaratinam S, Karloukovski V, Ahmed IA, Aslam Z, Grigg J (2021) Evidence for the presence of air pollution nanoparticles in placental tissue cells. Sci Total Environ 751. https://doi.org/10.1016/J.SCITOTENV.2020.142235
45. Ltifi M, Abichou T, Tisot JP (2014) Effects of soil aging on mechanical and hydraulic properties of a silty soil. Geotech Geol Eng 32(4):1101–1108. https://doi.org/10.1007/S10706-014-9784-1
46. Maher BA, González-Maciel A, Reynoso-Robles R, Torres-Jardón R, Calderón-Garcidueñas L (2020) Iron-rich air pollution nanoparticles: an unrecognised environmental risk factor for myocardial mitochondrial dysfunction and cardiac oxidative stress. Environ Res 188. https://doi.org/10.1016/J.ENVRES.2020.109816
47. Mandzhieva S, Chaplygin V, Chernikova N, Fedorenko A, Voloshina M, Minkina T, Rajput VD, Elinson M, Wong MH (2022) Responses of Spring Barley to Zn- and Cd-induced stress: morphometric analysis and cytotoxicity assay. Plants 11(23):3332. https://doi.org/10.3390/PLANTS11233332
48. Martin JL, Maris V, Simberloff DS (2016) The need to respect nature and its limits challenges society and conservation science. Proc Natl Acad Sci 113(22):6105–6112. https://doi.org/10.1073/PNAS.1525003113
49. Maurer-Jones MA, Christenson JR, Haynes CL (2012) TiO2 nanoparticle-induced ROS correlates with modulated immune cell function. J Nanopart Res 14(12):1291. https://doi.org/10.1007/S11051-012-1291-9
50. Maurer-Jones MA, Lin YS, Haynes CL (2010) Functional assessment of metal oxide nanoparticle toxicity in immune cells. ACS Nano 4(6):3363–3373. https://doi.org/10.1021/NN9018834
51. McLaughlin MJ, Smolders E (2001) Background zinc concentrations in soil affect the zinc sensitivity of soil microbial processes—a rationale for a metalloregion approach to risk assessments. Environ Toxicol Chem 20(11):2639–2643. https://doi.org/10.1002/ETC.5620201132
52. Medina-Pérez G, Fernández-Luqueño F, Vazquez-Nuñez E, López-Valdez F, Prieto-Mendez J, Madariaga-Navarrete A, Miranda-Arámbula M (2019) Remediating polluted soils using nanotechnologies: environmental benefits and risks. Polish J Environ Stud 28(3):1013–1030. https://doi.org/10.15244/PJOES/87099
53. Mehndiratta P, Jain A, Srivastava S, Gupta N (2013) Environmental pollution and nanotechnology. Environ Pollut 2(2). https://doi.org/10.5539/EP.V2N2P49
54. Mukhopadhyay S, Maiti SK (2010) Phytoremediation of metal mine waste. Appl Ecol Environ Res 8(3):207–222

55. Naderi Peikam E, Jalali M (2019) Application of three nanoparticles (Al_2O_3, SiO_2 and TiO_2) for metal-contaminated soil remediation (measuring and modeling). Int J Environ Sci Technol 16(11):7207–7220. https://doi.org/10.1007/S13762-018-2134-8
56. Pretty J, Bharucha ZP (2014) Sustainable intensification in agricultural systems. Ann Bot 114(8):1571–1596. https://doi.org/10.1093/AOB/MCU205
57. Qian Y, Qin C, Chen M, Lin S (2020) Nanotechnology in soil remediation—applications versus implications. Ecotoxicol Environ Saf 201. https://doi.org/10.1016/J.ECOENV.2020.110815
58. Qiao Y, Wu J, Xu Y, Fang Z, Zheng L, Cheng W, Tsang EP, Fang J, Zhao D (2017) Remediation of cadmium in soil by biochar-supported iron phosphate nanoparticles. Ecol Eng 106:515–522. https://doi.org/10.1016/J.ECOLENG.2017.06.023
59. Raffa CM, Chiampo F, Shanthakumar S (2021) Remediation of metal/metalloid-polluted soils: a short review. Appl Sci 11:4134; 11(9):4134. https://doi.org/10.3390/APP11094134
60. Rajput VD, Minkina T, Kumari A, Shende SS, Ranjan A, Faizan M, Barakvov A, Gromovik A, Gorbunova N, Rajput P, Singh A, Khabirov I, Nazarenko O, Sushkova S, Kızılkaya R (2022) A review on nanobioremediation approaches for restoration of contaminated soil. Eurasian J Soil Sci 11(1):43–60. https://doi.org/10.18393/EJSS.990605
61. Rasmussen LD, Sørensen SJ, Turner RR, Barkay T (2000) Application of a mer-lux biosensor for estimating bioavailable mercury in soil. Soil Biol Biochem 32(5):639–646. https://doi.org/10.1016/S0038-0717(99)00190-X
62. Reimann C, Birke M, Demetriades A, Filzmoser P, O'Connor P (2014) Chemistry of Europe's agricultural soils, Part B. Chem Eur Agric Soils 352
63. Rico CM, Majumdar S, Duarte-Gardea M, Peralta-Videa JR, Gardea-Torresdey JL (2011) Interaction of nanoparticles with edible plants and their possible implications in the food chain. https://doi.org/10.1021/JF104517J
64. Saha JK, Selladurai R, Coumar MV, Dotaniya ML, Kundu S, Patra AK (2017) Status of soil pollution in India 271–315. https://doi.org/10.1007/978-981-10-4274-4_11
65. Singh Y, Saxena MK (2022) Insights into the recent advances in nano-bioremediation of pesticides from the contaminated soil. Front Microbiol 13:3906. https://doi.org/10.3389/FMICB.2022.982611/BIBTEX
66. Song Y, Fang G, Zhu C, Zhu F, Wu S, Chen N, Wu T, Wang Y, Gao J, Zhou D (2009) Zero-valent iron activated persulfate remediation of polycyclic aromatic hydrocarbon-contaminated soils: An in situ pilot-scale study. Chem Eng J 355:65–75. https://doi.org/10.1016/J.CEJ.2018.08.126
67. Stater EP, Sonay AY, Hart C, Grimm J (2021) The ancillary effects of nanoparticles and their implications for nanomedicine. Nat Nanotechnol 16(11):1180–1194. https://doi.org/10.1038/s41565-021-01017-9
68. Stevenson F (1994) Humus chemistry: genesis, composition, reaction. Humus Chem 72(4):512
69. Struik PC, Kuyper TW (2017) Sustainable intensification in agriculture: the richer shade of green. A review. Agron Sustain Dev 37(5):1–15. https://doi.org/10.1007/S13593-017-0445-7/FIGURES/3
70. Sun RJ, Chen JH, Fan TT, Zhou DM, Wang YJ (2018) Effect of nanoparticle hydroxyapatite on the immobilization of Cu and Zn in polluted soil. Environ Sci Pollut Res Int 25(1):73–80. https://doi.org/10.1007/S11356-016-8063-5
71. Sunanda S, Misra M, Ghosh Sachan S (2022) Nanobioremediation of heavy metals: perspectives and challenges. J Basic Microbiol 62(3–4):428–443. https://doi.org/10.1002/JOBM.202100384
72. Taylor AF, Rylott EL, Anderson CWN, Bruce NC (2014) Investigating the toxicity, uptake, nanoparticle formation and genetic response of plants to gold. PLoS ONE 9(4):e93793. https://doi.org/10.1371/JOURNAL.PONE.0093793
73. Tian H, Liang Y, Yang D, Sun Y (2020) Characteristics of PVP-stabilised NZVI and application to dechlorination of soil-sorbed TCE with ionic surfactant. Chemosphere 239. https://doi.org/10.1016/J.CHEMOSPHERE.2019.124807

74. Tiede K, Boxall ABA, Tear SP, Lewis J, David H, Hassellöv M (2008) Detection and characterization of engineered nanoparticles in food and the environment. Food Addit Contam Part A, Chem Anal Control, Expo Risk Assess 25(7):795–821. https://doi.org/10.1080/02652030802007553
75. Usman M, Farooq M, Wakeel A, Nawaz A, Cheema SA, Rehman HU, Ashraf I, Sanaullah M (2020) Nanotechnology in agriculture: current status, challenges and future opportunities. Sci Total Environ 721:137778. https://doi.org/10.1016/J.SCITOTENV.2020.137778
76. Vittori Antisari L, Carbone S, Gatti A, Vianello G, Nannipieri P (2009) Toxicity of metal oxide (CeO_2, Fe_3O_4, SnO_2) engineered nanoparticles on soil microbial biomass and their distribution in soil. Soil Biol Biochem 60:87–94. https://doi.org/10.1016/J.SOILBIO.2013.01.016
77. Wang AN, Teng Y, Hu XF, Wu LH, Huang YJ, Luo YM, Christie P (2016) Diphenylarsinic acid contaminated soil remediation by titanium dioxide (P25) photocatalysis: degradation pathway, optimization of operating parameters and effects of soil properties. Sci Total Environ 541:348–355. https://doi.org/10.1016/J.SCITOTENV.2015.09.023
78. Wang L, Hu C, Shao L (2017) The antimicrobial activity of nanoparticles: present situation and prospects for the future. Int J Nanomed 12:1227. https://doi.org/10.2147/IJN.S121956
79. Xiao L, Cheng H, Cai H, Wei Y, Zan G, Feng X, Liu C, Li L, Huang L, Wang F, Chen X, Zou Y, Yang X (2022) Associations of heavy metals with activities of daily living disability: an epigenome-wide view of DNA methylation and mediation analysis. Environ Health Perspect 130(8):087009-1-087009–087011. https://doi.org/10.1289/EHP10602
80. Yang J, Teng Y, Wang J, Li J (2011) Vanadium uptake by alfalfa grown in V-Cd-contaminated soil by pot experiment. Biol Trace Elem Res 142(3):787–795. https://doi.org/10.1007/S12011-010-8777-Z
81. Zhang R, Zhang N, Fang Z (2018) In situ remediation of hexavalent chromium contaminated soil by CMC-stabilized nanoscale zero-valent iron composited with biochar. Water Sci Technol J Int Assoc Water Pollut Res 77(5–6):1622–1631. https://doi.org/10.2166/WST.2018.039
82. Zhang T, Lowry GV, Capiro NL, Chen J, Chen W, Chen Y, Dionysiou DD, Elliott DW, Ghoshal S, Hofmann T, Hsu-Kim H, Hughes J, Jiang C, Jiang G, Jing C, Kavanaugh M, Li Q, Liu S, Ma J, Alvarez PJJ (2019) In situ remediation of subsurface contamination: opportunities and challenges for nanotechnology and advanced materials. Environ Sci Nano 6(5):1283–1302. https://doi.org/10.1039/C9EN00143C
83. Zhang WX (2003) Nanoscale iron particles for environmental remediation: an overview. J Nanopart Res 5(3–4):323–332. https://doi.org/10.1023/A:1025520116015/METRICS
84. Zhu X, Wang J, Zhang X, Chang Y, Chen Y (2010) Trophic transfer of TiO(2) nanoparticles from Daphnia to zebrafish in a simplified freshwater food chain. Chemosphere 79(9):928–933. https://doi.org/10.1016/J.CHEMOSPHERE.2010.03.022

Development Strategies and Prospects of Carbon Nanotube as Heavy Metal Adsorbent

Shivani Tyagi, Pranchal Rajput, Aashna Sinha, Atreyi Pramanik, Kundan Kumar Chaubey, Sujata Jayaraman, Chetan Shrivastva, Ashok Kumar, Deepak Kumar Verma, Sapna Yadav, Deen Dayal, Versha Dixit, and Shiv Dayal Pandey

Abstract The carbon elements less than 100 nm are known as carbon nanoparticles. The family of carbon nanomaterials, which includes graphene (atomically flat carbon), graphene oxide, carbon nanoparticles, and carbon nanotubes, is expanding quickly. Many electrical, optical, and biological applications employ carbon nanoparticles. The processes of carbonization, heating, activation, and grinding are used to produce carbon nanoparticles. Its synthesis and activation procedure for various medicinal and biological uses and basic features are explained in this chapter. Such as the manufacturing of carbon-based nanomaterials, wet and dry synthesis, covalent and noncovalent binding, and there are some naturally occurring carbon nanotubes in the environment, and their potential as heavy metal absorbents is promising. This

Shivani Tyagi and Pranchal Rajput are equally contributed.

S. Tyagi · P. Rajput · A. Sinha · A. Pramanik · K. K. Chaubey (✉) · C. Shrivastva · S. Yadav
Division of Research and Innovation, School of Applied and Life Sciences, Uttaranchal University, Prem Nagar, P.O. Chandanwari, Arcadia Grant, Dehradun 248007, Uttarakhand, India
e-mail: kundan.chaubey@uttaranchaluniversity.ac.in

S. Tyagi · S. Jayaraman · D. Dayal
Department of Biotechnology, Institute of Applied Sciences and Humanities, GLA University, Mathura 281406, Uttar Pradesh, India

A. Kumar
Animal Health Division, ICAR-Central Institute for Research on Goats, Makhdoom, Farah, Mathura 281122, Uttar Pradesh, India

D. K. Verma
Department of Biotechnology, Sanskriti University, Chhata, Distt. Mathura, Uttar Pradesh, India

V. Dixit
Department of Life Sciences, Christ University, Bangalore 560029, Karnataka, India

S. D. Pandey
Uttaranchal Institute of Technology, Uttaranchal University, P.O. Chandanwari, Prem Nagar, Arcadia Grant, Dehradun 248007, Uttarakhand, India

R. K. Bachheti et al. (eds.), *Nanomaterials for Environmental and Agricultural Sectors*, Smart Nanomaterials Technology,
https://doi.org/10.1007/978-981-99-2874-3_4

study states that there are several techniques for creating carbon nanoparticles, which exhibit significant benefits as environmental heavy metal adsorbents.

Keywords Carbon nanotube · Heavy metal · Nanotechnology · Development · Strategies · Prospects

1 Introduction

The field of nanotechnology is emerging nowadays, it covers many fields of science and plays a frontier role in the transformation of various research areas like agriculture, food sector, health science, environmental remediation and so on, and resolving many dominating issues (Fig. 1) [1, 2]. The carbon elements known as carbon nanoparticles are created at the nanoscale using a number of procedures, including carbonization, heating, activation, and grinding. A typical diameter for all types of nanoparticles is less than 100 nm. Carbon nanoparticles are arranged electrically in a [He] $2s^2$ $2p^2$ configuration. Pure carbon particles have a molar mass of 12.01 g/mol and extraordinarily high melting and boiling temperatures that are both above 3500 °C. Several physical properties of carbon nanoparticles, such as the structure of the surface, are influenced by the manufacturing technique used to create the particles. When carbon is activated, it creates porous nanoparticles with a significant amount of surface area that may be employed for chemical reactions, binding, or absorption processes [3]. Due to their exceptional electrical, optical, thermal, chemical, and mechanical properties, carbon nanomaterials (CNMs) hold a special place in the field of nanoscience and have found use in a variety of fields, including composite materials, energy storage and conversion, sensors, drug delivery, medical devices, field emission devices, and nanoscale electronic components [4]. Human health faces great risk due to heavy metal contamination in water and soil, posing critical challenges. The "Environmental Protection Agency (EPA)", also known as the "Agency for Toxic Substances and Disease Registry (ATSDR)", situated in Atlanta, Georgia, U.S. Department of Health and Human Services stated the list for 2001 named the 20 topmost precarious substances, among them As, Pb, and Hg are ranked first, second, and third, respectively, while Cd is ranked seventh [5]. Besides these heavy metals, iron, zinc, copper, chromium, and manganese are the second most common heavy metals found in potable water [6] (Table 1). Even at low ion concentrations, almost all heavy metals are toxic to humans and other organisms [7] (Fig. 2).

2 Carbon Nanoparticles

Carbon nanoparticles are carbon elements that are produced at the nanoscale by a variety of processes, including carbonization, heating, activation, and grinding. Every form of nanoparticle typically has a diameter of less than 100 nm. The electrical

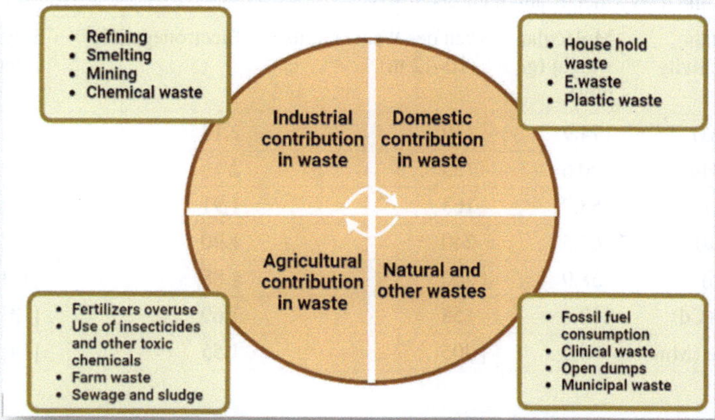

Fig. 1 Impurities from different sources

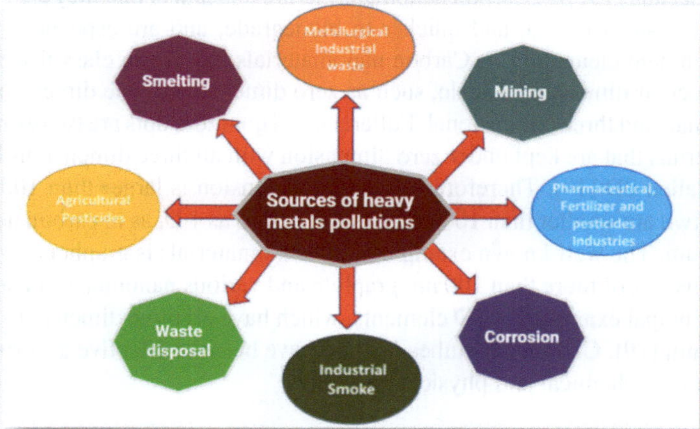

Fig. 2 Sources of heavy metal pollution

arrangement of carbon nanoparticles is [He] $2s^2$ $2p^2$. The molar mass of the pure carbon particles is 12.01 g/mol, and they have extremely high melting and boiling points that are both above 3500 °C. The manufacturing process used to manufacture the particles affects other physical characteristics of carbon nanoparticles, such as the surface's structure. When carbon is activated, it forms porous nanoparticles with a huge surface area that can be used for binding, chemical reactions, or absorption processes [3].

Table 1 Some heavy metals and their chemistry

Heavy metals electronegativity	Molecular weight (g/mol)	van der Waals radius (10–12 m)	Electronegativity	Key references
Arsenic (As)	74.9	119	2.18	[8]
Mercury (Hg)	200.6	155	2	[9]
Nickel (Ni)	58.7	163	1.91	[10]
Copper (Cu)	63.5	240	1.90	[11]
Cobalt (Co)	58.9	200	1.88	[12]
Cadmium (Cd)	112.4	158	1.69	[13]
Manganese (Mn)	54.9	205	1.55	[14]

3 Types of Carbon Nanomaterials

Material science is essential for realizing the objective of a clean environment, and this technology, particularly nanomaterials, has advanced at a breakneck pace in the previous decade [15, 16]. Carbon nanomaterials are unusual in that they are nontoxic, have a large surface area, and quicker to biodegrade, and are especially valuable for environment cleanup [17]. Carbon nanomaterials have been classified based on their molecular dimensional scale, such as zero dimensional, one dimensional, two dimensional, and three dimensional. Fullerene and quantum dots are two examples of nanomaterials that are kept under zero dimension with all three dimensions less than 100 nm called 0D [18]. Therefore, when one dimension is larger than 100 nm and the other two are smaller than 100 nm, it is referred to as 1 D, as in carbon nanotubes and titanium. The well-known example of 2D nanomaterials is graphene, which has two dimensions of more than 100 nm graphite and various nanomaterial composites are the principal examples of 3D elements, which have all three dimensions of more than 100 nm [19]. Carbon nanotubes (CNTs) have become effective adsorbents due to their unique chemical and physical properties.

4 Characteristics of Carbon Nanotubes

In 1999, Lijima revealed carbon nanotubes the first time and detailed them in single-wall (SWCNTs) and multi-wall CNTs (MWCNTs) (Viduet et al. 2014). This classification is according to the graphene layers involved in synthesizing CNTs as single-walled CNTs made up of monolayer of graphene and multi-walled CNTs with multiple layers of graphene [20]. Few-walled carbon nanotubes (FWCNTs), a type of MWCN, have been found as a transitional form between SWCNTs and MWNTs (Wang et al. 2009). A graphene layer is made up of sp2-hybridized hexagonal arrangement of carbon atoms with in-plane s-bonds and p-orbital out-of-plane. Some properties of CNTs make them an excellent nanomaterial for heavy metals

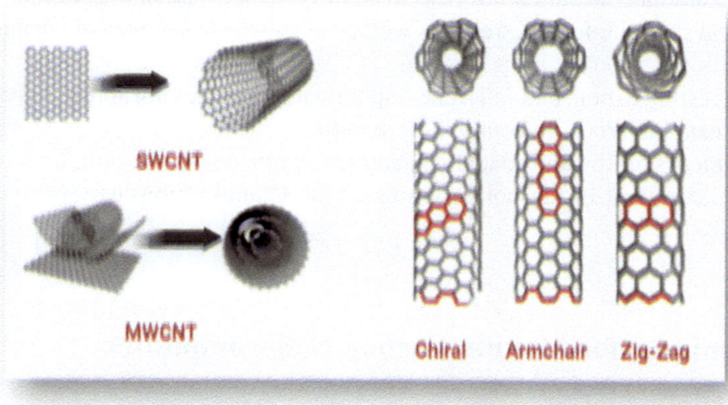

Fig. 3 Single and multi-walled CNPs, patterns of synthesis (single-walled); (zigzag nanotube, armchair nanotube, chiral nanotube

absorption like tiny size (approx. 1 nm to several nm), large surface area, and nature of mesopores.

Furthermore, zigzag, armchair, and chiral nanotubes are three other forms of CNTs depending on the two-dimensional form of their sheets [21] (Fig. 3b). The majority of the active sites found in carbon nanotubes have faulty sections, for example, pentagons aligned next to a tube body made up entirely of six-sided polygon, it's conceivable that this is what provides CNTs their extraordinary ability to collaborate with other molecules [22]. In order to make carbon nanotubes more responsive to various contaminants, their surfaces must be stimulated and functionalized with extra materials. The most frequent method of functionalization is the oxidation process. Oxidizing CNTs accomplishes it by heating them with acids such as nitric acid, H_2SO_4, and a combination of these [23].

5 Naturally Occurring Carbon Nanomaterials

Natural carbon-based nanoparticles have also been found, in addition to manufactured nanomaterials [24, 25]. According to Velasco-Santos et al. [25], carbon nanotubes were found in a coal-petroleum mixture (Chen et al. 2011). Employed metal-oxide-containing volcanic lava as a substrate and catalyst for the CVD technique of synthesizing SWCNTs. The scientists hypothesized that this mechanism would show that nanotubes can grow in the wild under situations of tremendous heat, such as during volcanic eruptions. There is proof that fullerenes may also be found in geological materials in addition to carbon nanotubes. In addition to carbon nanotubes, fullerenes have also been found in geological materials, according to the data. Little amounts of fullerenes (2% w/w) have been found in the naturally occurring mineral shungite from Karelia, as well as in meteorite samples of cosmic origin [26].

It's interesting to note that fullerenes' spherical shape does not appear to be limited to carbon nanomaterials. Recently, fullerene-like structures with potential functions for mechanical stability and adaptable qualities significant for the pollination process have been identified in the pollen grains of the Chinese hibiscus (*Hibiscus rosa-sinensis*) [27].

6 Techniques for Making Carbon Nanocomposites

6.1 Synthesis of Carbon-Based Nanomaterials in Industries

Fullerenes with the discovery of carbon-based nanomaterials, many synthesis techniques have been developed and their remarkable characteristics have undergone extensive study. Carbon vapours are the primary ingredients used in the creation of carbon nanomaterials. By evaporating graphite electrodes in a helium environment, W. Krätschmer and D. R. Huffman created fullerenes for the first time in 1990 (Kratschmer et al. 1990; Kratschmer 2011). Eventually, by creating an electric arc between two graphite electrodes, a reactor was altered. The resultant soot collects and is treated in boiling toluene, benzene, xylene, or other organic solvents after condensing on the reactor's cool surface. A black condensate is produced after the solvents have evaporated, and it contains minor quantities of higher fullerenes as well as roughly 10–15% of C60 and C70 fullerenes. The ratio of C60 to C70 fullerenes varies depending on the synthesis settings, although normally C60 is the predominate proportion. The broad family of plasma procedures, which are the most well-known and often applied as compared to other techniques, includes the described arc-discharge approach (Churilov 2010). Unfortunately, the high prices and low productivity of the present technologies for their synthesis limit the practical application of fullerenes.

Nanotubes of carbon, the fundamental techniques for CNT production, include arc discharge, laser ablation, and chemical vapour deposition (CVD) [28]. At the moment, CVD is one of the most studied and often applied methods for making CNTs [29]. The manufacturing of CNTs on a wide scale may be accomplished via CVD synthesis as opposed to the other two techniques (arc discharge and laser ablation), which need more complex equipment and harsher environmental conditions [30]. The CVD method of synthesis relies on the breakdown of hydrocarbons to carbon, followed by the synthesis of carbon nanostructures on a variety of substrates that contain catalysts for the growth of the nanotubes. The diameter of the nanotubes that are generated using metal-based nanoparticle catalysts (0.5–5 nm for SWCNTs, 8–100 nm for MWCNTs) closely corresponds with their size. SWCNT and MWCNT synthesis is often catalysed by nickel, cobalt, or iron nanoparticles.

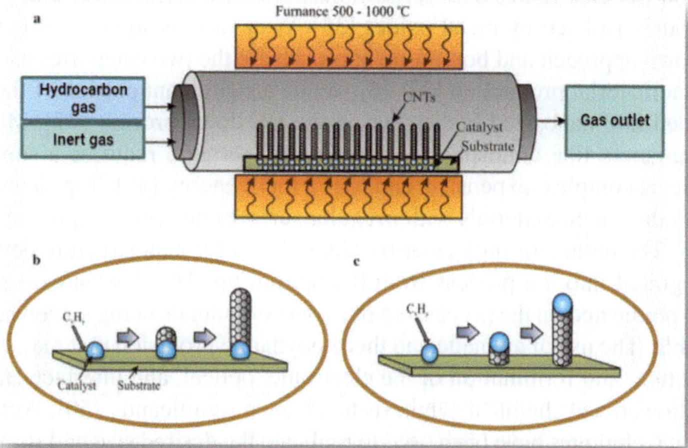

Fig. 4 Synthesis of carbon-based nanomaterials in industries

Inert gas and hydrocarbon-filled tubes are commonly seen in reaction chambers of reactors used for CVD synthesis (Fig. 4a). For the synthesis of SWCNTs, methane is typically employed, whereas MWCNTs are made with ethylene or acetylene. According to a condensed process description, the substrate is heated to 850–1000 °C for the creation of SWCNT and to 550–700 °C for the production of MWCNT. Hydrocarbons thermally decompose to produce carbon, which is then dissolved by the metal nanoparticle catalyst. As a beginning structure for the formation of a cylindrical shell nanotube, created by a continuous flow of carbon from the hydrocarbon source to the catalyst particle, it creates a semi-fullerene cap after reaching a particular threshold concentration of carbon (Fig. 4b, c). To produce CNTs of a better grade, the final removal of the catalysts from the nanotube tips and further purification are still being developed and optimized [31, 32].

6.2 *Wet Synthesis*

An important class of nanomaterials called metal oxide nanoparticles has many uses in both science and technology. Wet chemical propagation may be used to selectively design the surface, step, size, and shape of metal oxide nanoparticles, which helps to gather the required attributes [33]. Metal oxide nanoparticles and nanocomposites are becoming more common in applied ecology, particularly as adsorbents and photo-catalysts, as well as a resource for the fabrication of environmental monitoring systems [34] and catalytic and energy storage industries [35]. As a result, developments in the synthesis of metal oxide nanoparticles are noteworthy due to their use in the electronics and optics industries. The metal oxide charges can be spontaneously isolated by metal oxides [36]. With various significant applications, the top-down approach and bottom-up approach are the two categories used to categorize nanomaterial production [37–46]. While a significant portion of the material gets divided into nanosized bodies during the top-down process [47]. Maintaining exact parameters like atmosphere, heat, and temperature requires a sophisticated system that is complex, expensive, and uses a lot of energy [48]. Top-down methodologies produce nanomaterials with irregular surfaces that prevent practical deployment [49]. The atomic or molecular building blocks for nanostructure development were integrated into the process from the bottom up. The downstream strategy is primarily predicated on the processing of moist compounds being somewhat flexible and scalable. The use of a kinetic and thermodynamic provision that may change the size, structure, and formulation of the electronic, optical, and interface characteristics has allowed wet chemical synthesis to advance significantly [50]. Wet chemical processing techniques have been used to replicate the desired size and form of metal oxide nanoparticles. Wet chemical processes including sol–gel, co-precipitation, and hydrothermal synthesis were created to produce ultrafine, very homogeneous, and high purity powders. A deeper knowledge of the fundamental processes, the cycle by which the precursor, the surface stabilizing factor, and the reagent of the system are transformed, as well as its interactions with the rate of proliferation and nucleation, allows for the regulation of shape and size. Nanoparticles with a total scaling capacity of 10 mg/m may be produced using a typical traditional batch process. To comprehend the lucrative operation of colloidal metal oxide, the synthesis process should be flexible in terms of product quality and quantity to match industrial criteria. The method for synthesizing nanocomposites using chemical, physical, and biological processes is shown in Fig. 5.

6.3 Dry Synthesis

It is well known that the dry synthesis is a very good and suitable procedure for the propagation of carbon nanocomposite materials. This strategy's key advantages are its adaptability, improved adherence, and the advantages of having less design variables [51]. The technique most suited for carbonizing metal nanoparticles has been referred to as dry synthesis [51]. The functional oxygen groups [52] can act as a link between carbon resources and metal nanoparticles. It is also possible to treat carbon compounds with no useful surface groups consistently [53]. For the synthesis of carbon nanocomposite, Yogeshwaran et al. [54] have developed a rapid and less dry solvent synthesis technique. Dry mixing of the metal precursor salt and carbon materials (CNTs or graphene oxide (GO)) is the first stage of a two-phase direct process, which is followed by inert atmospheric heating [55]. The mechanochemical cycle does not require a solvent, extra reducing agents, or apply electric current [56]. According to studies, the mechanochemical approach often works with CNTs as well as other carbon-based materials that have a high heat conductivity, such as graphene,

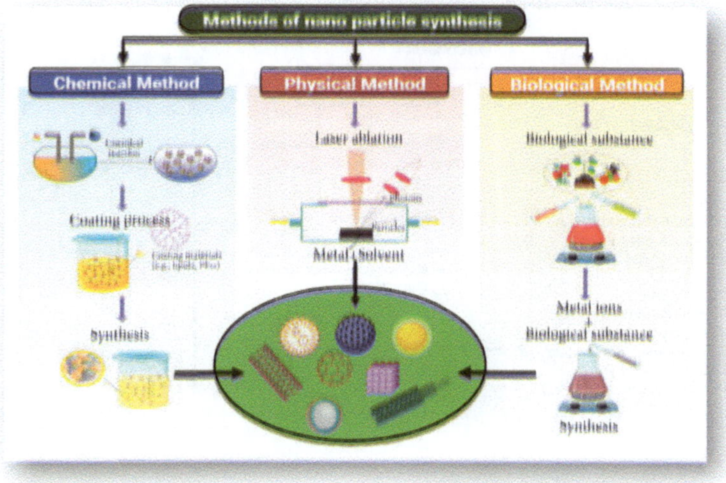

Fig. 5 Methods of nanoparticle synthesis

GO, and activated carbon [57]. The mechanochemical method is thought to be quick, flexible, and eventually customizable, enabling it to be employed in a variety of applications [58] for additional use. As it is similar to carbon nanocoating and is shown in Fig. 4 [59], it covers potential applications in domains including anticorrosion, anti-wear, super-hydrophobic area, self-cleaning, antifouling, antibacterial area, and electronics.

6.4 Functionalization or Modification of CNTs

The methods for modifying carbon nanotubes can be divided into four categories: covalent functionalization, noncovalent functionalization, exterior embellishment using inorganic elements, and endohedral filling (Fig. 6). The CNTs' surface has a quite hydrophobic character. Noncovalent and covalent functionalization methodologies have attempted different surface modification techniques to address this issue in raw CNTs.

In past studies, various chemistries have been created to alter the surface of CNTs. By subjecting CNTs to elemental fluorine, Mickelson et al. were able to functionalize the sidewalls of CNTs [60]. Grignard-reagent nucleophilic substitution or a reaction with alkyl lithium precursors were used to accomplish sidewall alkylation of the nanotubes [61]. Reactive alkyl oxycarbonyl nitrenes made from alkoxy carbonyl azides were used to add nitrenes to the sidewall of CNTs [62].

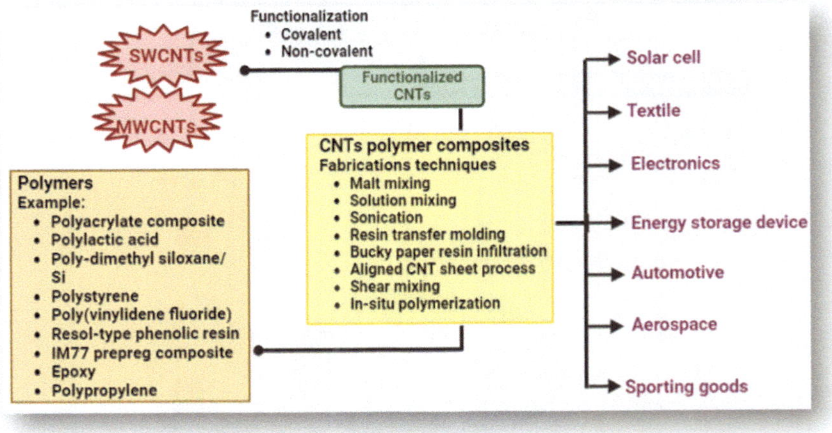

Fig. 6 Types of functionalization of CNTs

6.5 *Covalent Functionalization*

Covalent functionality involves attaching functional components chemically to the nanotube's carbon scaffold. Depending on where the contact occurs, it can take one of two forms: covalent sidewall functionalization or defect functionalization. A change in hybridization from sp2 to sp3 and conjugation loss are implied by the way the direct covalent sidewall functions [63]. Functionalization of faults is based on the site alterations that have previously been made [64]. The open ends and sidewall holes ended in hexagonal graphene frames, for example, by functional groups and flaws in stone/wales, may defect sites (5–7 faults). In oxidative purification, oxygenated sites that are created as defects are also included. SWCNTs are presented as bundles and exhibit low dispersibility [65]. In this case, it is assured that a highly reactive reagent will be used to form covalent bonds between the walls. It is impossible to predict in advance whether such extra reactions would occur more frequently at complete hexagonal portions of the edge or at faulty spots [66]. Covalent functionalization has been accomplished using a variety of covalent techniques, including oxidative purification, amidation, esterification, thiolation, halogenation, hydrogenation, and electrochemical functionality [67].

6.6 *Noncovalent Functionalization*

It is typical in both organic and inorganic structures for materials to be subjected to attractive or repulsive interactions, and this phenomenon is known as noncovalent functionalization or supramolecular correlations [68]. While it involves merging several relations, it is theoretically beneficial to attribute faults or abnormalities resulting from the change from sp2 to sp3 carbon in the case of graphene [69]. There is a strong correlation between graphene materials and energy dissociation of less than 50 kJ mol^{-1} [70]. Solvation and hydrophobic effects brought on by different interactions are a significant issue that has to be addressed in graphene and graphene oxide (GO) systems because they impact both their dispersibility and the discovery of associations that can be used to categorize them [71]. The compatibility of graphene with other nanomaterials or stimulants between its electron-rich and electron-poor areas is influenced by two types of interactions [72]. Both face-to-face and edge-to-face arrangements frequently exhibit this [73]. It is also found through electron interactions in physiologically significant compounds including DNA, RNA, and porphyrins. Also present in tiny molecules, these interactions are sent to the GO and G systems where they may be exploited to process and change characteristics [74]. By noncovalent functionalization with amine-packed and completely packed polymers, the effective dispersion of graphene was achieved [75]. Noncovalent grafting using a PS-NH2 polymer with an end-functionality might be used to disseminate reduced graphene in non-solvents [76]. Several non-solvents for reduced graphene, including benzene, hexane, oxylene, and dichloromethane, which are immiscible with the aqueous process, were utilized for noncovalent functionalization [77]. After the chemical reduction of graphene oxide, the residual carboxylate groups successfully served as the places for noncovalent functionalization of the protonated amine terminals of end-functional polymers [78]. Via straightforward sonication, the noncovalent functionality encouraged the transfer of graphene sheets from the aqueous phase to the organic phase [79].

7 Methods for Removal of Heavy Metals

Heavy metals can be removed from water using a variety of methods, including flocculation, coagulation, ion exchange, precipitation, reverse osmosis, oxidation, flotation, and photocatalysis [80–84]. Each of these strategies, however, has some disadvantages. The volume of sludge generated makes the flocculation and coagulation procedure ineffective, while the absence of recyclability is a disadvantage of the effective ion exchange technique. Hazardous waste production is the inefficiency of precipitation which further needs to be treated. The osmosis method's key constraints are the issues of generation, expense, and disposal of remaining materials. The disadvantage of the photocatalytic approach is the long time it takes [85].

Absorption is another suitable method that has been used for the removal of heavy materials from water with the help of various types of modified absorbents like activated carbon, landfill clay, extracellular polymeric substances, manganese oxides, granular biomass, and chitosan [86, 87], Hawari et al. 2006). Nanotechnology's revolution has opened up many possibilities for new adsorption methods.

8 Carbon Nanotubes and Their Derivatives: Heavy Metal Remediation

Photocatalysis, a semiconductor-based wastewater treatment technology, is one of the most advanced technologies in use [United Nations Children's Fund (UNICEF), 2000–2017]. Several researchers have developed an ultrathin network photocatalyst based on TiO_2-functionalized SWCNTs, effectively purifying oil from oil [88]. Alijani et al. utilized CNTs to synthesize nanocomposites using magnetite cobalt sulphide, and the produced nanocomposites were used to remove mercury (Hg); adsorption was more than 99.56 percent achieved in less than 7 min [89]. SWCNTs and their functionalized equivalents, such as SWCNTs complex with OH, NH_2, and COOH, have been employed for heavy metal ion adsorption including Cd^{2+}, Cu^{2+}, Pb^{2+}, and Hg^{2+} from aqueous systems [90]. Functionalization of SWCNT with L-cysteine is used to create nanocomposites for mercury removal from wastewater [91]. Gupta et al. [92] developed a film of SWCNTs-polysulfone nanocomposite for heavy metal removal. SWCNTs were used by Dehghani et al. [93] to remove Cr^{6+} ions from water. Using hydroxyl multi-wall CNTs and a PbO_2 complex, [94] developed photo-catalysts that successfully removed pyridine from water. The removal of Cr^{6+} ions by nanocomposites of oxidized MWCNTs with manganese oxide/iron (III) oxide has been observed, with an increase in the absorbent dose of 186 mg/g [95]. Nanotube fabrication using titanium oxide and manganese dioxide using a plasma-oxidized approach has also been described to eliminate lead ions from water [96]. The functionalization of MWCNTs with hydroxyl quinoline and their use for the elimination of copper (Cu), lead (Pb), cadmium (Cd), and other hazardous ions has also been described [97]. CNT-based nanocomposites were also used to remove iron and manganese from water in another experiment [98]. MWCNTs with magnetic grafting have a high absorption capacity, making them ideal for magnetic separation and heavy metal recovery [99].

The removal of arsenic from water is accomplished by employing CNTs as an adsorbent that has been changed by oxidation with an HNO_3 and H_2SO_4 combination [100]. The removal of Pb2 from CNTs treated with HNO_3 was also investigated [101]. With an adsorption capacity of 55.74 mg/g, synthetic Zeolite CNTs were employed as a novel technique for removing Pb_2 molecules. When compared to aminothiol-CNTs, small-wall CNTs functionalized with SH achieved an adsorption efficacy of 91% which is fivefold increased removal efficiency for Hg_2 [102]. Another study combined magnetite/thiol-functionalized MWCNTs with mercaptopropyl triethoxysilane to generate MPTS-CNTs/Fe_3O_4 nanocomposites to remove Pb_2 and Hg_2 [103]. MWCNTs synthesized with the Fe-Pt and Fe-Ti catalyst had a viability range of 95.9 to 99.8 percent viability compared with non-functionalized MWCNTs (Zhou et al. 2017).

9 Future Prospects of Carbon Nanoparticles as Heavy Metal Adsorbents

NMs are functionalized by joining with biomolecules, carbon, polymers, and inorganic substances to make separation easier and increase their capacity for adsorption. The tremendous potential of carbon nanoparticles as the primary agent for heavy metal removal has been revealed by lab-scale investigations, but most will not find practical applications in the context of larger scale remediation projects. Cost consideration is a major issue because the majority of these carbon nanoparticles which take the form of carbon nanotubes are expensive to produce in large amounts [104] (Table 2).

10 Conclusion

Water shortage, particularly in arid and semi-arid regions of the world, is placing extreme strain on supplies and needs providing adequate water for human consumption in addition to other purposes. Reusing and recycling water has shown to be effective and encouraging for delivering water that is consistent. Due to this, the focus is being made on the efficient treatment of other water sources, such as wastewater, saltwater, storm water, and industrial effluent. These additional water sources include wastewater that has been treated for sewage purposes. Carbon nanotubes (CNTs) are referred to be twenty-first-century technology. Because of their distinct physical and chemical characteristics, CNTs are now often utilized for the adsorption of heavy metals from water and waste. In this chapter, we looked at the various processes for making carbon nanoparticles as well as the uses of carbon nanomaterials for environmental and particular heavy metal purification from heavy metal pollutants. So, it is concluded that there are so many methods available for the production and functionalization of carbon nanoparticles which shows high impacts as absorbents of heavy metals present in the environment.

Table 2 Past, present, and possible future application of nanoparticle in contaminated soil remediation

S. no	Development strategies and prospects of carbon nanotube as heavy metal adsorbent	Key references
Past		
1	Strong oxidizers modify CNTs, such as $KMnO_4$, HNO_3, and H_2O_2, are more efficient at removing Cd^{2+} from aqueous solution	[105]
2	Effect of various morphologies of carbon nanotubes on Pb removal, after oxidation with nitric acid. CNTs with more defects introduce more functional groups onto their surfaces, having higher lead adsorption capacities	[106]
3	By increasing the agitation rate, solution pH, and nickel concentration, oxidized CNTs can absorb Ni^{2+} at a higher rate	[107]
4	In a batch reactor, the capacity of MWCNT treated with HNO_3 to remove Ni^{2+} cations from aqueous solutions was investigated	[7]
5	Surfaces of CNTs altered with alcohol and carboxyl groups to affect the aggregation state and the sites that are open for Cu^{2+} adsorption	[108]
6	In order to remove uranium from nuclear industrial effluents, amidoxime-grafted MWCNTs (AO-g-MWCNTs) was produced	[109]
7	Using magnetic MWCNT/iron oxide composite (Fe_3O_4/MWCNT) helps in removal of U(VI)	[110]
8	MWCNTs with chitosan-functionalization and magnetic $CoFe_2O_4$ nanoparticles (MNP-CTS) applied for Pb removal	[111]
9	Functionalized CNT-polydopamine-polyethylenimine (CNT-PDA-PEI) used for treatment of Cu	[112]
10	Use of MWCNTs in the removal of inorganic mercury (Hg^{+2}) or methylmercury. Absorption rate is affected by variables like pH (from 3 to 7), the concentration of mercury, and interaction time	[113]
11	Utilizing oxidized multi-walled carbon nanotubes to remove V(V) ions from aqueous solutions	[114]
12	Using potato peel waste MWCNTs was prepared from activated carbon by mixing it with the melamine (nitrogen-based material) and iron (III) oxalate hexahydrate used for heavy metal removal mainly lead from water	[115]
13	Purified and polyhydroxybutyrate functionalized carbon nanotube adsorbents for selected heavy metal (Pb, Cr, Cd, Ni, Cu, Fe, and Zn) removal from electroplating wastewater	[116]
Present		
1	Novel buckypaper membranes fabricated by MWCNTs and two biopolymers Chitosan and i-carrageenan. Used in removal of Co^{2+}, Ni^{2+}, Cu^{2+}, Cd^{2+}, Ba^{2+}, and Pb^{2+} ions	[117]
2	In this paper, MWCNTs were first partially oxidized to form 6O-MWCNTs, and then combined with magnetic ions to give O-MWCNTs@Fe_3O_4 and enhanced the adsorption capacities of CNTs for heavy metal ions from wastewater	[99]

(continued)

Table 2 (continued)

S. no	Development strategies and prospects of carbon nanotube as heavy metal adsorbent	Key references
3	Benzothiazole ionic liquid (N-butyl benzothiazole tetrafluoroborate (C_4Bth][PF_6])) was loaded into MWCNTs by impregnation method and then mixed with the derived cellulose to form a composite tablet, which was applied to remove tetracyclines and metal ions mainly Cr^{6+} and Cu^{2+} from water	[118]
4	Oxidized multi-walled carbon nanotubes (oxMWCNTs) were used for removing low levels of Cd(II) and Pb(II) from aqueous media. The studies were performed under specific conditions like low m/V ratio, high ionic strength, low and narrow C0 range performed good absorption	[119]
5	MWCNTs were used as adsorbents for Cr(VI), and the influence of operating parameters, on adsorption process, such as pH, concentration, and interaction time. Adsorption was inversely proportional with pH value	[120]
6	CNTs functionalized with amine groups and conjugated with two low-symmetry porphyrin derivatives to increase their active sites and showing adsorption of methylene blue and toxic metal ions from industrial wastewaters. They displayed 100% removal of pollutants	[121]
7	β-cyclodextrin was covalently grafted onto the surface of magnetic MWCNTs to synthesize a novel composite material (denoted as β-CD@Fe_3O_4/MWCNT) to remove nickel ions from aquatic solution. The adsorption amount of nickel ions (Ni^{2+}) was 103 mg g^{-1}	[122]
8	Application of MWCNT on the phytoremediation efficiency of hyper-accumulator plant, marigold (*Calendula officinalis* L.) and antioxidant defence under Pb and Cd stresses	[123]
10	Thiol-functionalized multi-walled carbon nanotubes (MWCNT-SH) were produced using 2-mercaptoethanol and MWCNT-COOH as a precursor. It effectively removed Pb when used as an adsorbent (II)	[124]
11	To enhance the hydrophilic surface of MWCNTs-KOH, Ni nanoparticles were added which is employed to remove Pb(II), As(V), and Cd(II) contamination	[125]
12	For the adsorption of metal ions, danthron-saturated carbon nanotubes were used with solid phase extraction towards a few selected metal ions, including lead, copper, cadmium, zinc, cadmium, cobalt, and nickel	[126]
13	Lead is removed from wastewater using an environmentally friendly substance called an N-doped carbon nanotube embellished with molybdenum disulphide nanosheet that resembles fish scales	[127]
14	For the first time, a carbon nanotube modified with tetrahydrofuran was created to remove copper from effluent	[128]
15	Without any pre-treatment, a new form of effective and affordable absorbent was created using steel slag, a cheap by-product of the steel industry, as a substrate medium for carbon nanotube (CNT) development by chemical vapour deposition (CVD) for Pb and Cu removal	[129]

(continued)

Table 2 (continued)

S. no	Development strategies and prospects of carbon nanotube as heavy metal adsorbent	Key references
Future		
1	Strong hydrophobicity and a low affinity for heavy metals are two properties of CNTs. To increase the adsorption effectiveness, CNTs must be further modified	[125]
2	CNTs have been reported to be more expensive to purchase than activated carbon. Making and exploiting lower grade carbon nanoparticles offers a potential solution to the cost barrier. More emphasis should be placed on using agriculture and biomass wastes as feedstock to create carbon-based nanomaterials	[129]
3	CNTs can be recycled and used again for the variety of applications. Cross-linking between CNTs can be synthesized and break to allow the recycling of CNTs for further use	[130]
4	MWCNT application on the phytoremediation efficacy of hyper-accumulator plants was discussed by Sharifi et al. (2021) provided new insights into the special consequences of MWCNTs in the phytoremediation of HMs-contaminated soils	[123]
5	Future prospective studies on CNT may take into account cutting-edge multivariate modelling techniques like artificial neural networks (ANNs)	[93]
6	Deep eutectic solvents (DES), a frequently employed functionalization agent, were discovered to have low toxic impacts, an easy synthesis procedure, high levels of material availability, and eco-friendly qualities. Future research is necessary to examine the various DES kinds and the ideal optimal ratios for improving CNTs' adsorption ability	[131]

References

1. Gupta N, Rai DB, Jangid AK, Kulhari H (2019) Use of nanotechnology in antimicrobial therapy. Methods Microbiol 46:143–172
2. Thangadurai D, Sangeetha J, Prasad R, Editors (2020) Nanotechnology for food, agriculture, and environment. Springer
3. Walker K (2015) Working with carbon nanoparticles. https://www.azonano.com/article.aspx?ArticleID=4149
4. Ray SC, Jana NR (2017) Carbon nanomaterials for biological and medical applications. Elsevier 2017:1–41
5. Sethy K, Pati S, Jena D, Mishra CK (2020) Heavy metal toxicity in animals: a review. Pharma Innov J 9:134–137
6. Izah SC, Chakrabarty N, Srivastav AL (2016) A review on heavy metal concentration in potable water sources in Nigeria: human health effects and mitigating measures. Expos Health 8(2):285–304
7. Mobasherpour I, Salahi E, Ebrahimi M (2012) Removal of divalent nickel cations from aqueous solution by multi-walled carbon nano tubes: equilibrium and kinetic processes. Res Chem Intermed 38(9):2205–2222
8. Shaniuk TJ (2005) Arsenic removal media. Google Patents
9. Wallace GT, Seibert DL, Holzknecht SM, Thomas WH (1982) The biogeochemical fate and toxicity of mercury in Controlled Experimental Ecosystems. Estuar Coast Shelf Sci 15 (2):151–182

10. Reesink BH, van Gasteren N (2003) US Patent 6,524,994
11. Chaignon V, Sanchez-Neira I, Herrmann P, Jaillard B, Hinsinger P (2003) Copper bioavailability and extractability as related to chemical properties of contaminated soils from a vine-growing area. Environ Pollut 123(2):229–238
12. Barceloux DG, Barceloux D (1999) Molybdenum. J Toxicol: Clin Toxicol 37(2):231–237
13. Elinder CG, Kjellström T, Hogstedt C, Andersson K, Spång G (1985) Cancermortality of cadmium workers. Occup Environmen Med 42(10):651–655
14. Fang H, Wu Y, Zhao J, Zhu J (2006) Silver catalysis in the fabrication of silicon nanowire arrays. Nanotechnology 17(15):3768
15. Mauter MS, Elimelech M (2008) Environmental applications of carbon-based nanomaterials. Environ Sci Technol 42(16):5843–5859
16. Wu Y, Pang H, Liu Y, Wang X, Yu S, Fu D, Chen J, Wang X (2019) Environmental remediation of heavy metal ions by novel-nanomaterials: a review. Environ pollut 246:608–620
17. Baby R, Saifullah B, Hussein MZ (2019) Carbon nanomaterials for the treatment of heavy metal-contaminated water and environmental remediation. Nanoscale res lett 14(1):1–17
18. Han X, Li S, Peng Z, Al-Yuobi ARO, Bashammakh ASO, Leblanc RM (2016) Interactions between carbon nanomaterials and biomolecules. J oleo sci ess15248
19. Lee K, Mazare A, Schmuki P (2014) One-dimensional titanium dioxide nanomaterials: nanotubes. Chem Rev 114(19):9385–9454
20. Obitayo W, Liu T (2012) A review: carbon nanotube-based piezoresistive strain sensors. J Sens
21. Baughman RH, Zakhidov AA, De Heer WA (2002) Carbon nanotubes–the route toward applications. Sci 297(5582):787–792
22. Abbasi M (2017) Synthesis and characterization of magnetic nanocomposite of chitosan/SiO_2/carbon nanotubes and its application for dyes removal. J Clean Prod 145:105–113
23. Esumi K, Ishigami M, Nakajima A, Sawada K, Honda H (1996) Chemical treatment of carbon nanotubes. Carbon (New York, NY) 34(2):279–281
24. MacKenzie KJ, See CH, Dunens OM, Harris AT (2008) Do single-walled carbon nanotubes occur naturally? Nat Nano technol 3(6):310–310
25. Velasco-Santos C, Martínez-Hernández AL, Consultchi A, Rodríguez R, Castaño VM (2003) Naturally produced carbon nanotubes. Chem Phys Lett 373(3–4):272–276
26. Becker L, Bada JL, Winans RE, Bunch TE (1994) Fullerenes in allende meteorite. Nature 372(6506):507
27. Andrade K, Guerra S, Debut A (2014) Fullerene-based symmetry in hibiscus rosa-sinensis pollen. PLoS ONE 9(7):e102123
28. Gore JP, Sane A (2011) Flame synthesis of carbon nanotubes. Carbon Nanotubes-Synth, Charact Appl 1:16801
29. Kumar M, Ando Y (2010) Chemical vapor deposition of carbon nanotubes: a review on growth mechanism and mass production. J Nano sci Nano technol 10(6):3739–3758
30. Zhang Q, Huang JQ, Zhao MQ, Qian WZ, Wei F (2011) Carbon nanotube mass production: principles and processes. Chem Sus Chem 4(7):864–889
31. Matsuzawa Y, Takada Y, Kodaira T, Kihara H, Kataura H, Yoshida M (2014) Effective nondestructive purification of single-walled carbon nanotubes based on high-speed centrifugation with a photochemically removable dispersant. J Phys Chem C 118(9):5013–5019
32. Morsy M, Helal M, El-Okr M, Ibrahim M (2014) Preparation, purification and characterization of high purity multi-wall carbon nanotube. Spectrochim Acta Part A Mol Biomol Spectrosc 132:594–598
33. Shameli K, Bin AM, Yunus WZW, Ibrahim NA, Darroudi M (2010) Synthesis and characterization of silver/talc nanocomposites using the wet chemical reduction method. Int J Nanomed 5(1):743–751
34. Chae C, Kim KW, Kim SJ, Lee D, Jo Y, Yun YJ, Moon J, Choi Y, Lee SS, Choi S, Jeong S (2015) 3D intra-stacked CoO/carbon nanocomposites welded by Ag nanoparticles for high-capacity, reversible lithium storage. Nanoscale 7(23):10368–10376

35. Dong J, Liu W, Li H, Su X, Tang X, Uher C (2013) In situ synthesis and thermoelectric properties of PbTe–graphene nanocomposites by utilizing a facile and novel wet chemical method. J Mater Chem A 1(40):12503–12511
36. Du WC, Yin YX, Zeng XX, Shi JL, Zhang SF, Wan LJ, Guo YG (2016) Wet chemistry synthesis of multidimensional nanocarbon-sulfur hybrid materials with ultrahigh sulfur loading for lithium-sulfur batteries. ACS Appl Mater Interfaces 8(6):3584–3590
37. Husen A, Iqbal M (2019b) Nanomaterials and plant potential: an overview. In: Husen A, Iqbal M (eds) Nanomaterials and plant potential, vol 11. Springer International Publishing AG, Gewerbestrassep, Cham, pp 3–29 (https://doi.org/10.1007/978-3-030-05569-1_1
38. Husen A, Siddiqi KS (2014) Phytosynthesis of nanoparticles: concept, controversy and application. Nanoscale Res Lett 9(229):1–24
39. Husen A (2023) Smart nanomaterials from agricultural and horticultural products. Springer Nature Singapore Pte Ltd., 152 Beach Road, #21-01/04 Gateway East, Singapore 189721, Singapore
40. Husen A (2023) Secondary metabolites based green synthesis of nanomaterials and their applications. Springer Nature Singapore Pte Ltd., 152 Beach Road, #21–01/04 Gateway East, Singapore 189721, Singapore
41. Husen A, Siddiqi KS (2023) Advances in smart nanomaterials and their applications. Elsevier Inc., 50 Hampshire St., 5th Floor, Cambridge, MA 02139, USA
42. Husen, A (2022) Engineered nanomaterials for sustainable agricultural production, soil improvement and stress management. Elsevier Inc., 50 Hampshire St., 5th Floor, Cambridge, MA 02139, USA
43. Husen A, Iqbal M (2019a) Nanomaterials and plant potential, vol 11. Springer International Publishing AG, Gewerbestrasse. Cham, Switzerland. https://doi.org/10.1007/978-3-030-05569-1
44. Jin-Chul K, Madhusudhan A, Husen A (2021) Smart nanomaterials in biomedical applications, vol 11. Springer Nature Switzerland AG, Gewerbestrasse. Cham, Switzerland, p 6330
45. Taghiyari HR, Morrell JJ, Husen A (2022) Emerging nanomaterials (Opportunities and Challenges in Forestry Sectors), 11. Springer Nature Switzerland AG, Gewerbestrasse, Cham, Switzerland, p 6330 https://doi.org/10.1007/978-3-031-17378-3
46. Yogeshwaran S, Natrayan L, Udhayakumar G, Godwin G, Yuvaraj L (2021) Effect of waste tyre particles reinforcement on mechanical properties of jute and abaca fiber- epoxy hybrid composites with pre-treatment. Mater Today Proc 37(Part 2):1377–80.
47. Li S, Wang Y, Lai C, Qiu J, Ling M, Martens W et al (2014) Directional synthesis of tin oxide graphene nanocomposites via a one-step up-scalable wet-mechanochemical route for lithium ion batteries. J Mater Chem A 2(26):10211–10217
48. Ramachandran R, Saranya M, Grace AN, Wang F (2017) MnS nanocomposites based on doped graphene: simple synthesis by a wet chemical route and improved electrochemical properties as an electrode material for supercapacitors. RSC Adv 7(4):2249–2257
49. Gich M, Fernández-Sánchez C, Cotet LC, Niu P, Roig A (2013) Facile synthesis of porous bismuth–carbon nanocomposites for the sensitive detection of heavy metals. J Mater Chem A 1(37):11410–11418
50. Mirzaee SA, Jaafarzadeh N, Gomes HT, Jorfi S, Ahmadi M (2019) Magnetic titanium/carbon nanotube nanocomposite catalyst for oxidative degradation of Bisphenol A from high saline polycarbonate plant effluent using catalytic wet peroxide oxidation. Chem Eng J 370:372–386
51. Ko TH, Radhakrishnan S, Seo MK, Khil MS, Kim HY, Kim BS (2017) A green and scalable dry synthesis of $NiCo_2O_4$/graphene nanohybrids for high-performance supercapacitor and enzymeless glucose biosensor applications. J Alloys Compd 696:193–200
52. Kalathil S, Khan MM, Banerjee AN, Lee J, Cho MH (2012) A simple biogenic route to rapid synthesis of Au@TiO_2 nanocomposites by electrochemically active biofilms. J Nanopart Res 14(8):1–9
53. Kellici S, Acord J, Vaughn A, Power NP, Morgan DJ, Heil T et al (2016) Calixarene assisted rapid synthesis of silver-graphene nanocomposites with enhanced antibacterial activity. ACS Appl Mater Interfaces 8(29):19038–19046

54. Yogeshwaran S, Natrayan L, Rajaraman S, Parthasarathi S, Nestro S (2021) Experimental investigation on mechanical properties of Epoxy/graphene/fish scale and fermented spinach hybrid bio composite by hand lay-up technique. Mater Today Proc 37(Part 2):1578–1583.
55. Vergaro V, Pisano I, Grisorio R, Baldassarre F, Mallamaci R, Santoro A et al (2019) $CaCO_3$ as an environmentally friendly renewable material for drug delivery systems: uptake of HSA-$CaCO_3$ nanocrystals conjugates in cancer cell lines. Mater 12(9):1481
56. da Silva LMG, Lemos HG, Santos SF, Antunes RA, Venancio EC (2018) Polyaniline/Carbon black nanocomposites: the role of synthesis conditions on the morphology and properties. Mater Today Commun. 16:14–21
57. Wu KH, Chang YC, Yang CC, Gung YJ, Yang FC (2009) Synthesis, infrared stealth and corrosion resistance of organically modified silicate–polyaniline/carbon black hybrid coatings. Eur Polym J 45(10):2821–2829
58. Esmizadeh E, Naderi G, Paran SMR (2017) Preparation and characterization of hybrid nanocomposites based on NBR/Nanoclay/Carbon black. Polym Compos 38:E181–E188
59. Wu KH, Ting TH, Wang GP, Ho WD, Shih CC (2008) Effect of carbon black content on electrical and microwave absorbing properties of polyaniline/carbon black nanocomposites. Polym Degrad Stab 93(2):483–488
60. Mickelson ET, Chiang IW, Zimmerman JL, Boul PJ, Lozano J, Liu J, Margrave JL (1999) Solvation of fluorinated single-wall carbon nanotubes in alcohol solvents. J Phy Chem B 103(21):4318–4322
61. Herrero MA, Prato M (2008) Recent advances in the covalent functionalization of carbon nanotubes. Mol Cryst Liq Cryst 483(1):21–32
62. Holzinger M, Abraham J, Whelan P, Graupner R, Ley L, Hennrich F, Hirsch A (2003) Functionalization of single-walled carbon nanotubes with (R-)oxycarbonylnitrenes. J Am Chem Soc 125(28):8566–8580
63. Chen RJ, Bangsaruntip S, Drouvalakis KA, Wong Shi Kam N, Shim M, Li Y, Kim W, Utz PJ, Dai H (2003) Noncovalent functionalization of carbon nanotubes for highly specific electronic biosensors. Proc Natl Acad Sci 100(9):4984–4989
64. Zhao YL, Stoddart JF (2009) Noncovalent functionalization of single-walled carbon nanotubes. Acc Chem Res 42(8):1161–1171
65. Zhao J, Lu JP, Han J, Yang CK (2003) Noncovalent functionalization of carbon nanotubes by aromatic organic molecules. Appl Phys Lett 82(21):3746
66. Georgakilas V, Tiwari JN, Kemp KC, Perman JA, Bourlinos AB, Kim KS et al (2016) Noncovalent functionalization of graphene and graphene oxide for energy materials, biosensing, catalytic, and biomedical applications. Chem Rev 116(9):5464–5519
67. Nakayama-Ratchford N, Bangsaruntip S, Sun X, Welsher K, Dai H (2007) Noncovalent functionalization of carbon nanotubes by fluorescein-polyethylene glycol: supramolecular conjugates with pH-dependent absorbance and fluorescence. J Am Chem Soc 129(9):2448–2449
68. Chen J, Collier CP (2005) Noncovalent functionalization of single-walled carbon nanotubes with water-soluble porphyrins. J Phys Chem B 109(16):7605–7609
69. Petrov P, Stassin F, Pagnoulle C, Jérôme R (2003) Noncovalent functionalization of multi-walled carbon nanotubes by pyrene containing polymers. Chem Commun 3(23):2904–2905
70. Ghosh A, Rao KV, George SJ, Rao CNR (2010) Noncovalent functionalization, exfoliation, and solubilization of graphene in water by employing a fluorescent coronene carboxylate. Chem A Eur J 16(9):2700–2704
71. Tian R, Jia X, Yang J, Li Y, Song H (2020) Large-scale, green, and high-efficiency exfoliation and noncovalent functionalization of fluorinated graphene by ionic liquid crystal. Chem Eng 395:125104
72. Kim J, Cha J, Chung B, Ryu S, Hong SH (2020) Fabrication and mechanical properties of carbon fiber/epoxy nanocomposites containing high loadings of noncovalently functionalized graphene nanoplatelets. Compos Sci Technol 192:108101
73. Alzate-Carvajal N, Bolivar-Pineda LM, Meza-Laguna V, Basiuk VA, Basiuk EV, Baranova EA (2020) Oxygen evolution reaction on single-walled carbon nanotubes noncovalently functionalized with metal phthalocyanines. Chem Electro Chem 7(2):428–436

74. Ju Z, Yao X, Luo Z, Cao M, Xiao W (2020) Theoretical studies on the noncovalent interaction of fructose and functionalized ionic liquids. Carbohydr Res 487:107882
75. Karimi M, Solati N, Amiri M, Mirshekari H, Mohamed E, Taheri M et al (2015) Carbon nanotubes part I: preparation of a novel and versatile drug-delivery vehicle. Expert Opin Drug Deliv 12(7):1071–1087
76. Ribeiro RS, Silva AMT, Figueiredo JL, Faria JL, Gomes HT (2016) Catalytic wet peroxide oxidation: a route towards the application of hybrid magnetic carbon nanocomposites for the degradation of organic pollutants. A review. Appl Catal B Environ 187:428–460
77. Zhou X, Dai Z, Bao J, Guo YG (2013) Wet milled synthesis of an Sb/MWCNT nanocomposite for improved sodium storage. J Mater Chem A 1(44):13727–13731
78. Zein SHS, Yeoh LC, Chai SP, Mohamed AR, Mahayuddin MEM (2007) Synthesis of manganese oxide/carbon nanotube nanocomposites using wet chemical method. J Mater Process Technol 190(1–3):402–405
79. Rumyantseva MN, Kovalenko VV, Gaskov AM, Pagnier T, Machon D, Arbiol J, Morante JR (2005) Nanocomposites SnO_2/Fe_2O_3: wet chemical synthesis and nanostructure characterization. Sens Actuators B Chem 109(1):64–74
80. Bissen M, Frimmel FH (2003) Arsenic- a review. Part II: oxidation of arsenic and its removal in water treatment. Acta Hydrochim Hydrobiol 31(2):97–107
81. Gihring TM, Druschel GK, McCleskey RB, Hamers RJ, Banfield JF (2001) Rapid arsenite oxidation by Thermus aquaticus and Thermus thermophilus: field and laboratory investigations. Environ sci technol 35(19):3857–3862
82. Kim J, Benjamin MM (2004) Modeling a novel ion exchange process for arsenic and nitrate removal. Water Res 38(8):2053–2062
83. Kumar PR, Chaudhari S, Khilar KC, Mahajan SP (2004) Removal of arsenic from water by electrocoagulation. Chemosphere 55(9):1245–1252
84. Ning RY (2002) Arsenic removal by reverse osmosis. Desalination 143(3):237–241
85. Abbas A, Al-Amer AM, Laoui T, Al-Marri MJ, Nasser MS, Khraisheh M, Atieh MA (2016) Heavy metal removal from aqueous solution by advanced carbon nanotubes: critical review of adsorption applications. Sep Purif Technol 157:141–161
86. Baskar AV, Bolan N, Hoang SA, Sooriyakumar P, Kumar M, Singh L, Jasemizad T, Padhye LP, Singh G, Vinu A, Sarkar B, Kirkham MB, Rinklebe J, Wang S, Wang H, Balasubramanian R, Siddique KHM (2022) Recovery, regeneration and sustainable management of spent adsorbents from wastewater treatment streams: a review. Sci Total Environ 822:153555
87. Kim EJ, Lee CS, Chang YY, Chang YS (2013) Hierarchically structured manganese oxide-coated magnetic nanocomposites for the efficient removal of heavy metal ions from aqueous systems. ACS Appl Mater Interfaces 5(19):9628–9634
88. Gupta RK, Dunderdale GJ, England MW, Hozumi A (2017) Oil/water separation techniques: a review of recent progresses and future directions. J Mat Chem A 5(31):16025–16058
89. Alijani H, Shariatinia Z (2018) Synthesis of high growth rate SWCNTs and their magnetite cobalt sulfidenanohybrid as super-adsorbent for mercury removal. Chem Eng Res Des 129:132–149
90. Anitha K, Namsani S, Singh JK (2015) Removal of heavy metal ions using a functionalized single-walled carbon nanotube: a molecular dynamics study. J Phys Chem 119(30):8349–8358
91. Zazouli MA, Yousefi Z, Yazdani Cherati J, Tabarinia H, Tabarinia F, Akbari Adergani B (2014) Evaluation of L-Cysteine functionalized single-walled carbon nanotubes on mercury removal from aqueous solutions. J Maz Univ Med Sci 24(111):10–21
92. Gupta S, Bhatiya D, Murthy CN (2015) Metal removal studies by composite membrane of polysulfone and functionalized single-walled carbon nanotubes. Sep Sci Technol 50(3):421–429
93. Dehghani MH, Yetilmezsoy K, Salari M, Heidarinejad Z, Yousefi M, Sillanpää M (2020) Adsorptive removal of cobalt (II) from aqueous solutions using multi-walled carbon nanotubes and γ-alumina as novel adsorbents: modelling and optimization based on response surface methodology and artificial neural network. J Mol Liq 299:112154

94. Xu Z, Liu H, Niu J, Zhou Y, Wang C, Wang Y (2017) Hydroxyl multi-walled carbon nanotube-modified nanocrystalline PbO_2 anode for removal of pyridine from wastewater. J Hazard Mater 327:144–152
95. Li H, Ha CS, Kim I (2009) Fabrication of carbon nanotube/SiO_2 and carbon nanotube/SiO_2/Ag nanoparticles hybrids by using plasma treatment. Nanoscale Res Lett 4(11):1384–1388
96. Zhao X, Jia Q, Song N, Zhou W, Li Y (2010) Adsorption of Pb (II) from an aqueous solution by titanium dioxide/carbon nanotube nanocomposites: kinetics, thermodynamics, and isotherms. J Chem Eng Data 55(10):4428–4433.
97. Saifullah B, Chrzastek A, Maitra A, Naeemullah B, Fakurazi S, Bhakta S, Hussein MZ (2017) Novel anti-tuberculosis nanodelivery formulation of ethambutol with graphene oxide. Molecules 22(10):1560
98. Elsehly EM, Chechenin NG, Bukunov KA, Makunin AV, Priselkova AB, Vorobyeva EA, Motaweh HA (2016) Removal of iron and manganese from aqueous solutions using carbon nanotube filters. Water Sci Technol Water Supply 16(2):347–353
99. Wang Z, Xu W, Jie F, Zhao Z, Zhou K, Liu H (2021) The selective adsorption performance and mechanism of multiwall magnetic carbon nanotubes for heavy metals in wastewater. Sci Rep 11(1):16878
100. Veličković ZS, Marinković AD, Bajić ZJ, Marković JM, Perić-Grujić AA, Uskokovic PS, Ristic MD (2013) Oxidized and ethylenediamine-functionalized multi-walled carbon nanotubes for the separation of low concentration arsenate from water. Sep Sci Technol 48(13):2047–2058
101. Tofighy MA, Mohammadi T (2011) Adsorption of divalent heavy metal ions from water using carbon nanotube sheets. J Hazard Mater 185(1):140–147
102. Bandaru NM, Reta N, Dalal H, Ellis AV, Shapter J, Voelcker NH (2013) Enhanced adsorption of mercury ions on thiol derivatized single wall carbon nanotubes. J Hazard Mater 261:534–541
103. Zhang Q, Vigier KD, Royer S, Jérôme F (2012) Deep eutectic solvents: syntheses, properties and applications. Chem Soc Rev 41(21):7108–7146
104. Mahesh N, Balakumar S, Shyamalagowri S, Manjunathan J, Pavithra MKS, Babu PS, Kamaraj M, Govarthanan M (2022) Carbon-based adsorbents as proficient tools for the removal of heavy metals from aqueous solution: a state of art-review emphasizing recent progress and prospects. Environ Res 213:113723
105. Li ZX, Li XH, Kinny PD, Wang J, Zhang S, Zhou H (2003) Geochronology of Neoproterozoic syn-rift magmatism in the Yangtze Craton, South China and correlations with other continents: evidence for a mantle superplume that broke up Rodinia. Precambr Res 122(1–4):85–109
106. Li J, Zhang G, Qi S, Li X, Peng X (2006) Concentrations, enantiomeric compositions, and sources of HCH, DDT and chlordane in soils from the Pearl River Delta, South China. Sci Total Environ 372(1):215–224
107. Lu G, Ocola LE, Chen J (2009) Reduced graphene oxide for roomtemperature gas sensors. Nanotechnology 20(44):445502
108. Rosenzweig C, Jones JW, Hatfield JL, Ruane AC, Boote KJ, Thorburn P, Antle JM, Nelson GC, Porter C, Janssen S, Asseng S (2013) The agricultural model intercomparison and improvement project (AgMIP): protocols and pilot studies. Agric For Meteorol 170:166–182
109. Wang Y, Jodoin PM, Porikli F, Konrad J, Benezeth Y, Ishwar P (2014) CDnet 2014: An expanded change detection benchmark dataset. In: Proceedings of the IEEE conference on computer vision and pattern recognition workshops, pp 387–394
110. Zong P, Gou J (2014) Rapid and economical synthesis of magnetic multiwalled carbon nanotube/iron oxide composite and its application in preconcentration of U (VI). J Mol Liq 195:92–98
111. Zhou J (2014) Multicatalyst system in asymmetric catalysis. John Wiley & Sons
112. Xie W, Lv X, Ye L, Zhou P, Yu H (2015) Construction of lycopeneoverproducing Saccharomyces cerevisiae by combining directed evolution and metabolic engineering. Metab Eng 30:69–78

113. Yaghmaeian K, Khosravi Mashizi R, Nasseri S, Mahvi AH, Alimohammadi M, Nazmara S (2015) Removal of inorganic mercury from aquatic environments by multi-walled carbon nanotubes. J Environ Health Sci Eng 13:1–9
114. Sobhanardakani S, Zandipak R (2015) 2, 4-Dinitrophenylhydrazine functionalized sodium dodecyl sulfate-coated magnetite nanoparticles for effective removal of Cd (II) and Ni (II) ions from water samples. Environ Monit Assess 187:1–4
115. Al Osman M, Yang F, Massey IY (2019) Exposure routes and health effects of heavy metals on children. Biometals 32:563–573
116. Bankole MT, Abdulkareem AS, Mohammed IA, Ochigbo SS, Tijani JO, Abubakre OK, Roos WD (2019) Selected heavy metals removal from electroplating wastewater by purified and polyhydroxylbutyrate functionalized carbon nanotubes adsorbents. Sci Rep 9(1):4475
117. Alshahrani B, Olarinoye IO, Mutuwong C, Sriwunkum C, Yakout HA, Tekin HO, Al-Buriahi MS (2021) Amorphous alloys with high Fe content for radiation shielding applications. Radiat Phys Chem 183:109386
118. Chen C, Feng X, Yao S (2021) Ionic liquid-multi walled carbon nanotubes composite tablet for continuous adsorption of tetracyclines and heavy metals. J Cleaner Prod 286:124937
119. Šolić M, Maletić S, Isakovski MK, Nikić J, Watson M, Kónya Z, Rončević S (2021) Removing low levels of Cd (II) and Pb (II) by adsorption on two types of oxidized multiwalled carbon nanotubes. J Environ Chem Eng 9(4):105402
120. Mpouras T, Polydera A, Dermatas D, Verdone N, Vilardi G (2021) Multi wall carbon nanotubes application for treatment of Cr (VI)-contaminated groundwater; Modeling of batch & column experiments. Chemosphere 269:128749
121. Saleh TA, Elsharif AM, Bin-Dahman OA (2021) Synthesis of amine functionalization carbon nanotube-low symmetry porphyrin derivatives conjugates toward dye and metal ions removal. J Mol Liq 340:117024
122. Lin Y, Huang R, Sun X, Yu X, Xiao Y, Wang L, Hu W, Zhong T (2021) The p-Anisaldehyde/β-cyclodextrin inclusion complexes as fumigation agent for control of postharvest decay and quality of strawberry. Food Control 130:108346
123. Sharifi P, Bidabadi SS, Zaid A, Latef AA (2021) Efficacy of multi-walled carbon nanotubes in regulating growth performance, total glutathione and redox state of Calendula officinalis L. cultivated on Pb and Cd polluted soil. Ecotoxicol Environmen Saf 213:112051
124. Qu G, Bai Y, Zhang Y, Jia Q, Zhang W, Yan B (2009) The effect of multiwalled carbon nanotube agglomeration on their accumulation in and damage to organs in mice. Carbon 47(8):2060–2069
125. Egbosiuba TC, Egwunyenga MC, Tijani JO, Mustapha S, Abdulkareem AS, Kovo AS, Krikstolaityte V, Veksha A, Wagner M, Lisak G (2022) Activated multi-walled carbon nanotubes decorated with zero valent nickel nanoparticles for arsenic, cadmium and lead adsorption from wastewater in a batch and continuous flow modes. J Hazard Mat 423:126993
126. Shahryari T, Singh P, Raizada P, Davidyants A, Thangavelu L, Sivamani S, Naseri A, Vahidipour F, Ivanets A, Hosseini-Bandegharaei A (2022) Adsorption properties of Danthron-impregnated carbon nanotubes and their usage for solid phase extraction of heavy metal ions. Colloids Surf A: Physicochem Eng Aspects 641:128528
127. Zhang L, Liang J, Yue L, Xu Z, Dong K, Liu Q, Luo Y, Li T, Cheng X, Cui G, Tang B (2022) N-doped carbon nanotubes supported CoSe2 nanoparticles: a highly efficient and stable catalyst for H2O2 electrosynthesis in acidic media. Nano Res 15(1):304–309
128. Ghanavati B, Bozorgian A, Esfeh HK (2022) Thermodynamic and kinetic study of adsorption of cobalt II using adsorbent of magnesium oxide nano-particles deposited on chitosan. Prog Chem Biochem Res 5(2)
129. Hoang AT, Nižetić S, Cheng CK, Luque R, Thomas S, Banh TL, Nguyen XP (2022) Heavy metal removal by biomass-derived carbon nanotubes as a greener environmental remediation: a comprehensive review. Chemosphere 287:131959
130. Wang Q, Feng X, Liu Y, Cui W, Sun Y, Zhang S, Wang F (2022) Effects of microplastics and carbon nanotubes on soil geochemical properties and bacterial communities. J Hazard Mater 433:128826

131. AlOmar MK, Alsaadi MA, Hayyan M, Akib S, Ibrahim M, Hashim MA (2017) Allyl triphenyl phosphonium bromide based DES-functionalized carbon nanotubes for the removal of mercury from water. Chemosphere 167:44–52
132. Churilov GN (2008) Synthesis of fullerenes and other nanomaterials in arc discharge. Fuller, Nanotub Carbon Nanostruct 16(5–6):395–403
133. Hawari AH, Mulligan CN (2006) Biosorption of lead (II), cadmium (II), copper (II) and nickel (II) by anaerobic granular biomass. Bioresour Technol 97(4):692–700
134. Krätschmer W (2011) The story of making fullerenes. Nanoscale 3(6):2485–2489
135. Krätschmer W, Lamb LD, Fostiropoulos K, Huffman DR (1990) Solid C60: a new form of carbon. Nature 347(6291):354–358
136. Mracek JD, Fagan RM, Stengelin R, Hesjedal T (2011) Are carbon nanotubes a naturally occurring material? hints from methane CVD using lava as a catalyst. Curr Nano sci 7(3):294–296
137. Vidu R, Rahman M, Mahmoudi M, Enachescu M, Poteca TD, Opris I (2014) Nanostructures: a platform for brain repair and augmentation. Front Syst Neurosci 8:91

Recent Development and Importance of Nanoparticles in Disinfection and Pathogen Control

Deepak Kumar Verma, Aishwarya Sharma, Laxmi Awasthi, Himanshi Singh, Pankaj Kumar, Pranchal Rajput, Aashna Sinha, Kundan Kumar Chaubey, Anil Kumar, Nishant Rai, and Rakesh Kumar Bachheti

Abstract Antibiotic resistance has become alarmingly common in recent years, posing a severe threat to public health globally as it causes millions of deaths worldwide. Multidrug-resistant (MDR) microorganisms have emerged due to widespread dissemination of resistance and sharing of resistance genes, exacerbated by biofilms. This has inspired global attempts to create new and enhanced antimicrobial drugs as well as creative and effective methods for administering and directing antibiotics. To manage or prevent diseases brought on by microorganisms, researchers are looking for substitutes for currently used antimicrobial agents. Effective antibacterial drugs are developed using a variety of techniques, and nanoparticles (NPs) are unquestionably promising candidates Additionally, the different sizes and physical properties of NPs make it possible for them to target biofilms and treat diseases that

---

D. K. Verma (✉) · L. Awasthi · H. Singh
Department of Biotechnology, Sanskriti University, 28 K.M. Stone, Mathura-Delhi Highway, Chhata, Mathura, U.P., India
e-mail: deepakvermabiotek@gmail.com

A. Sharma
Division of Zoology, Department of Bio-Sciences, Career Point University, Hamirpur, H.P., India

P. Kumar
Department of Physics, Career Point Univeristy, Hamirpur, H.P., India

P. Rajput · A. Sinha · K. K. Chaubey
Division of Research and Innovation, Uttaranchal University, Dehradun, U.K., India

A. Kumar
Department of Microbiology, Karnali Academy of Health Sciences, Jumla, Karnali, Nepal

N. Rai
Department of Biotechnology, Graphic Era (Deemed to be) University, Clement Town, Dehradun, U.K., India

R. K. Bachheti
Department of Industrial Chemistry, College of Applied Science, Addis Ababa Science & Technology University, P.O. Box 16417, Addis Ababa, Ethiopia

R. K. Bachheti et al. (eds.), *Nanomaterials for Environmental and Agricultural Sectors*, Smart Nanomaterials Technology,
https://doi.org/10.1007/978-981-99-2874-3_5

are resistant. Pathogenic germs' cell membranes can be penetrated by NPs, which then create special antimicrobial mechanisms by interfering with crucial molecular processes. NPs can be used with antibiotics to reduce bacterial resistance. NPs work by evading bacterial defenses against drugs and preventing the growth of biofilms or other critical virulence-related activities. Bacterial cell walls and membranes can be penetrated by NPs, which disturb crucial molecular processes. NPs may exhibit synergy when used in conjunction with the proper antibiotics and aid in halting the growing global epidemic of bacterial resistance. The most widely employed types of nanoparticles worldwide are those based on Ag, Au, Cu, and Zn. The antimicrobial and disinfectant characteristics of nanoparticles eliminate the undesired microorganisms that develop resistance to different antimicrobials. Silver and zinc nanoparticles are often utilized in the pharmaceutical, agricultural, biomedical, and food packaging industries. This chapter aims to list down different types of NPs used as disinfectant, their importance and applications.

Keywords Nanoparticles · Infection · Disinfection · Pathogens · Control · Antimicrobials · Chemotherapy

1 Introduction

Nanotechnology (NT) is a Greek word meaning 'dwarf.' A nanometer is around 100,000th of hair width and is referred to as a nanometer. The size range between 0.1–100 nm is considered nanoparticles, and the biomolecule shape depends upon the compound structure [86]. Nanoparticles (NPs) fall in the range of about 1–100 nm [25]. NT consists of small-sized particles that have applications in different fields of science, such as chemistry, biology, physics, and neuroscience. NT has a wide range of applications and technological advancements in genetics, agriculture, biological, chemical, and pharmaceutical sciences [37, 41–45, 48]. NT has resolved problems related to pathogen and disinfection applications [32]. In the last decades, the use of NT has increased and become a promising concept [8, 9]. NPs impact the microbes, including the breakdown of cell structure, loss of functional activity, and production of neutralized variants. These particles possess antimicrobial action due to their wide surface area and volume ratios. Disinfectants possess various levels of physical variables like osmotic pressure, pH, temperature, and surface tension, while their action also depends upon parameters like concentration and time of application [46]. Disease detection and treatment of infectious diseases exist as a severe threat in the present world. The spread of infection with context to the health concept has been influenced by environmental pollution [6]. Various types of disinfectants possess characteristic features like antibacterial and antimicrobial applications and help remove harmful bacterial consequences [23]. Properties like allergy symptoms and germ resistance have been developed through the use of disinfectants. Harmful and resistant pathogens have increased the demand for disinfection strategies with context to antimicrobial constituents [83].

The application of NT can prevent biofilm formation in most medical devices, foods, and cosmetics. Nano disinfection applications include interactions occurring in the nanoparticles. About 99% of the mortality rate and long-term effects are known to be harmful to many bacteria and viruses [72]. Until now, no symptoms have been found regarding microbial resistance. Most of the nano-products used for disinfection are non-toxic and capable of influencing various microorganisms, which are efficient in various useful applications [13]. The toxic consequences of nano-products are being kept negligible or moderately low, and the most commonly used metallic nanoparticles in bacterial studies include zinc, silver, copper, and titanium [79, 97, 100]. In this chapter, we will discuss the various types of NPs used for treatment of bacterial growth, their applications, and their importance towards the therapy and disinfection of pathogens.

2 Mechanism of Action of Nanoparticles as Antibacterials

Nanomaterials can use a number of bactericidal techniques to fight bacteria, including the generation of reactive oxygen species (ROS), the rupturing of cell walls and membranes, the transfer of drugs via membrane fusion, and interaction with intra-cellular components (e.g., DNA and ribosomes). It is stated that Van der Waals forces, receptor-ligand interactions, hydrophobic interactions, and electrostatic attractions dictate the unique physicochemical properties of nanomaterials, specifically their chelation with bacterial cells [35].

2.1 Destruction of Cell Wall and Cell Membrane

Pits begin to form in the bacterial cell wall as a result of NPs implementation. Because of this, they have the ability to infiltrate cells, modify cell membranes, and cause structural damage and cell death. Positively charged ions produced by NPs have a stronger bactericidal impact by interacting with the negative charges on the bacterial surfaces (through carboxyl or phosphate groups). However, it is more challenging for NPs to enter gram-positive bacteria due to their thick peptidoglycan covering. As a result, NPs only engage the surface of bacteria [77].

2.2 Production of Reactive Oxygen Species (ROS)

One important mechanism of nanotoxicity has been identified as the production of ROS, which causes oxidative stress in cells and microbial death. The need for new antibacterial with novel modes of action may be met by NPs as more research demonstrates that NP-induced oxidative stress can be used to kill a variety of microorganisms [39].

ROS is a byproduct that occurs from oxidative metabolism. They have an impact on the growth, signaling, survival, and demise of cells. They also have a high positive redox potential. Superoxide radicals (O_2), singlet oxygen (O_2), hydroxyl radicals (OH), and hydrogen peroxide are examples of ROS (H_2O_2). Various NPs produce distinct ROS combinations with various antibacterial properties. For instance, while Ag and Cu NPs produce all types of ROS, MgO NPs and ZnO NPs, respectively, only produce O_2 radicals and mixtures of (H_2O_2) and OH radicals. ROS are produced when the respiratory chain is disrupted or when NPs are present. Normally, scavenger molecules like reduced glutathione keep them at a low level [77].

2.3 Binding to and Damaging Intracellular Component and Inhibition of RNA and Protein Synthesis Through the Induction of Intracellular Effects

Cellular homeostasis and intracellular signaling pathways are essential for the survival and function of bacteria. When nanoparticles are engineered to impede these routes, cell death may follow. Among these disruptions are modifications in protein synthesis, DNA damage, and alterations in gene expression [77].

2.4 Destruction of Biofilm Architecture

NPs' strong penetrating power enables them to destroy the biofilm. They pierce thick biofilm (extracellular polysaccharide matrix) layers, and after doing so, interact with the bacteria and have bactericidal effects. The interaction and penetration of biofilm are significantly influenced by the charge of NPs. For instance, cationic or positively charged NPs are interacting with the anionic matrix [38].

3 Type of Nanoparticles Used as Disinfectants

3.1 Silver (Ag) Nanoparticles

Silver (Ag)-based NPs are used in antimicrobial as well as anticancer therapy. They are also employed in wound healing. The biological mechanism of Ag-NPs action mainly involves the release of silver ions (Ag^+), active oxygen species (AOS) generation, and destruction of membrane structure. Ag-NPs possess antibacterial effects and are thus used to control bacterial growth. They play an important role in providing disinfection with context to dental work, surgical applications, and wound treatment. Ag-NPs are known to hold catalytic redox properties for biological agents like dyes and chemical entities, which make them useful in the degradation process reducing disinfection chances. Ag-NPs are also used in food packaging and as sterilizing agents in medical equipment. Ag^+ has been known for their antimicrobial action and medicinal importance for many centuries. It acts as an essential constituent in Ayurveda and medicine. Ag^+ has versatile functions due to which it shows a great response with regard to the antimicrobial activity, offering high surface exposure towards the microbes. Silver is known as a significant noble metal because of its antibacterial properties and wide range of applications [5, 33]. Ag^+ has an affinity to interact with the negatively charged nucleic acid and the plasma membrane, leading to the breakdown of DNA structure and making the cell membrane unstable [93].

Ag^+ ions can bind with the -SH group, also known as the thiol group, present at the active sites of the enzymes, resulting in the stable complex formation Ag–S. When a blockage takes in the enzyme, cell death occurs due to a decreased respiration rate [14]. Biocides and traditional antibiotics manifest biochemical destruction of cells leading to resistance. Metallic nanoparticles can trigger bacterial cells by physical or chemical means [92]. Silver nanoparticles can be regarded as the best and most efficient alternatives for cleaning and disinfecting equipment with multidrug-resistant species of bacteria. They are known to improve human health and the environment as well. Silver nanoparticles have various applications as disinfectants in different sectors, as shown in Fig. 1.

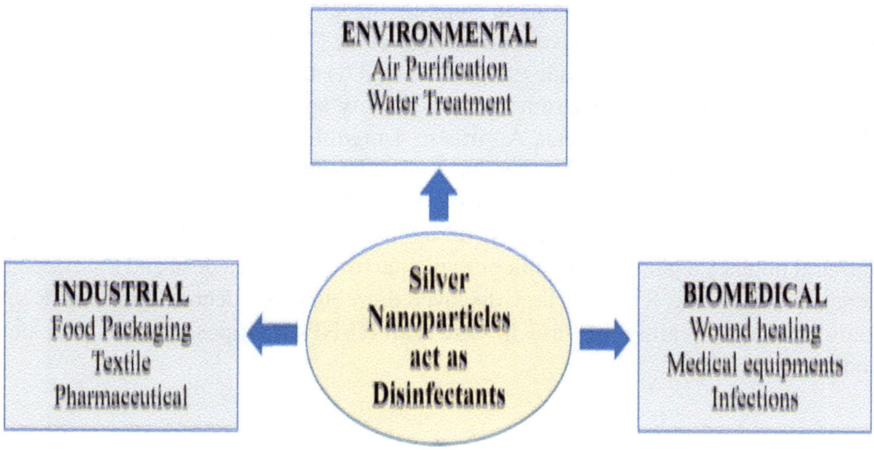

Fig. 1 Role of Silver nanoparticles as disinfectants

3.2 Zinc Nanoparticles

Zinc oxide is known to be a safe antibacterial agent because of its stability. It shows antibacterial action toward various microbes [30]. Zinc oxide nanoparticles have an excellent disinfectant application as they possess significant action even at very low concentrations [3]. Antimicrobial action is the mechanism that is known to suppress the development of microbes and ZnO nanoparticles are known to show great antifungal as well as antiviral activities, which make them harmful to some microbes [68]. ZnO nanoparticles exhibit UV absorption capacity and transparency towards visible light, making them good antibacterial agents against various microbes. They are used in products like varnishes and cosmetics [40].

3.3 Iron Oxide Nanoparticles

Due to their great biocompatibility and magnetic characteristics, iron oxide nanoparticles (IO-NPs) are a popular nanomaterial utilized in biomedical applications such as cancer nanotherapy, biosensors, and contrast agents for imaging. It is noteworthy that IO-NPs may be altered by an external magnetic field and utilized to target drugs or to cause localized hyperthermia to treat tumors [15]. The US Food and Drug Administration (FDA) and several European Union agencies have approved IO-NPs for their demonstrated safety in the treatment of anemia, which when combined with their antibacterial and antiviral activity against influenza subtypes, rotavirus, and Dengue virus may make them useful tools in stopping the spread of viral infections [2]. By a variety of in vitro and in vivo studies, Qin et al. [74] investigated the effectiveness of large (200 nm) iron oxide (magnetite, Fe_3O_4) nanoparticles against influenza A and discovered broad-spectrum antiviral activity against 12 virus subtypes (H1–H12). Pre-exposure of influenza A virus to 4 mg/mL iron oxide nanoparticles for two hours effectively prevented virus-induced cell death in MDCK cells, and 14 days after infection, animal models demonstrated a 100% survival rate. Similar to this, Kumar et al. [56] found that small, 10–15 nm diameter nanoparticles had significant antiviral potency against H1N1 influenza plaque formation in a dose and time dependent manner. Using RT-PCR, they discovered an eightfold decrease in viral RNA transcripts after treating the virus in vitro with IO-NPs as opposed to the untreated control.

3.4 Copper Based Nanoparticles

The Smith Papyrus, an Egyptian medical document written between 2600 and 2200 B.C., contains the earliest known usage of copper for its antibacterial qualities. It describes how humans came to realize copper's use in preserving clean drinking water and sterilizing chest wounds. The discovery that copper miners were significantly less vulnerable to cholera in the 1832 and subsequent epidemics in Paris, France, led to a greater understanding of copper's medicinal potential in the nineteenth century. Since then, scientific study of copper has allowed for a more thorough knowledge of the process behind copper's function as a natural microbicide. As a result, copper has developed into a significant metal with several applications, including sterile touch surfaces and medications [22]. Copper is a required trace element for normal metabolism in the human body, making it safe in the small doses that may be employed in Cu nanoparticle-based surface coatings, in contrast to other nanoparticle materials like gold, silver, and silica. Given their adaptability, accessibility, and comparatively low toxicity, copper and a few of its compounds are now intriguing possibilities in the field of nanomaterials research. This section highlights a number of significant investigations exploring the effectiveness of copper-based nanoparticles as antivirals.

3.5 Gold Based Nanoparticles

Because of its stability, inert biocompatibility, and propensity to form complexes with proteins, gold is a metal that has the potential to be used safely for sterilization. These characteristics also contribute to its antiviral capabilities. One of the main ways that gold, a nanoparticle, limits the propagation of viruses is by interacting with and inhibiting the binding receptors on viral membranes, which stops the virus from adhering to and fusing with host cells. For instance, using 11 nm gold nanoparticles created by the Turkevich process [51], Melendez-Villanueva et al. discover an antiviral mechanism against the measles virus (MeV) which modified with chloroauric acid and garlic extract (Allium sativum) as reducing agents (HAuCl4) [61].

Direct interaction with viral surfaces, proteins, and genetic material occurs through the diffusion of nanoparticles and metal ions to damage viral integrity and obstruct functions like protein synthesis and genome replication.

4 Importance and Applications of Nanoparticles

Nanoparticles have a wide applications as shown in Fig. 2 in various fields as they can generate a wide range of materials with context to the industrial and medical aspects. Nanoparticles such as Ag, Au, Cu, and Zn, are the most common types of nanoparticles being used worldwide. Nanoparticles are also employed in treatment processes like chemotherapy. At the same time, the latter group has fullerenes, gold nanoparticles, and quantum dots. Nanoparticles occur naturally in an environment and help remove environmental contamination by interacting with all life forms, such as bacteria, fungi, algae, plants, and other animals.

Fig. 2 Applications of nanotechnology

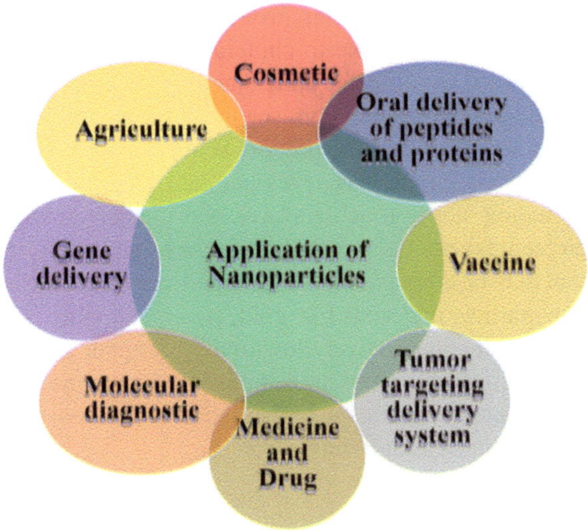

4.1 Use of ZnO-PVP Nanoparticles in COD Reduction

Nanoparticles are highly important due to their efficiency in removing environmental factors such as COD (chemical oxygen demand) and phosphate. Scientists have evaluated that the modified form of zinc, ZnO-PVP (Zinc oxide Poly pyrrolidone), helps to reduce chemical oxygen demand and remove phosphate from the wastewater. The influence of single ZnO and ZnO-PVP is compared by using various doses of the nanoparticles (1, 1.5,2 and 2.5 g/L). It is also proved that the use of ZnO-PVP shows improvements in the photocatalytic COD treatment (63–83%) and phosphate about (68–87%), when compared to the individual ZnO. Maximum of about 95–97% of COD and phosphate removal efficiency can occur with the 1.5 g/L dosage of ZnO-PVP nanoparticles having a UV intensity of about 2293 Mw/cm^2 as shown in Fig. 3.

Researchers are showing their interest in the context of the ZnO-PVP nanoparticle applications employed in the treatment process of wastewater plants. The nanoparticles are used in the photocatalytic degradation of the dye mixture and real textile wastewater. It has been estimated that between 75 and 94.1% of the COD is removed in the case of real textile wastewater and dye mixtures. According to studies, the photocatalytic performance of nanoparticles is affected by the intensity of UV light and the initial dose of the nanoparticles. It has been proven that maximum UV radiation can produce more hydroxyl radicals and increase the reaction rate, resulting in greater efficiency. On the other hand, the higher ZnO-PVP dosage can lead to an increased total surface area, increasing the free radicals that facilitate the organic pollutant degradation rate.

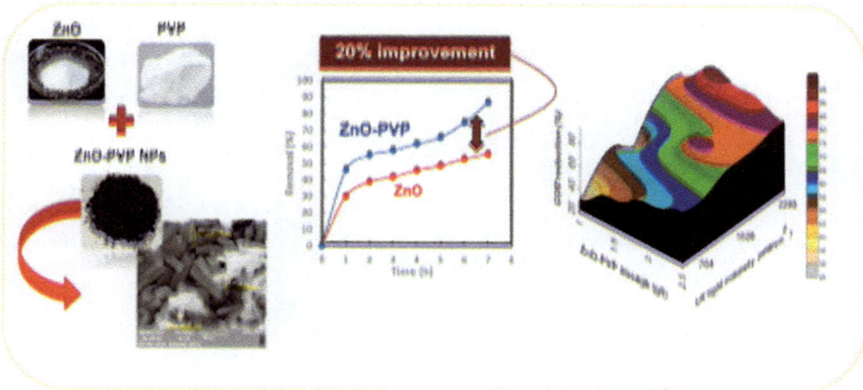

Fig. 3 ZnO-PVP nanoparticles are used in COD reduction

4.2 *Importance of Nanoparticles in the Agricultural Sector*

In agricultural pest control, nanotechnology holds a very promising aspect. It helps provide ecofriendly agriculture and also helps in reducing the continuous use of chemical fertilizers and pesticides and increases the smart and targeted use of nanoparticles depending upon the chemical pesticides and fertilizers delivery. They are also used to monitor pest infestation, soil properties, plant growth, and crop growth regulation. Silver (Ag) nanoparticles have been widely used in the agricultural sector due to their antimicrobial and insecticidal properties. Besides this, silver nanoparticles have a promising future with context to the agricultural industry as shown in Fig. 4 and the progress is still under development compared to others. NPs not only promote plant growth but also shield it from abiotic stress. Due to the NP's huge surface area and small size, toxic metal binds to it, decreasing its availability. Drought, salinity, alkalinity, temperature swings, and metal and mineral toxicity are examples of abiotic stress. NPs can operate as nano-enzymes that can scavenge from oxidative stress and mimic the activities of antioxidant enzymes [81]. Although photosynthesis is a very vulnerable process and a crucial component of plant metabolism, its proper operation can be preserved by reducing oxidative and osmotic stress. In stressful circumstances, photosystem II (PS II), rubisco, and ATP are the main targets in the photosynthetic apparatus (e.g., nutrient deficiency, salinity, water, drought, and heat) [65].

The usual hexagonal form of the nanotubes' two sides is embellished with additional atomic groups to create a hollow carbon tube. Nanotubes are used in medical tools, alumina, sports equipment, and food processing machinery because of their great temperature resistance, flexibility, and strength. Food packaging with carbon nanotubes has increased the mechanical qualities. Certain polymers have had their tensile strength improved using carbon nanotubes and polyamides. Due to their quick detection, easy procedure, and lower cost, carbon nanotube-based biosensors have been used to identify microorganisms, hazardous substances, and other metabolites

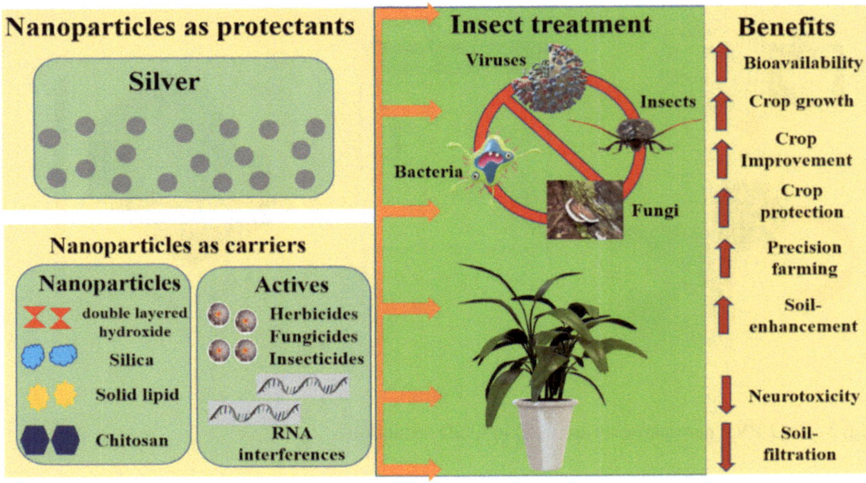

Fig. 4 Role of nanoparticles in agriculture

in beverages and food. It has been observed that carbon nanotubes based on TiO_2 have increased disinfection power against *Bacillus cereus* spores. TiO_2 nanoparticles coated with silver had greater antibacterial activity against *E. coli*. Carbon nanotubes' potent antibacterial properties killed them when they came into touch with them [69].

Microalgae can produce nanoparticles by involving the action of phenolics, amines, amides, alkaloids, and reducing agents. These nanoparticles help in controlling agricultural pathogens. Micro-algae can transform harmful metals into non-toxic variants using chelation method, leading to the formation of nanoparticles. Silver nanoparticles can also be synthesized from microalgae. The increased surface area of silver results in greater efficiency [55]. Silver nanoparticles and microalgal biomass are obtained, which show an excellent alternative for pathogen control that help in disease management in the agricultural sector [90].

4.3 Nanoparticles in Disinfection

Current research is only beginning to scratch the surface of the variety of advantages and prospective uses that nanoparticles can provide in the fusion of medicines and new materials. The use of inorganic nanoparticles in antiseptic coatings to stop the spread of infection-causing pathogens is one such application that has shown encouraging advancements. Importantly, the powerful inactivation of viruses by metal-based nanoparticles is made possible by the high reactive surface area to volume ratio and special chemical characteristics of these particles. Via processes such as suppression of virus-cell receptor binding, reactive oxygen species oxidation, and damaging displacement bonding with essential viral structures, nanoparticles carry out their

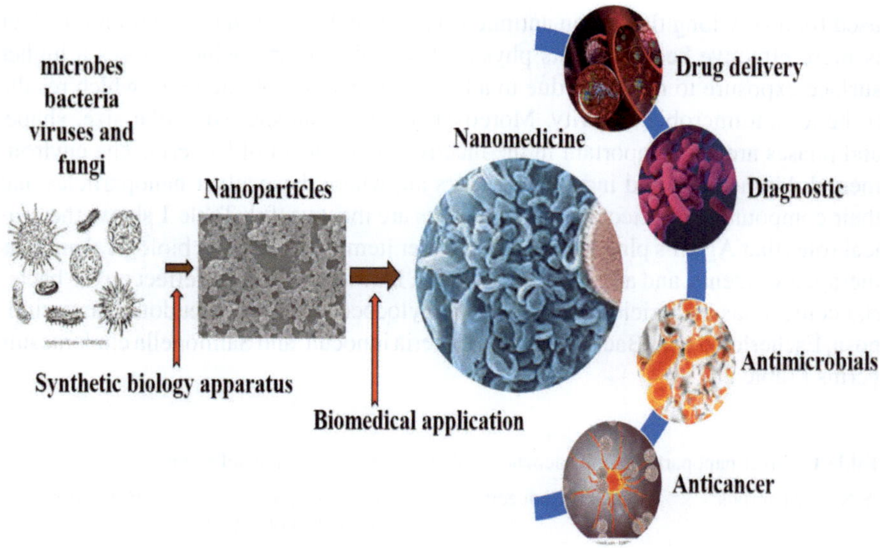

Fig. 5 Role of nanoparticles in disinfection

virucidal activity. The relevance of research efforts to produce nanoparticles for preventative antiviral applications is emphasised by the fact that preventing viral epidemics is one of the biggest issues facing medical science today [58]. Due to their large specific surface area and distinctive chemical characteristics, metal and metal-based nanoparticles (NPs) are at the forefront of nanomaterials research and are potential for applications ranging from disease treatments to antiviral surface coatings (Fig. 5). Importantly, effective antiviral action is made possible by nanoparticles' high surface area to volume ratio even when only modest levels of metal are present. Also, their size makes it simple to incorporate them into pharmaceuticals, polymers, and other surfaces [28]. Metals like copper, silver, and gold display the oligodynamic effect, which has well-known biocidal properties that still hold true at the nanoscale [10].

4.3.1 An Efficient Disinfectant: Silver Nanoparticles (AgNPs)

AgNPs in particular have drawn a lot of interest because of their unique optical, chemical, electrical, and catalytic properties that can be tailored with surface nature, size, and shape, etc. These crystals have been used in a variety of fields, including catalysis, sensor, electronic components, and antimicrobial agents in the health industry, among others. AgNPs-based disinfectants have drawn the most interest among them because of their useful uses in our daily lives. As a result, AgNPs have been employed in a variety of industries, including silver-based textile, animal husbandry, biomedical, and food packaging [20]. As is the standard practice, silver nitrate has been used for a very long time as an antimicrobial agent, but nowadays nano-based silver is more effective because of its physicochemical property, which causes a higher surface exposure to microbes due to a larger surface-to-volume ratio, which results in better antimicrobial activity. Moreover, unique characteristics like size, shape, and phases are very important in the inactivation or death of bacteria. The environmental, biomedical, and industrial sectors are where these silver nanoparticles and their compounds' physicochemical qualities are most useful. Table 1 shows the critical roles that Ag NPs play in textile consumer items, wound care, biological areas as therapeutic agents, and air/water purification. Due to a more toxic effect on the bacterial cells, it has bactericide effects on Staphylococcus aureus, Pseudomonas aeruginosa, Escherichia coli, Bacillus cereus, Listeria innocua, and Salmonella choleraesuis germs (Table 2).

Table 1 Silver nanoparticles' physicochemical characteristics as a disinfectant

S.No	Materials	Disinfectant activity	Physicochemical properties of Ag NPs	References
1	Ag@ZnO	Photo-oxidative killing of bacteria	O_2^-, OH	[21, 60]
2	Ag/BC	Wound healing ability	Porosity	[70]
3	Ag/TiO_2	Visible active antibacterial activity	Reduction in band gap leads to visible active material	[19, 63]
4	Ag polyamide	Sustainable release of Ag^+ ion for antibacterial effect	Size 40–60 nm	[103]
5	AgNP@SiO_2	Prompt and synergic antibacterial activity of air filter	Spherical morphology	[53]
6	Ag@Co-NPs	Water purification	Magnetic and antibacterial	[4]
7	PLA/ZnO:Cu/Ag bio nanocomposites	Enhance shelf life of food	Mechanical/structural, antibacterial and barrier property to UV light	[95]
8	Ag NPs/Chitosan	Wound healing ability	Porosity, moisture retention capability, blood-clotting capability	[57]
9	Ag NPs	Elimination of aerosolized bacteriophage MS2 virus particles	Anti-viral	[49]
10	Ag NPs	Photocatalytic, self-cleaning bacterial inactivation	Surface functionalization	[7, 60, 98, 98]

Table 2 Nanoparticles with context to the bacterial strains

Testing with context to bacterial strains	Shape	Size	Method used	Conclusion	References
ZnONPs					
S. aureus *E. coli* *B. subtilis*	Spherical	5–38 nm	Disk diffusion and MIC	Decreased size of the particles exhibits good antibacterial action and the size influences	[62]
S. aureus *E. col*	Spherical	20.2, 27.1 and 36.8 nm	Disk diffusion, MIC and MBC	Smaller size leads to efficient antibacterial action	[34]
E. coli	Spherical	25–300 nm	Disk diffusion and MIC	Size defects and modification in the surface of ZnO NPs play crucial role in the toxicological action of the ZnO nanoparticles	[82]
P. aeruginosa	Spherical	12 nm 45 nm 2 μm	Disk diffusion, MIC and MBC	The maximum antibacterial action was seen in case of ZnO suspension of range 12 nm	[1]
S.aureus; *S. epidermidis*; *S. pyogenes*; *E. faecalis*; *B. subtilis*; *B. cereus*; *E. coli*, *P. vulgaris*, *S. typhimurium*, *S. flexneri*, *P. alcaligenes*, and *E. aerogenes*	(i) Flower-like (ii) spherical	12–212 nm	Turbidity	Smaller NPs gave good results	[54]

(continued)

Table 2 (continued)

Testing with context to bacterial strains	Shape	Size	Method used	Conclusion	References
C. albicans *A. braseliensis* *E. coli* *P. aeruginosa* S. aureas	(i) Platelet (ii) Platelet (iii) Rodlik	(i) 14.7 nm (ii) 17.5 nm (iii) 76.2 nm	Disk diffusion and MIC	Increased antimicrobial action was reported for NPs having a small crystal size and large porosity	[47]
S. aureus and *E. coli*	(i) Flower-like (ii) Hexagonal-rod (iii) Spherical	(i) 45 nm (ii) 76 nm (iii) 65 nm	Colony count method	Better action was seen in NPs having small size	[66]
S. aureus and *E. coli*	Prism; Ellipse. Spherical	(a) Length = 1um Diameter = 100 nm (b) Length = 500 to 600 nm Diameter = 100 nm (c) 30 nm	Colony count method	Better results were observed for nanoparticles having small size along with the spherical shape	[59]
Au-NPs					
B. cereus	Spherical	10	Disk diffusion, MIC and MBC	Superior antimicrobial activity was initiated for NPs	[18]
Serratia nematodiphila	Spherical; face-centered cubic	10–31	Disk diffusion, MIC and MBC	Improved activity remained experimental for NPs	[88]
E. coli S. Typhimurium	Spherical; Face-centered cubic	10–100	Colony count method	Improved results were found with slighter NPs	[26]
B. flexus, *B. pseudomycoides* *C. runiversalis* *K. rosea*	Spherical and triangular shape; Face centered cubic	12–65	Disk diffusion, MIC	Small size exhibits good antibacterial action	[50]

(continued)

Table 2 (continued)

Testing with context to bacterial strains	Shape	Size	Method used	Conclusion	References
B. subtilis	Crystalline	5–15	Colony count method	Toxicological action of ZnO Nanoparticles	[84]
Klebsiella pneumonia, Enterobacter cloacae	Spherical	28.2–122	Disk diffusion, MIC and MBC	Slighter size exhibits improved antibacterial activity	[99]
Escherichia coli, Micrococcus luteus	Spherical and Triangular	5–50	Disk diffusion, MIC and MBC	The antibacterial activity was higher for ZnO	[94]

4.3.2 Antiviral Potential of Metals and Metal-Based Nanoparticles

Due to their large specific surface area and distinctive chemical characteristics, metal and metal-based nanoparticles (NPs) are at the forefront of nanomaterials research and are potential for applications ranging from disease treatments to antiviral surface coatings. Importantly, effective antiviral action is made possible by nanoparticles' high surface area to volume ratio even when only modest levels of metal are present. Also, their size makes it simple to incorporate them into pharmaceuticals, polymers, and other surfaces [29]. Metals like copper, silver, and gold display the oligodynamic effect, which has well-known biocidal properties that still hold true at the nanoscale [73, 75]. To understand their antiviral characteristics and processes, these metals and a number of their oxides have been extensively researched. The most common ways that metal nanoparticles have been found to stop viruses are: [11, 101]

- Preventing adhesion and entrance into host cells by binding to or disabling viral surface features (such as spike glycoproteins).
- The generation of metal ions and reactive oxygen species (ROS) through chemical mechanisms that cling to and break down viral components such as the glycan shield, lipid envelope, protein capsid, and nucleic acids.
- Cleavage of disulfide bonds between cysteine amino acid residues to denature and disable viral glycoproteins.

Figure 6 presents a summary of them. The precise processes by which individual nanoparticles interact with various viruses vary greatly, depending on both the reactivity of the nanoparticles and the architecture of the viruses. The sections that follow will examine several metal-based nanoparticle types and give an overview of the existing research on their antiviral capabilities.

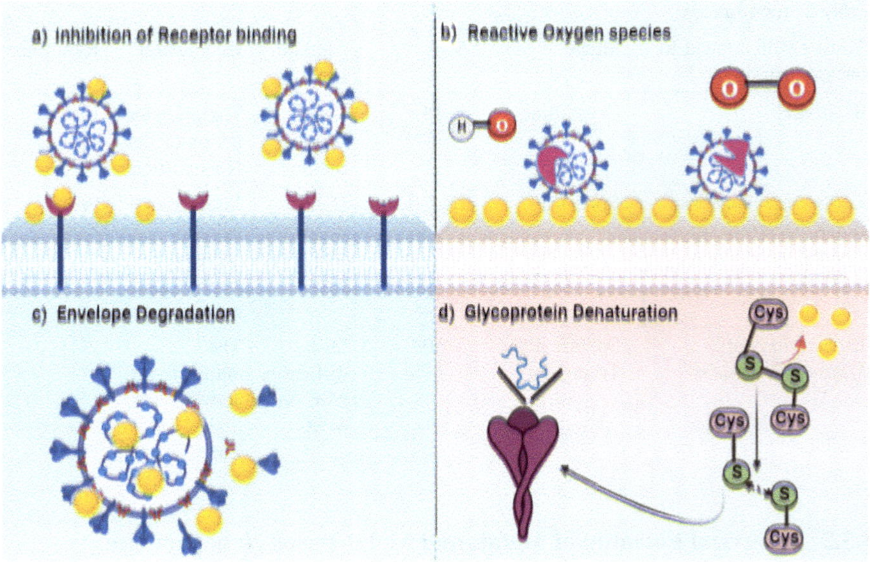

Fig. 6 Diagram showing potential metal and metal-based nanoparticle antiviral mechanisms. **a** To prevent virus-host cell adhesion and entrance, nanoparticles interact electrostatically and establish complex bonds with viral surface proteins. **b** Nanoparticles go through processes like UV photocatalysis to produce reactive oxygen species, which easily interact with viral biomolecules and cause harm to them. **c** Viral capsid or envelope structure is weakened as a result of direct interactions between nanoparticles and generated metal ions and their passage through them. **d** Viral glycoproteins' disulfide bonds are broken by processes involving nanoparticles, which denaturize the proteins and lessen their infectious potential by preventing the binding of the virus to its target receptors on host cells

4.4 *Nanoparticles in COVID-19 Disinfection*

COVID-19 virus was known to spread havoc all over the world. It is crown-shaped, as shown in Fig. 7. It forms a transport system for a nanoparticle that provides various ways for controlling antibacterial and anti-intranasal aspects [91]. Therapeutic drug particles are included in the theragnostic nanoparticle systems and act as an effective drug for seizing COVID-19 infection transmission. Silver nanoparticles are known to cause inhibition in the SARS-CoV-2 offering antiviral characteristics.

Nanoparticles may rupture and infiltrate the membrane in a way akin to the bacterial inactivation process as enveloped viruses like SARS-CoV-2 also have an outer membrane made of a lipid bilayer [12, 80]. In reality, El-Megharbel et al. found this to be the case. It is widely known that amphiphilic soap efficiently eliminates SARS-CoV-2 virions by dissolving their lipid membrane. This lends credence to the idea that nanoparticles like ZnO may achieve viral inactivation by concentrating on the same structures. References [16, 64] proposed the ZnO micro-nano structure shape by electrostatically targeting virions designed for HSV that may be modified to mimic other viral targets, such as that of SARS-CoV-2. The cleavage of oligosaccharides

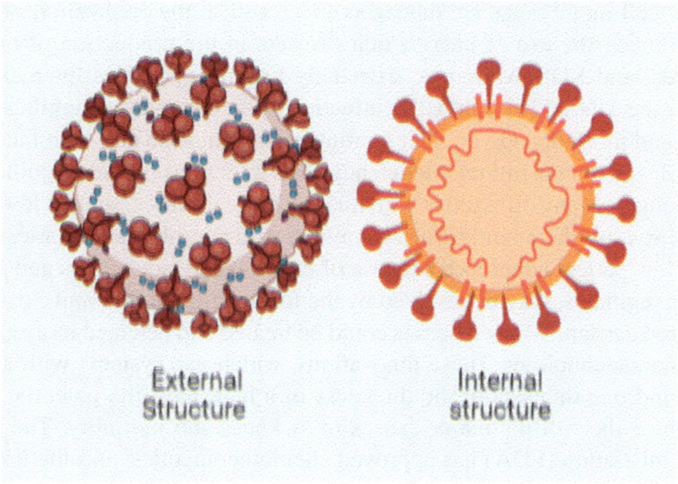

Fig. 7 COVID-19 structure

from the glycan shield by hydroxyl radicals and the ROS-mediated deterioration of viral integrity are further inactivation mechanisms that ZnO and TiO_2 may share [102], nevertheless, owing to ZnO-NPs' significant cytotoxicity.

4.5 Nanoparticles in Pathogen Control

Many plant viruses and pests attack and damage many crops, and millions of dollars are spent on their management. Initially, it appeared that using chemical pesticides would be the solution to the issue, but their frequent and haphazard use has caused serious environmental damage and health risks, forcing us to consider other options. The present application of nanotechnology in agri-science is promising. Nanotechnology is a developing field in bioscience and medical science. The foundation of nanotechnology is the use of nano-particles (size 109 mm) made of metals like gold, silver, copper, silica, zinc, etc. The creation of the so-called "Nano pesticides" is being fueled by the unique uses of nanotechnology for plant protection [31]. Gold nanoparticles provide a diverse range of emerging technologies, including optical and chemical properties. These nanoparticles improve sensitivity and speed for the detection and diagnosis of bacteria. They are also used as photothermal, stabilizing, or antimicrobial agents [71]. Hydrophilic, biocompatible, and very stable inorganic nanoparticles. Calcium phosphate nanoparticles, gold nanoparticles, iron oxide nanoparticles, zinc oxide nanoparticles, silver nanoparticles, and others are examples of inorganic nanoparticles. Ions, which are antibacterial agents, are present in inorganic nanoparticles. ROS are produced by zinc and copper, which kill the pathogen that has been engulfed. At low concentrations, gold and silver are extremely poisonous to bacteria.

Pathogens' cell membranes are damaged as a result of the application of nanoparticles. Moreover, the use of nanoparticles results in the production of metal ions, free radicals, and ROS. Moreover, extremely tiny nanoparticles intercalate themselves into the DNA [36]. Globally, infectious illnesses play a significant role in morbidity and mortality. Given their continued transmission and high fatality rates, malaria, TB, and human immunodeficiency virus infections provide significant treatment challenges. Many difficulties with treating these diseases, such as low on-target bioavailability, sub-therapeutic drug accumulation in microbial sanctuaries and reservoirs, and low patient adherence because of drug-related side effects and prolonged therapeutic regimens, may be resolved by the formulation of new and existing drugs in nano-sized carriers. Many diseases could be treated and detected more effectively, thanks to nanotechnology. These innovations, which use systems with a diameter that is around one thousandth the thickness of a hair, have the potential to significantly reduce the world's major causes of sickness and mortality. The Food and Drug Administration (FDA) has approved chemotherapeutics, anesthetics, imaging agents, nutritional supplements, and other products, thanks to decades of extensive study on Nano systems [52]. The creation of suitable components for sensor systems that can detect and prevent diseases is another use of nanomaterials in the medical industry. The tests that communicate with the outside of the patient's body show internal body ailments like heart attack, tumor, or localized infections are completed with environmental sensors that are constructed on a very small chip. Currently, a drop of blood can be used to diagnose a number of diseases using laser systems that operate in the infrared, visible, and ultraviolet frequency ranges. The development of nanotechnology in the biological sciences and medical field is also demonstrated by new methods for creating DNA-based nanoscale instruments. Several disorders can now be identified and treated right away, thanks to the usage of these novel medicines. One of the biggest obstacles to adopting nanotechnology in medicine is the ability of defense systems to kill nanoparticles. Within minutes, the administered nanoparticles are successfully caught and expelled from the body. Nanoparticles with cell membrane coatings can last in the body for several hours without suffering any harm. Among these particles, protein nanoparticles are of interest due to a variety of advantages, including simple resource access, renewable resources, affordable price, biocompatibility, biodegradability. The presence of multiple functional groups to carry high doses of the drug, and the capacity to link simultaneously targeting groups to target nanoparticles to specific cells or tissues [78].

5 Conclusion

Multidrug resistant bacteria have demonstrated to be a severe health problem that needs to be addressed on a global level as they have emerged and grown. Nanomaterials present an original, "out of the box" approach to the treatment of persistent MDR planktonic bacterial and biofilm infections. To enhance their therapeutic potential and lessen host risk, it is crucial to optimize their physical characteristics, especially size and surface charge. Current concerns about the long-term effects of nanoparticles on humans and systemic safety are limiting clinical application. Future research should include a thorough understanding of the NP influence on gene expression due to concerns with metabolism, toxicity, stability, and gene-level mechanisms.

References

1. Abdo AM, Fouda A, Eid AM, Fahmy NM, Elsayed AM, Khalil AMA, Alzahrani OM, Ahmed AF, Soliman AM (2021) Green synthesis of Zinc Oxide Nanoparticles (ZnO-NPs) by Pseudomonas aeruginosa and their activity against pathogenic microbes and common house mosquito, Culex pipiens. Mater 14(22):6983
2. Abo-zeid Y, Ismail NS, McLean GR, Hamdy NM (2020) A molecular docking study repurposes FDA approved iron oxide nanoparticles to treat and control COVID-19 infection. Eur J Pharm Sci 153:105465
3. Ahmad I, Alshahrani MY, Wahab S, Al-Harbi AI, Nisar N, Alraey Y, Alqahtani A, Mir MA, Irfan S Saeed M (2022) Zinc oxide nanoparticle: an effective antibacterial agent against pathogenic bacterial isolates. J King Saud Univ Sci 102110
4. Alonso A, Muñoz-Berbel X, Vigués N, Rodríguez-Rodríguez R, MacAnás J, Muñoz M, Mas J, Muraviev DN (2013) Superparamagnetic AgCo-nanocomposites on granulated cation exchange polymeric matrices with enhanced antibacterial activity for the environmentally safe purification of water. Adv Funct Mater 23(19):2450–2458
5. Arshad H, Saleem M, Pasha U, Sadaf S (2022) Synthesis of Aloe vera-conjugated silver nanoparticles for use against multidrug-resistant microorganisms. Electron J Biotechnol 55:55–64
6. Ashok A, Nayar SA, Shankar P, Das A, Mohanan PV (2022) Sterilization for biological products. Biomed Prod Mater Eva 725–742
7. Atiyeh BS, Costagliola M, Hayek SN, Dibo SA (2007) Effect of silver on burn wound infection control and healing: review of the literature. Burns 33(2):139–148
8. Bachheti RK, Abate L, Bachheti A, Madhusudhan A, Husen A (2021) Algae-, fungi-, and yeast-mediated biological synthesis of nanoparticles and their various biomedical applications. In: Handbook of greener synthesis of nanomaterials and compounds, pp 701–734. Elsevier, Amsterdam. https://doi.org/10.1016/B978-0-12-821938-6.00022-0
9. Bachheti A, Bachheti RK, Abate L, Husen A (2022) Current status of Aloe-based nanoparticle fabrication, characterization and their application in some cutting-edge areas. S Afr J Bot 147:1058–1069
10. Bamal D, Singh A, Chaudhary G, Kumar M, Singh M, Rani N, Mundlia P, Sehrawat AR (2021) Silver nanoparticles biosynthesis, characterization, antimicrobial activities, applications, cytotoxicity and safety issues: an updated review. Nanomater 11(8)
11. Bawage SS, Tiwari PM, Singh A, Dixit S, Pillai SR, Dennis VA (2016) Gold nanorods inhibit respiratory syncytial virus by stimulating the innate immune response. Nanomed Nanotechnol, Biol Med 12(8):2299–2310
12. Bianchi M, Benvenuto D, Giovanetti M, Angeletti S, Ciccozzi M, Pascarella S (2020) Sars-CoV-2 Envelope and membrane proteins: structural differences linked to virus characteristics? BioMed Res Int :4389089
13. Caldara M, Belgiovine C, Secchi E, Rusconi R (2022) Environmental, microbiological, and immunological features of bacterial biofilms associated with implanted medical devices. Clin Microbiol Rev 35(2):e00221-e320
14. Campo-Beleño C, Villamizar-Gallardo RA, López-Jácome LE, González EE, Muñoz-Carranza S, Franco B, Morales-Espinosa R, Coria-Jimenez R, Franco-Cendejas R, Hernández-Durán M, Lara-Martínez R (2022) Biologically synthesized silver nanoparticles as potent

antibacterial effective against multidrug-resistant Pseudomonas aeruginosa. Lett Appl Microbiol 75(3):680–688
15. Chang D, Lim M, Goos JACM, Qiao R, Ng YY, Mansfeld FM (2018) Biologically targeted magnetic hyperthermia: potential and limitations. Front Pharmacol 9:831
16. Chin AWH, Chu JTS, Perera MRA, Hui KPY, Yen HL, Chan MCW, Peiris M, Poon LLM (2020) Stability of SARS-CoV-2 in different environmental conditions. Lancet 1(1):e10
17. Danish M, Shahid M, Zeyad MT, Bukhari NA, Al-Khattaf FS, Hatamleh AA Ali S (2022) Bacillus mojavensis, a Metal-Tolerant Plant Growth-Promoting Bacterium, Improves Growth, Photosynthetic Attributes, Gas Exchange Parameters, and Alkalo-Polyphenol Contents in Silver Nanoparticle (Ag-NP)-Treated WithaniasomniferaL.(Ashwagandha). ACS omega 7(16):3878–13893
18. Danish P, Ali Q, Hafeez MM, Malik A (2020) Antifungal and antibacterial activity of aloe vera plant extract. Biol Clin Sci Res J
19. Deshmukh SP, Mullani SB, Koli VB, Patil SM, Kasabe PJ, Dandge PB, Pawar SA, Delekar SD (2018) Ag nanoparticles connected to the surface of TiO_2 electrostatically for antibacterial photoinactivation studies. Photochem Photobiol 94(6):1249–1262
20. Deshmukh SP, Patil SM, Delekar MSB, SD, (2019) Silver nanoparticles as an effective disinfectant: a review. Mat Sci Eng C 97:954–965
21. Dunnill CW, Page K, Aiken ZA, Noimark S, Hyett G, Kafizas A, Pratten J, Wilson M, Parkin IP (2011) Nanoparticulate silver coated-titania thin films—Photo-oxidative destruction of stearic acid under different light sources and antimicrobial effects under hospital lighting conditions. J Photochem Photobiol A 220(2–3):113–123
22. Ermini ML, Voliani V (2021) Antimicrobial nano-agents: the copper age. ACS Nano 15(4):6008–6029
23. Fang Z, Chen R, Zheng J, Khan AI, Neilson KM, Geiger SJ, Callahan DM, Moebius MG, Saxena A, Chen ME, Rios C (2022) Ultra-low-energy programmable non-volatile silicon photonics based on phase-change materials with graphene heaters. Nat Nanotechnol 17(8):842–848
24. Feng F, Wen Z, Chen J, Yuan Y, Wang C, Sun C (2022) Strategies to develop a mucosa-targeting vaccine against emerging infectious diseases. Viruses 14(3):520
25. Feynman RP (1959) Plenty of room at the bottom. In APS annual meeting
26. Gabrielyan L, Badalyan H, Gevorgyan V, Trchounian A (2020) Comparable antibacterial effects and action mechanisms of silver and iron oxide nanoparticles on Escherichia coli and Salmonella typhimurium. Sci Rep 10(1):13145
27. Gabrielyan L, Badalyan H, Gevorgyan V, Trchounian A (2021) Author correction: comparable antibacterial effects and action mechanisms of silver and iron oxide nanoparticles on Escherichia coli and Salmonella typhimurium. Sci Rep 11(1):1–3
28. Galdiero S, Falanga A, Vitiello M, Cantisani M, Marra V, Galdiero M (2011) Silver nanoparticles as potential antiviral agents. Mol 16(10):8894
29. Galdiero S, Falanga A, Vitiello M, Cantisani M, Marra V, Galdiero M (2011) Silver nanoparticles as potential antiviral agents. Mol 10:8894–8918
30. Gandhi RT, Malani PN, Del Rio C (2022) COVID-19 therapeutics for nonhospitalized patients. JAMA 327(7):617–618
31. Ghosh SK, Bera T (2021) Unraveling the mechanism of nanoparticles for controlling plant pathogens and pests. In: Advances in nano-fertilizers and nano-pesticides in agriculture, pp 415–436
32. Gokce C, Gurcan C, Besbinar O, Unal MA, Yilmazer A (2022) Emerging 2D materials for antimicrobial applications in the pre-and post-pandemic era. Nanoscale 14(2):239–249
33. Gonfa YH, Gelagle AA, Hailegnaw B, Kabeto SA, Workeneh GA, Tessema FB, Tadesse MG, Wabaidur SM, Dahlous KA, Abou Fayssal S, Kumar P (2023) Optimization, characterization, and biological applications of silver nanoparticles synthesized using essential oil of aerial part of laggera tomentosa. Sustainability 15(1):797
34. Gudkov SV, Burmistrov DE, Serov DA, Rebezov MB, Semenova AA, Lisitsyn AB (2021) A mini review of antibacterial properties of ZnO nanoparticles. Front Phy 9:641481

35. Gupta A, Mumtaz S, Li CH, Hussain I, Rotello VM (2019) Combatting antibiotic-resistant bacteria using nanomaterials. Chem Soc Rev 48(2):415–427
36. Hamid A, Salee S (2022) Role of nanoparticles in management of plant pathogens and scope in plant transgenics for imparting disease resistance 58(3)
37. Hamdi SS, Al-Kayiem HH, Alsabah MS, Muhsan AS (2022) A comparative study of dispersed and grafted nanofluids of graphene nanoplatelets with natural polymer in high salinity brine for enhanced oil recovery. J Petrol Sci Eng 210:110004
38. Hosnedlova B, Kabanov D, Kepinska M, Narayanan VHB, Parikesi AA, Fernandez C, Bjørklund G, Nguyen HV, Farid A, Sochor J, Pholosi A, Baron M, Jakubek M, Kizek R (2022) Effect of biosynthesized silver nanoparticles on bacterial biofilm changes in S. aureus and E. coli. Nanomater 12(13):2183
39. Hung YP, Chen YF, Tsai PJ, Huang IH, Ko WC, Jan JS, (2021) Advances in the application of nanomaterials as treatments for bacterial infectious diseases. Pharmaceutics 13(11):1913
40. Husain FM, Qais FA, Ahmad I, Hakeem MJ, Baig MH, Masood Khan J, Al-Shabib NA (2022) Biosynthesized zinc oxide nanoparticles disrupt established biofilms of pathogenic bacteria. Appl Sci 12(2):710
41. Husen A (2022) Engineered nanomaterials for sustainable agricultural production, soil improvement and stress management. Elsevier Inc., 50 Hampshire St., 5th Floor, Cambridge, MA 02139, USA
42. Husen A (2023) Secondary metabolites based green synthesis of nanomaterials and their applications. Springer Nature Singapore Pte Ltd., 152 Beach Road, #21–01/04 Gateway East, Singapore 189721, Singapore
43. Husen A (2023) Smart nanomaterials from agricultural and horticultural products. Springer Nature Singapore Pte Ltd., 152 Beach Road, #21–01/04 Gateway East, Singapore 189721, Singapore
44. Husen A, Iqbal M (2019) Nanomaterials and plant potential. Springer International Publishing AG, Gewerbestrasse 11, 6330 Cham, Switzerland. https://doi.org/10.1007/978-3-030-05569-1
45. Husen A, Siddiqi KS (2023) Advances in smart nanomaterials and their applications. Elsevier Inc., 50 Hampshire St., 5th Floor, Cambridge, MA 02139, USA
46. Jadoun S, Chauhan NPS, Zarrintaj P, Barani M, Varma RS, Chinnam S, Rahdar A (2022) Synthesis of nanoparticles using microorganisms and their applications: a review. Environ Chem Lett 20:1–45
47. Jin SE, Jin HE (2021) Antimicrobial activity of zinc oxide nano/microparticles and their combinations against pathogenic microorganisms for biomedical applications: from physicochemical characteristics to pharmacological aspects. Nanomater 11(2):263
48. Jin-Chul K, Madhusudhan A, Husen A (2021) Smart nanomaterials in biomedical applications. Springer Nature Switzerland AG, Gewerbestrasse 11, 6330 Cham, Switzerland
49. Joe YH, Park DH, Hwang J (2016) Evaluation of Ag nanoparticle coated air filter against aerosolized virus: anti-viral efficiency with dust loading. J Hazard Mater 301:547–553
50. Kanikireddy V, Varaprasad K, Rani MS, Venkataswamy P, Reddy BJ, Vithal M (2020) Biosynthesis of CMC Guar gum-Ag0 nanocomposites for inactivation of food pathogenic microbes and its effect on the shelf life of strawberries. Carbohydr Polym 236:116053
51. Kimling J, Maier M, Okenve B, Kotaidis V, Ballot H, Plech A (2006) Turkevich method for gold nanoparticle synthesis revisited. J Phys Chem B 110(32):15700–15707
52. Kirtane AR, Verma M, Karandikar P, Furin J, Langer R, Traverso G (2021) Nanotechnology approaches for global infectious diseases. Nat Nanotechnol 16(4):369–384
53. Ko YS, Joe YH, Seo M, Lim K, Hwang J, Woo K (2014) Prompt and synergistic antibacterial activity of silver nanoparticle-decorated silica hybrid particles on air filtration. J Mater Chem B 2(39):6714–6722
54. Koul B, Poonia AK, Jin YD, JO, (2021) Microbe-mediated biosynthesis of nanoparticles: applications and future prospects. Biomol 11(6):886
55. Krishnaraj C, Jagan EG, Rajasekar S, Selvakumar P, Kalaichelvan PT, Mohan NJCSBB (2010) Synthesis of silver nanoparticles using Acalypha indica leaf extracts and its antibacterial activity against water borne pathogens. Colloids Surf B 76(1):50–56

56. Kumar R, Nayak M, Sahoo GC, Pandey K, Sarkar MC, Ansari Y (2019) Iron oxide nanoparticles based antiviral activity of H1N1 influenza A virus. J Infect Chemother 25(5):325–329
57. Liang D, Lu Z, Yang H, Gao J, Chen R (2016) Novel asymmetric wettable AgNPs/chitosan wound dressing: in vitro and in vivo evaluation. ACS Appl Mater Interfaces 8(6):3958–3968
58. Lin N, Verma D, Saini N, Arbi R, Munir M, Jovic M, Turak A (2021) Antiviral nanoparticles for sanitizing surfaces: a roadmap to self-sterilizing against COVID-19. Nano Today 40:101267
59. Mahmoud UT, Abdel-Mohsein HS, Mahmoud MA, Amen OA, Hassan RI, Abd-El-Malek AM, Rageb SM, Waly HS, Osman AA, Osman MA (2020) Effect of zinc oxide nanoparticles on broilers' performance and health status. Trop Anim Health Prod 52(4):2043–2054
60. Manna J, Goswami S, Shilpa N, Sahu N, Rana RK (2015) Biomimetic method to assemble nanostructured AgZnO on cotton fabrics: application as self-cleaning flexible materials with visible-light photocatalysis and antibacterial activities. ACS Appl Mater Interfaces 7(15):8076–8082
61. Meléndez-Villanueva MA, Morán-Santibañez K, Martínez-Sanmiguel JJ, Rangel-López R, Garza-Navarro MA, Rodríguez-Padilla C (2019) Virucidal activity of gold nanoparticles synthesized by green chemistry using garlic extract. Vir 11(12):1111
62. Mendes CR, Dilarri G, Forsan CF, Sapata VDMR, Lopes PRM, de Moraes PB, Montagnolli RN, Ferreira H, Bidoia ED (2022) Antibacterial action and target mechanisms of zinc oxide nanoparticles against bacterial pathogens. Sci Rep 12(1):1–10
63. Mihaly Cozmuta A, Peter A, Mihaly Cozmuta L, Nicula C, Crisan L, Baia L, Turila A (2015) Active packaging system based on Ag/TiO_2 nanocomposite used for extending the shelf life of bread. Chemical and microbiological investigations. Pack Tech Sci 28(4):271–284
64. Mishra YK, Adelung R, Röhl C, Shukla D, Spors F, Tiwari V (2011) Virostatic potential of micro-nano filopodia-like ZnO structures against herpes simplex virus-1. Antivir Res 92(2):305–312
65. Mittal D et al (2021) Nanoparticle-based sustainable agriculture and food science: recent advances and future outlook. Front Nanotechnol 2:10
66. Mohd Yusof H, Abdul Rahman NA, Mohamad R, Hasanah Zaidan U, Samsudin AA (2021) Antibacterial potential of biosynthesized zinc oxide nanoparticles against poultry-associated foodborne pathogens: an in vitro study. Anim 11(7):2093
67. Morones JR, Elechiguerra JL, Camacho A, Holt K, Kouri JB, Ramírez JT, Yacaman MJ (2005) The bactericidal effect of silver nanoparticles. Nanotech 16(10):2346
68. Nazoori ES, Kariminik A (2018) In vitro evaluation of antibacterial properties of zinc oxide nanoparticles on pathogenic prokaryotes. J Appl Biotech Rep 5(4):162–165
69. Neme K et al (2021) Application of nanotechnology in agriculture. postharvest loss reduction and food processing: food security implication and challenges 7(12):e08539
70. Pal S, Nisi R, Stoppa M, Licciulli A (2017) Silver-functionalized bacterial cellulose as antibacterial membrane for wound-healing applications. ACS Omega 2(7):3632–3639
71. Pissuwan D, Cortie CH, Valenzuela SM, Cortie MB (2010) Functionalised gold nanoparticles for controlling pathogenic bacteria. Trends Biotechnol 28(4):207–213
72. Prakash J, Krishna SBN, Kumar P, Kumar V, Ghosh KS, Swart HC, Bellucci S, Cho J (2022) Recent advances on metal oxide-based nano-photocatalysts as potential antibacterial and antiviral agents. Catal 12(9):1047
73. Prasher P, Singh M, Mudila H (2018) Oligodynamic effect of silver nanoparticles: a review. BioNanoScience 8(4):951–962
74. Qin L, Zeng Z, Zeng G, Lai C, Duan A, Xiao R, Huang D, Fu Y, Yi H, Li B, Liu X (2019) Cooperative catalytic performance of bimetallic Ni-Au nanocatalyst for highly efficient hydrogenation of nitroaromatics and corresponding mechanism insight. Appl Catal B: Environ 259:118035
75. Rai M, Deshmukh SD, Ingle AP, Gupta IR, Galdiero M, Galdiero S (2014) Metal nanoparticles: the protective nanoshield against virus infection. Crit Rev Microbiol 42(1):46–56
76. Rajivgandhi G, Gnanamangai BM, Prabha TH, Poornima S, Maruthupandy M, Alharbi NS, Kadaikunnan S, Li WJ (2022) Biosynthesized zinc oxide nanoparticles (ZnO NPs) using

actinomycetes enhance the antibacterial efficacy against K. Pneumoniae. J King Saud Univ Sci 34(1):101731

77. Ramadan YFN, Al-Harbi AIA, Ahmed E, Battah A, Ellah NH, Zanetti S, Hetta HF, Ramadan YN, Al-Harbi AI, Ahmed EA, Battah B, Ellah NHA, Zanetti S, Donadu MG (2023) Nanotechnology as a promising approach to combat multidrug resistant bacteria: a comprehensive review and future perspectives. Biomed 11(2):413
78. Rezaei R, Safaei M, Mozaffari HR, Moradpoor H, Karami S, Golshah A, Salimi B, Karami H (2019) The role of nanomaterials in the treatment of diseases and their effects on the immune system. 14;7(11):1884–1890
79. Sánchez-López AL, Perfecto-Avalos Y, Sanchez-Martinez A, Ceballos-Sanchez O, Sepulveda-Villegas M, Rincón-Enríquez G, Rodríguez-González V, Garcia-Varela R, Lozano LM, Navarro-López DE, Sanchez-Ante G (2022) Influence of erbium doping on zinc oxide nanoparticles: structural, optical and antimicrobial activity. Appl Surf Sci 575:151764
80. Schoeman D, Fielding BC (2019) Coronavirus envelope protein: current knowledge. Virol J 16(1):1–22
81. Sharifi A, Khavarian-Garmsir AR (2020) The COVID-19 pandemic: Impacts on cities and major lessons for urban planning, design, and management. Sci Total Environ 749:142391
82. Siddiqi KS, Husen A (2018) Properties of zinc oxide nanoparticles and their activity against microbes. Nanoscale res lett 13(1):1–13
83. Singh MP, Yadav SK, Khan MM, Ahmad S, Khan R, Khan AQ, Haque R, Raza SS (2022) Brain infectious diseases and nano-therapy. Nanotechnol Infect Dis 575–602
84. Singla A (2022) Review of biological treatment solutions and role of nanoparticles in the treatment of wastewater generated by diverse industries. Nanotechnol Environ Eng 7(3):699–711
85. Singla S, Jana A, Thakur R, Kumari C, Goyal S, Pradhan J (2022) Green synthesis of silver nanoparticles using Oxalis griffithii extract and assessing their antimicrobial activity. OpenNano 7:100047
86. Soltani M, Kashkooli FM, Fini MA, Gharapetian D, Nathwani J, Dusseault MB (2022) A review of nanotechnology fluid applications in geothermal energy systems. Renew Sustain Energy Rev 167:112729
87. Tariq M, Mohammad KN, Ahmed B, Siddiqui MA, Lee J (2022) Biological synthesis of silver nanoparticles and prospects in plant disease management. Mol 27(15):4754
88. Tariq SL, Ali HM, Akram MA, Janjua MM, Ahmadlouydarab M (2020) Nanoparticles enhanced phase change materials (NePCMs)-A recent review. Appl Therm Eng 176:115305
89. Taghiyari HR, Morrell JF, Husen A (2022) Emerging nanomaterials (Opportunities and Challenges in Forestry Sectors). Springer Nature Switzerland AG, Gewerbestrasse 11, 6330 Cham, Switzerland https://doi.org/10.1007/978-3-031-17378-3
90. Terra ALM, Kosinski RDC, Moreira JB, Costa JAV, Morais MGD (2019) Microalgae biosynthesis of silver nanoparticles for application in the control of agricultural pathogens. J Environ Sci Health B 54(8):709–716
91. Thakur N, Thakur N, Chauhan P, Kumar K, Jeet K, Kumar A, Sharma DP (2022) Futuristic role of nanoparticles for treatment of COVID-19. Biomater Pol Horizon 1(2)
92. Uddin I, Parimi DS, Bollu TK, Bhatt CS, Suresh AK (2022) Silver nanoparticles as potent multidrug-resistant incorporants in biomedicine. Emerg Modal Mitigat Antimicrob Res 475–488
93. Umapathi A, Madhyastha H, Navya PN, Singh M, Madhyastha R, Daima HK (2022) Surface chemistry driven selective anticancer potential of functional silver nanoparticles toward lung cancer cells. Colloids Surf, A Physicochem Eng Asp 652:129809
94. Urnukhsaikhan E, Bold BE, Gunbileg A, Sukhbaatar N, Mishig-Ochir T (2021) Antibacterial activity and characteristics of silver nanoparticles biosynthesized from Carduus crispus. Sci Rep 11(1):1–12
95. Vasile C, Râpă M, Ştefan M, Stan M, Macavei S, Darie-Niţă RN, Barbu-Tudoran L, Vodnar DC, Popa EE, Ştefan R, Borodi G, Brebu M (2017) New PLA/ZnO:Cu/Ag bionanocomposites for food packaging EXPRESS Polym Lett 11(7):531–544

96. Vimala RTV, Rajivgandhi G, Sridharan S, Jayapriya M, Ramachandran G, Kanisha CC, Manoharan N, Li WJ (2022) Biosynthesis and characterization of silver nanoparticles from actinobacteria. Methods Actinobacteriol 709–712
97. Virto L, Simões-Martins D, Sánchez MC, Encinas A, Sanz M, Herrera D (2022) Antimicrobial effects of a new brushing solution concept on a multispecies in vitro biofilm model growing on titanium surfaces. Clin Oral Implants Res 33(2):209–220
98. Wagener S, Dommershausen N, Jungnickel H, Laux P, Mitrano D, Nowack B, Schneider G, Luch A (2016) Textile functionalization and its effects on the release of silver nanoparticles into artificial sweat. Environ Sci Technol 50(11):5927–5934
99. Wang X, Zhang Y, Li C, Li G, Wu D, Li T, Qu Y, Deng W, He Y, Penttinen P, Zhang H (2022) Antimicrobial resistance of Escherichia coli, Enterobacter spp., Klebsiella pneumoniae and Enterococcus spp. isolated from the feces of giant panda. BMC microbiol 22(1):1–11
100. Warsi MF, Chaudhary K, Zulfiqar S, Rahman A, Al Safari IA, Zeeshan HM, Agboola PO, Shahid M, Suleman M (2022) Copper and silver substituted MnO_2 nanostructures with superior photocatalytic and antimicrobial activity. Ceram Int 48(4):4930–4939
101. Watanabe Y, Allen JD, Wrapp D, McLellan JS, Crispin M (2020) Site-specific glycan analysis of the SARS-CoV-2 spike. Sci 369(6501):330–333
102. Yasuda J, Eguchi H, Fujiwara N, Ookawara T, Kojima S, Yamaguchi Y, Nishimura M, Fujimoto J, Suzuki K (2006) Reactive oxygen species modify oligosaccharides of glycoproteins in vivo: a study of a spontaneous acute hepatitis model rat (LEC rat). Biochem Biophy Res Comm 342(1):127–134
103. Zille A, Fernandes MM, Francesko A, Tzanov T, Fernandes M, Oliveira FR, Almeida L, Amorim T, Carneiro N, Esteves MF, Souto AP (2015) Size and aging effects on antimicrobial efficiency of silver nanoparticles coated on polyamide fabrics activated by atmospheric DBD plasma. ACS Appl Mater Interfaces 7(25):13731–13744

Recent Advances in Nanoparticles for Environmental Monitoring and Sensing: An Overview

Addisu Tamir Wassie, Rakesh Kumar Bachheti, and Archana Bachheti

Abstract Environmental monitoring and sensing are two of the first issues to address in order to keep our environment safe. Nanoparticle materials have tremendous potential to combat environmental pollution because of their novel chemical and physical properties. A pollution sensor was created using nanotechnology. It is becoming more common to use a variety of materials in their nano form to clean the environment, purify drinking water, and sanitize soil. These materials include iron oxide, manganese dioxide, indium phosphide nanowire, platinum, rhodium, zinc oxide, titanium dioxide, silica, and graphene. Sensible materials are being improved with the use of nanotechnology research. Nanomaterials should be monitored to ensure that they do not encourage more environmental degradation while we are using them to clean up environmental contaminants. This chapter primarily concentrates on the uses of nanoparticles such as capped nanoparticles, iron-based nanoparticles, gold nanoparticles, metal oxide nanoparticles, and metal nanoparticles in the defensible expansion, with a focus on environmental monitoring and sensing. Additionally, the paper highlights current advancements in several types of nanoparticles focused on monitoring, resolving, and sensing pollution.

Keywords Nanoparticles · Nanosensors · Pollutants · Capping · Controlling

A. T. Wassie (✉) · R. K. Bachheti
Department of Industrial Chemistry, College of Applied Sciences, Addis Ababa Science and Technology University, P.O.Box: 16417, Addis Ababa, Ethiopia
e-mail: addisutkm@gmail.com

R. K. Bachheti
Department of Allied Sciences, Graphic Era Hill University (G.E.H.U), Society Area, Clement Town, Dehradun 248002, India

A. Bachheti
Department of Environment Science, Graphic Era (Deemed to Be University), Dehradun, Uttarakhand 248002, India

R. K. Bachheti et al. (eds.), *Nanomaterials for Environmental and Agricultural Sectors*, Smart Nanomaterials Technology,
https://doi.org/10.1007/978-981-99-2874-3_6

1 Introduction

Environmental monitoring for human health, detecting gases, checking chemical dispensation, supervising the quality of water and air, also infrared radiation are all necessary for our safe and productive living on this planet [66]. Environmental pollution has grown to be one of the world's major problems since the start of the industrial revolution. The three main environmental issues brought on by multiple developing pollutants are air, water, and soil contamination [27]. The environment and human health are seriously endangered by contaminants in water and air, such as dangerous heavy metals, organic or inorganic chemicals, deadly gases, pesticides, bacteria, and antibiotics [27, 66]. The discharge of such dangerous pollutants into the ecosystem is a major issue, as is the growing environmental contamination brought on by the emergence of such harmful agents. The main drawbacks of conventional approaches in environmental monitoring are their inability to identify low sample concentrations and their lack of selectivity and sensitivity. A lengthy and intricate sample pretreatment procedure takes time as well. Given their global impact, rapid and repeatable identification of these contaminants is consequently required. Designing and developing monitoring techniques that would provide more efficacy and precision to detect a wide range of different contaminants is absolutely necessary [66]. In this situation, we require a system that can detect, diminish, avoid, and treat environmental pollution. Nanotechnology has the highest potential to offer a long-term solution to the problems with soil, water, and air pollution that exist around the world. Nanotechnology enables the atomic and molecular level design and manipulation of materials. These materials can be designed to have certain capabilities that can identify a specific contaminant in a combination [17]. Given their high surface area to volume ratio and average size range of 1–100 nm dimension, nanoparticles are distinguished by their distinct shapes and characteristics such as mechanical, optical, and electrical [2, 34].

Nanoparticles are essentially tiny objects that function as a single entity based on their attributes and modes of travel. They may also be created to enhance the medicinal and pharmacological benefits of the medications. Additionally, due to their large surface areas and the ability for numerous functional groups to cling to them, they have the ability to bind to tumor cells. They have been shown to be an excellent alternative to chemotherapy and radiation because they can quickly assemble in the tumor's microenvironment [20]. Nanoparticles (NPs) can be detected with extreme sensitivity due to their small size. These unique characteristics of NPs will make it possible to create highly precise, sensitive, and miniaturized pollution-monitoring systems to find contaminants in water and the air [17, 32, 42].

Device sensitivity increases in the presence of ambient noise because it interferes with signal demand and causes a sharp decline in the signal to noise ratio. Any sensor's capacity to determine a signal and the progression of it accurately depends heavily on the characteristics of the material that makes up the sensor and the sophistication of the manufacturing process. Safety for users and financial gain for the company are prerequisites for producing these devices on an industrial basis. Additionally, industrial processing speed must match the demands of the ever expanding market [42].

Single photon detectors have replaced noisy photomultiplier tubes in photon detecting technologies. This has been accomplished by collaborative and competitive research amongst several domains, primarily comprising microscopic material research and production technology (macroscopic). Since they are frequently employed in intricate surroundings, sensing devices must be lightweight, robust, and simple to use. There have been several developments in the field of sensing, including plasma mass spectrometry and surface-enhanced Raman scattering [66], chemiluminescence, fluorescence, electrochemistry, and many more [20, 66]. Although the majority of approaches have high accuracy and trustworthy repeatability as benefits, their primary drawbacks are expensive fabrication costs, complicated operation, and extensive sample preparation. As a result, the practical applications of these devices are severely constrained. Traditional semiconductor-based sensors have been researched and tested for a very long period to achieve the performance goal. However, the working conditions of these devices, such as their high operating temperatures (200–500 °C), fluctuations in accuracy and performance with humidity, and their prolonged operation hours, frequently limited their efficacy [66]. Therefore, rather than depending on improvements in preparation technology, it is projected that the best way forward for the development of the next generation of detecting technology will be the development of sensing materials with excellent all-around performance. In general, scale-up production and decreased amount of dimensions of the detecting material allow for the natural amplification of their response, whether it be in terms of optical, electrical, or chemical capabilities, or all three at once. According to their physical and chemical characteristics, the system of ordered assembly of low dimensional material in a layer-by-layer sequence or array has been studied for its capacity to identify environmental issues [43, 66].

Most zero-dimensional (0D) formations are composed of spherical particles with diameters ranging from nanometers to angstroms. The most prevalent zero-dimensional material is quantum dots, which may be built into layered structures using a variety of deposition processes. When it comes to quantum dot material, metallic nanoparticles dominate the bottom shelf, while other organic materials rise to the top in a distinctive fashion. Another carbon allotrope that meets the requirements to be a 0D material is fullerene, also referred to as "bucky ball" since it has a spherical structure made of 60 carbon atoms. The most prevalent examples of artificial allotropes that are similar to C60 in terms of behavior and physical characteristics are C70 and C540.

Nanomaterials are used in a wide range of industries, including electronics, magnets, optoelectronics, biomedicine, pharmaceuticals, cosmetics, and catalytic processes, as well as energy, environmental, and other activities. Nowadays, nanotechnologies are widely believed to have the potential to benefit a variety of industries, including medicine, water purification, and the development of stronger and lighter materials [58]. Over time, nanotechnology has attracted a lot of attention.

The nanoscale platform is utilized in practically every industry, including environmental science, health sciences, electronics, industrial separation, small- and large-scale portable water treatment systems, catalysts, and energy generation and storage [20]. The aim of this chapter is to highlight the recent development in nanoparticles for environmental monitoring and sensing.

2 Nanoparticles for Environmental Remediation

Nanoparticles are an excellent alternative today for use in environmental cleanup and boosting performance in the renewable energy sector because of their special chemical and physical characteristics. Nanoparticles are prevalent in nature, and some of them have been proven to have positive effects on the environment. Environmental remediation using nanoparticles, also known as nanoremediation, has been used for more than 10 years to treat or decontaminate the air, water, and soil. The area that is most frequently employed is in industrial stacks to either fully eliminate or reduce the pollutant level to permitted levels that reduce air pollution [20, 58].

Nanotechnology is also employed as an environmental technology to safeguard the environment by preventing, treating, and eradicating pollutants. Nanoremediation has the potential to not only lower the total costs of cleaning up large-scale polluted sites, but also to shorten clean-up times, do away with the need to treat and dispose of polluted soil, and reduce some pollutant concentrations to almost nothing. In general, NPs have some inherent qualities that can be used in environmental rehabilitation. The NP surface expands exponentially with the same density when the diameter decreases as a result of its smaller size. Additionally, because of their small size, NPs have great mobility in solutions and can be used in small quantities to swiftly scan a whole volume [60]. In general, Table1 lists many types of nanoparticle synthesis and applications for cleaning up the environment.

2.1 Impact of Capped Nanoparticles in Environmental Remediation

As a result of the irreparable harm it causes, environmental pollution has risen steadily over time to rank among the most critical global threats facing civilization. By releasing harmful substances, smoke, and toxic gases into the environment, nonstop urbanization and also rapid industrial development process have upset an equilibrium environment's composition, which has deleterious consequences for all living things [12]. In addition, overuse of natural resources brought on by an overabundance of people, automobiles, and industry, as well as a host of other things, destroys the environment. Some examples of hazardous materials are pesticides, heavy metals, dangerous gases, herbicides, unused pharmaceuticals, fertilizers,

Table 1 Synthesis and applications of different Nanoparticles in environmental remediation

Type of nanoparticles	Synthesis	Applications	References
AgNPs	Laser ablation technique	• Dye removal	[64]
Au–AgNPs	Biological method	• Biosensor	[1]
AuNPs	Brust-Schiffrin	• H_2O_2 sensor • Enhance Raman scattering activity	[2]
Carbon nanotube	Catalytic Chemical Vapor Deposition (CCVD)	• Thermal conductivity • Thermal materials • Structural applications • Field emission • Energy storage • Fibers and fabrics • Conductive adhesive • Biomedical applications • Catalyst supports • Air & Water Filtration	[21, 36, 67]
Fe–Mn	Any green synthesis method	• Acetamiprid degradation	[19]
Iron oxide	Thermal decomposition	• Wastewater treatment	[1]
MnO_2	co-precipitation method	• Wastewater treatment	[33]
$NdVO_4$	Ultrasound assisted method	• Photocatalytic	[29]
NiNPs	Solvothermal	• Synthesis of amines • Reduction of 4-nitrophenol (4-NP)	[49]
Ni–Zn ferrite	Sol–gel auto combustion	• Gas sensor	[18]
Platinum nanoparticle	Reduction method	• Fuel cell	[4]
Platinum	Biogenic synthesis method	• Oxygen reduction	[59]
Pt–Rh	Wet-chemistry based method	• Dye remova	[70]
Rhodium	Wetchemical reduction	• Catalyst	[14]
Silica	Sol–gel process, and flame synthesis	• Treating grasserie disease of silkworm	[16, 41]
Titanium dioxide	Hydrothermal method	• Improved photovoltaic performance of solar cells • Photovoltaic cell • Removal of benzene and toluene • Degradation of dye • Photodecomposition of phenol	[17]
V_2O_5	Hydrothermal method	• Photocatalytic and antibacterial	[38]

(continued)

Table 1 (continued)

Type of nanoparticles	Synthesis	Applications	References
Zinc oxide	Refluxing	• Solar cell • Antifungal agent	[5, 25]
ZnO − Zn composite	Sol–gel method	• Dye removal	[62]
(ZnMn)-co-doped magnetite	Co-precipitation method	• Magnetic hyperthermia therapy	[58]
$(Zn, Fe)_3O_4$	Co-precipitation method	• Biosensor and magnetic hypothermia therapy	[56]

industry effluents, oil spills, sulfur-containing chemicals, pathogens, matter particles, organic compounds, sewage, etc. [6, 35]. The maintenance of clean water and air is currently the major problem given the current situation. Thankfully, new techniques for removing unwanted substances from the air, water, and soil are increasingly being used. Among these, nanotechnology shows significant promise for developing and employing cutting-edge and reasonably priced methods for environmental pollutants to be cleaned, monitored, and catalyzed into less harmful forms [35, 39]. In contrast to bulk materials, nano technological products offer innovative physical and chemical properties because of their smaller size (100 nm), increased surface to volume ratio, and effectiveness as catalysts. Recently developed nanomaterials such as carbon nanotubes, polymers, dendrimers, metallic oxide nanoparticles, and many more have been employed to filter water, air, and soil. However, scientists must ensure that nanoparticles used in environmental remediation do not contribute to environmental degradation [47].

Most frequently, nanomaterials have a propensity to aggregate, adsorb into bigger particles or surfaces, or even decompose, which can be hazardous to human health and the environment [35, 40]. In recent years, the field of environmental remediation has seen significant growth in the polymer functionalization of nanoparticles due to the integration of both nanoparticles and polymers under a single system. Polymer-coated nanoparticles are useful because of their pore space, excellent, long-lasting mechanical strength, and surface properties. Additionally, nanoparticles with polymer coatings retain their original characteristics and provide stability and biocompatibility, making them a better surface covering. Polymer-coated nanoparticles' primary modes of action include adsorption, catalytic breakdown of contaminants, and an antibacterial activity that is particularly useful in the filtration of drinking water [47, 57]. These nanoparticles propose a more environmentally benign route for environmental rehabilitation because they are target-specific, do not produce waste, and may be used without needing to be disposed away after treatment [31].

3 Treatment of Water with Help of Nanoparticles

Due to industrialization, pure, and clean water is becoming increasingly scarce, with the poorer countries particularly suffering from this problem. Significant risks to the surface and subsurface habitats have been created by the expansion of the metallurgical, mining, nuclear energy, and chemical manufacturing industries, which have released significant volumes of effluent into the environment, including water [46]. Organic materials, bacteria, viruses, colors, and ions of heavy metals that aren't biodegradable, like cadmium (Cd), lead (Pb), nickel (Ni), zinc (Zn), arsenic (As), chromium (Cr), and mercury (Hg), are all examples of water contaminants that pose a serious threat to human health. Anemia, encephalopathy, miscarriages, hepatitis, cancer, kidney damage, nephritic syndrome, anemia, and hepatitis are just a few of the harmful impacts that heavy metal ions can have [11]. In addition to acid lead battery manufacturing, the paper, glass, and polishing sectors are the main sources of lead ions emitted into the environment. Water discharged from electroplating processes used in the creation of batteries, solar cells, metallurgical processes, and textile industries frequently contains cadmium. When nickel ions come into contact with jewellery waste, such as zips, watches, coins, etc., they can lead to skin disorders. In addition to causing stomach discomfort, nephritis, and liver damage, chromium metal ions (VI) are a primary factor in the development of nasal mucous ulcers [10]. In order to clean up contaminated water, it is required to remove these ions of heavy metal [67]. The adsorption of heavy metal ions on the carbon nanomaterial (graphene) is shown in Fig. 1 and the use of nanomaterials for environmental cleanup, monitoring, and energy efficiency is shown in Fig. 2.

NPs are extremely flexible and can be used to purify industrial effluents and wastewater in both ex situ and in- situ forms. For instance, it is simple to use NPs

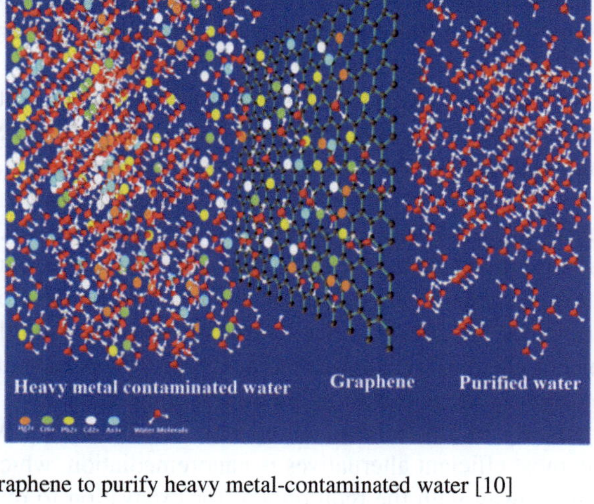

Fig. 1 Use of graphene to purify heavy metal-contaminated water [10]

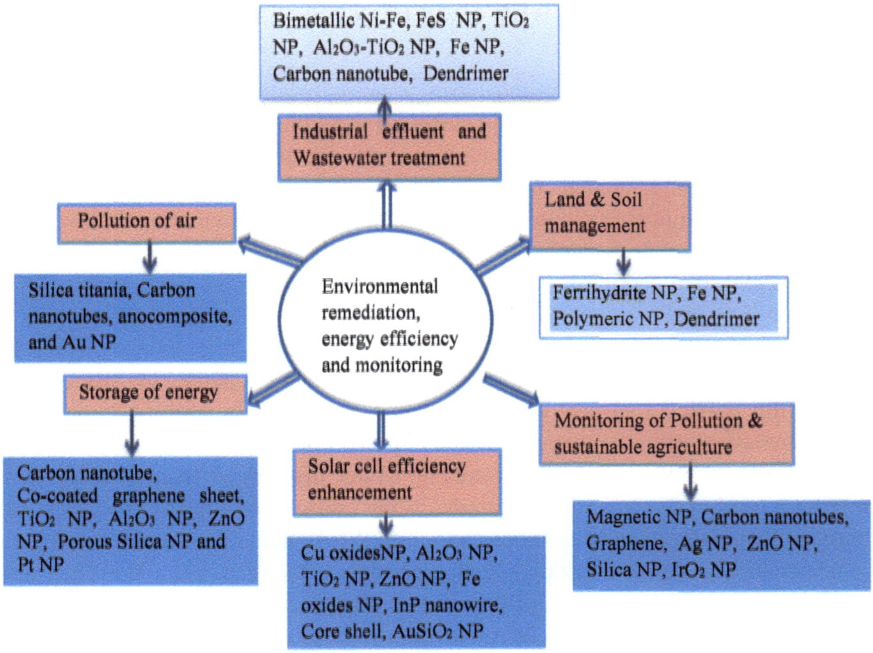

Fig. 2 The use of nanomaterials for environmental remediation, monitoring, and energy efficiency

in slurry reactors of ex-situ to remediate contaminated soils, sediments, and solid waste. For better waste water treatment, they can also be fixed to a solid matrix like zeolite, carbon, or a membrane. A successful use of nanotechnology-mediated environmental remediation is the use of zero-valente iron (Fe^0) nanoparticles to treat contaminated soil and groundwater. While oxidized iron rusts rapidly when exposed to air, contaminants like carbon tetrachloride (CCl_4), dioxins, and trichloroethylene (TCE) are broken down into simpler, considerably less hazardous carbon molecules when iron is oxidized in their presence [17].

Recent breakthroughs in nanotechnology and membrane separation have been identified as some practical and efficient methods for improving membrane enactment with combined effects for wastewater and water treatment. Due to their exceptional water-transport capabilities, high specific surface area, high mechanical strength, and excellent chemical inertness, carbon nanotubes (CNTs), specifically single-walled carbon nanotubes (SWCNTs) and multi-walled carbon nanotubes (MWCNTs), have attracted a great deal of attention in the development of new composite membranes for use in water treatment [24, 45]. Additionally, having the attractive electrochemical, catalytic, and adsorption properties of CNTs, which are useful for coupling these functions with membrane separation processes and enhancing the efficacy of CNTs-based composite membranes for water treatment [50]. One of the most efficient alternatives is nanoremediation, which offers in situ treatment and does away with the requirement for excavation to reach the desired location as well as ground water pumped out for treatment. When a nanoparticle is injected into a specific site, it travels with the flow of groundwater and cleans the water by immobilizing the impurities. Redox reactions are one of the main decontamination methods. Consequently, filtration, and disinfection, desalination are used to treat surface water with nanoparticles [20].

3.1 Iron-Contained Nanoparticles Applications in Wastewater Treatment

The interdisciplinary uses of iron-contained nanoparticles (IONPs) have been fantastic. It can be employed as a pigment in the manufacturing sector, the catalyst substantial, and an adsorbent in the treatment of water. Additionally, they are essential resources for the production of magnetic nature data storage devices, ion exchangers, magnetic resonance imaging, gas sensors, coatings, magnetic recording, devices for biological separation, and also magnetic resonance recording [8]. Currently, research is focused on environmentally friendly and sustainable trash rehabilitation and remediation methods [26]. Pathogenic microbes, poisonous organics, and inorganics are the three broad groups of contaminants that can be found in water and wastewater [28]. Many potent contaminants can be found in the wastewater produced by various sources. The most harmful substance to the environment, for instance, is the everlasting dye flow from the leather, paint, and textile industries [7, 51, 61]. Wastewater from certain sectors is permitted to include heavy metals. They may be poisonous and carcinogenic, and they may seriously harm aquatic ecosystems and people [52]. Therefore, it is essential to remediate these pollutants effectively. Currently, industrial effluent treatment and industrial water contamination control is the subject of extensive knowledge and a distinct scientific viewpoint. Today, many nanomaterials are used to combat environmental problems. In particular, iron oxides (such as magnetite, hematite, and maghemite) and iron oxhydroxides (such as goethite, akaganeite, and lepidocrocite) are being investigated for their possible application in wastewater treatment for the elimination of harmful metal ions [54]. The different type synthesis and applications of nanoparticles for environmental remediation are listed in Table 1.

4 Different Applications of Nanoparticles in Sensing Materials

For sensor technology and biomedical requests, a variety of noble metal nanoparticles are used. However, due to a variety of engaging qualities, gold nanoparticles (AuNPs) are primarily used. Some of these qualities, along with some of the most prevalent ones, are given in Table 2.

Table 2 Sensors based on gold nanoparticles (AuNPs)

Properties of Sensor	The AuNPs features that add value	References
Optical sensors that depend on variations in optical characteristics	Exacerbation of changes in refractive index	[69]
Unique optical properties, narrow size distributions and good biological affinity	Sensing analysis, catalytic, environmental monitoring, and disease therapy	[48]
Utilizing sensors of electrochemical to keep track of variations in electrical properties	Enhanced transferring of electron enhanced catalysis surface area	[3]
High conductivity	Electrochemical sensors for monitoring of bisphenol A in the application of Covalent-organic frameworks provided a significant guidance	[71]
Sensors for measuring gravity depend on variations in mass	Mass fluctuations are intensified an increased surface area	[30]

4.1 Application of Metal Nanoparticles (MNPs) for Collecting the Analyte

The magnetic nature of MNPs makes their use in sensing the most clear. This broad concept, however, can be more narrowly specified to provide MNPs particular features that enable quick, highly selective, and long-lasting sensors. For a variety of sensing applications, the MNPs with super paramagnetism, high saturation magnetization, surface belongings and constant magnetic, and also sensing-specific qualities work at their best [65]. Analytes in the trial solution diffuse to the sensing shallow in a distinctive sensor, somewhere the transducer, either of optical or supplementary produces a signal according to the degree of contact among the surface and analytes. In removing the sample's analytical component and bringing the analyte to the surface of the sensor while being controlled magnetically, MNPs have traditionally been employed to aid mass transport to the sensing surface [44]. The preconcentration of the sample allows for a decrease in response time and an increase in sensitivity, and it also enables the special and time-based parting of sample gathering from the transduction quantity of the sample, which are the two key advantages that an MNP sensor can offer over a traditional sensor by gathering the sample [23, 68].

4.2 Application of MNPs for Electrochemical Transduction

MNPs are a component of the transducing NP system because they can be utilized to connect certain goals to a NP sensor in addition to separating and preconcentrating the analyte. Despite not directly sensing the analyte, transduction is made possible by the MNPs' binding to or alteration of the analyte [55]. Other MNP characteristics, like ionic confrontation in the solution or conductivity of electric as soon as mixed through Au, have used MNPs directly in the analyte sensing in addition to their magnetic capabilities. As with preconcentration for electrochemical detection, straightforward transduction of the electrochemical using MNPs entails delivering of catalyzing materials to word electrode. This was demonstrated by Lin and colleagues using AuNPs and MNPs treated through urease and anti-Listeria polyclonal antibodies [15, 55]. A sandwich complex is created after Listeria cells are separated from MNPs by the improved AuNPs and Listeria-bound MNPs. This complex, when magnetically gathered at an electrode, catalyzes the conversion of urea into ammonium carbonate. Electrochemical impedance spectroscopy is used to measure the consequent rise in ionic strength. Similar to electrochemical sensors, MNPs can reduce the reaction period of optical sensors straightforward to the concentration method [23]. The refractive index of an optically active surface can be changed to make MNPs, which are not optically active, become optical transducers, or optically active materials can be coated onto MNPs to make them optically active [22].

4.3 Application of MNPs for Magnetic Transduction

The magnetic response can be used to detect a signal, which is an easy technique to use MNPs as transducing components in sensing. Simple MNPs can be utilized instead of complicated coatings or processing, which saves time, money, and perhaps increases toxicity for biological applications. This is especially helpful for sensors. With transduction of magnetic, a single parameter generates the signal (the magnetic moment or relaxation time). Weissleder and colleagues progressively labeled MNPs with various molecular markers, and then they measured accumulating magnetic signals to achieve multiplexed detection. When sequential labeling is successful, more MNP probes may be placed on target cells and more markers can be quantitatively detected on a single sample. This matters for a variety of diagnostic and imaging applications [63].

4.4 Application of Metal Oxide Nanoparticles (MONPs) in Gas Sensors

Gas sensor research is increasingly focusing on creating highly effective, consistently dependable gas sensors that can precisely measure at or near room temperature (RT). With the help of MONPs, signal transmission is improved, in-situ monitoring is made possible, and surface plasmonic resonance has the potential to be used to improve sensing performance. Wider operational parameters to be evaluated (quick changes in light intensity, polarization, wavelength modulation, etc.), increased selectivity, and simplicity of implementation are a few of these distinct characteristics [63]. The most popular sensing materials historically have been metal oxide semiconductors (MOS), which have benefits including a quick response, low cost, ease of handling, a wide range of target gases, and longer lives. Due to their large surface-to-volume ratios, high specific surface areas with extremely high surface reactivity and abundance of surface active sites, nano-metal oxides, are employed in gas sensing [53]. An MONP-based gas sensor typically comprises of conducting electrodes to measure resistance, a sensing film that changes resistance in response to exposure, and a heating layer or wire to reach the appropriate operating temperature (Fig. 3). When exposed to the target gas, conductometric gas sensors see a change in electrical conductivity. In contrast to the transducer function, which is dependent on the microstructure of the oxide and can be changed by mixing different oxides or by adding noble metals, the receptor function depends on the interactions between gases and solids. The amount of reactive surface sites and the amount of oxygen species that are absorbable, which rises the amount of reactive external sites, decide the gas sensing mechanism [9]. The surface stoichiometry has a significant impact on the surface conductivity of the metal oxide because oxygen vacancies increase surface conductivity while absorbed ions lower it. When molecules like O_2 or NO_2 bind to the vacancy sites of the oxide, the electrons flow out of the conduction band and reduce conductivity [37]. In contrast, when CO or H_2 in an oxygen-rich environment react with adsorbed O_2, the electrons are released and conductivity is increased. Since metal oxide serves as a transducer, interactions with the environment alter its conductivity, which allows electrical signals to be sent to the electrode. To alter the device's electrical conductivity, indirect heating is performed [13].

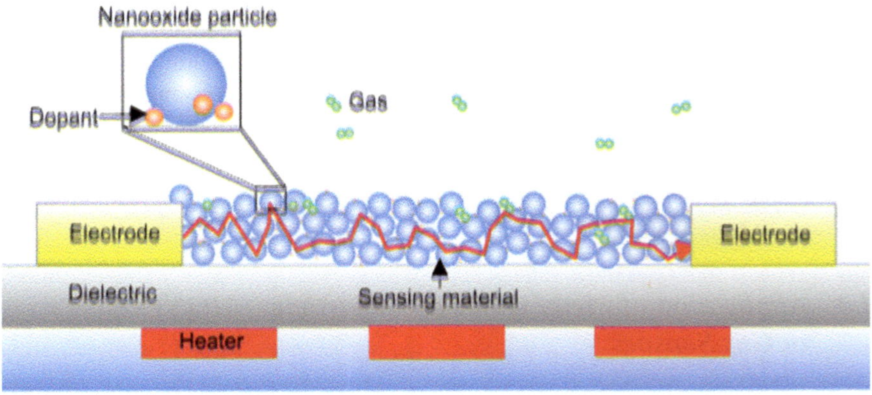

Fig. 3 A schematic of a conductometric gas sensor and its operating mechanism [13]

5 Conclusion

NPs have a significant potential to combat environmental contamination because of their novel physicochemical properties. The ability of a wide variety of nanomaterials to clean the environment is being researched. Their intriguing qualities, such as their large surface area, are what made the application effective. However, the majority of these environmental monitoring and treatment methods based on nanomaterials are still in the research stage. Some commercial goods are currently on the market and suitable for use on a large scale. It is projected that there will be a large number of nanoparticle technologies available for environmental monitoring and remediation in the years to come. However, it needs to be properly monitored to ensure that nanoparticles do not further harm the environment. Due to their unique chemical, magnetic, mechanical, optical, and physical capabilities, nanoparticle materials are primarily employed for creating biosensors with higher specificity and sensitivity of detection. This literature review's findings suggest that many nanoparticle materials offer intriguing physicochemical features and the dynamic potential to be used for environmental management, remediation, sensing, and water purification.

References

1. Abegunde SM, Idowu KS, Adejuwon OM, Adeyemi-Adejolu T (2020) A review on the influence of chemical modification on the performance of adsorbents. Resour Environ Sustain 1:1001
2. Aguilar-Pérez K, Heya M, Parra-Saldívar R, Iqbal HM (2020) Nano-biomaterials in-focus as sensing/detection cues for environmental pollutants. Case Stud Chem Environ Eng 2:1055
3. Ahmad R, Griffete N, Lamouri A, Felidj N, Chehimi MM, Mangeney C (2015) Nanocomposites of gold nanoparticles@ molecularly imprinted polymers: chemistry, processing, and applications in sensors. Chem Mater 27(16):5464–5478
4. Ahmed RK, Saad EM, Fahmy HM, El Nashar RM (2021) Design and application of molecularly imprinted Polypyrrole/Platinum nanoparticles modified platinum sensor for the electrochemical detection of Vardenafil. Microchem J 171:106771
5. Al-Juaid F, Merazga A (2013) Increasing dye-sensitized solar cell efficiency by ZnO spin-coating of the TiO_2 electrode: effect of ZnO amount. Energy Power Eng
6. Amiri P, Behin J (2021) Assessment of aqueous phase ozonation on aggregation of polyvinylpyrrolidone-capped silver nanoparticles. Environ Sci Pollut Res 28:34838–34851
7. Aragaw TA (2020) Recovery of iron hydroxides from electro-coagulated sludge for adsorption removals of dye wastewater: adsorption capacity and adsorbent characteristics. Surf Interfaces 18:100439
8. Aragaw TA, Bogale FM, Aragaw BA (2021) Iron-based nanoparticles in wastewater treatment: a review on synthesis methods, applications, and removal mechanisms. J Saudi Chem Soc 25(8):101280
9. Ayesh AI, Abu-Hani AF, Mahmoud ST, Haik Y (2016) Selective H2S sensor based on CuO nanoparticles embedded in organic membranes. Sens Actuators, B Chem 231:593–600
10. Baby R, Saifullah B, Hussein MZ (2019) Carbon nanomaterials for the treatment of heavy metal-contaminated water and environmental remediation. Nanoscale Res Lett 14(1):1–17
11. Baby Shaikh R, Saifullah B, Rehman FU (2018) Greener method for the removal of toxic metal ions from the wastewater by application of agricultural waste as an adsorbent. Water 10(10):1316

12. Bapat MS, Singh H, Shukla SK, Singh PP, Vo D-VN, Yadav A, Goyal A, Sharma A, Kumar D (2022) Evaluating green silver nanoparticles as prospective biopesticides: an environmental standpoint. Chemosphere 286:131761
13. Chavali MS, Nikolova MP (2019) Metal oxide nanoparticles and their applications in nanotechnology. SN Appl Sci 1(6):1–30
14. Chen C, Pan Y, Zhao H, Xu X, Xu J, Zhang Z, Xi S, Xu L, Li H (2018) A versatile rhodium (iii) catalyst for direct acyloxylation of aryl and alkenyl C–H bonds with carboxylic acids. Org Chem Front 5(3):415–422
15. Chen Q, Wang D, Cai G, Xiong Y, Li Y, Wang M, Huo H, Lin J (2016) Fast and sensitive detection of foodborne pathogen using electrochemical impedance analysis, urease catalysis and microfluidics. Biosens Bioelectron 86:770–776
16. Das S, Debnath N, Patra P, Datta A, Goswami A (2012) Nanoparticles influence on expression of cell cycle related genes in Drosophila: a microarray-based toxicogenomics study. Toxicol Environ Chem 94(5):952–957
17. Das S, Sen B, Debnath N (2015) Recent trends in nanomaterials applications in environmental monitoring and remediation. Environ Sci Pollut Res 22(23):18333–18344
18. de Mello LB, Varanda LC, Sigoli FA, Mazali IO (2019) Co-precipitation synthesis of (Zn-Mn)-co-doped magnetite nanoparticles and their application in magnetic hyperthermia. J Alloy Compd 779:698–705
19. Duan P, Ma T, Yue Y, Li Y, Zhang X, Shang Y, Gao B, Zhang Q, Yue Q, Xu X (2019) Fe/Mn nanoparticles encapsulated in nitrogen-doped carbon nanotubes as a peroxymonosulfate activator for acetamiprid degradation. Environ Sci Nano 6(6):1799–1811
20. Ealia SAM, Saravanakumar M (2017) A review on the classification, characterisation, synthesis of nanoparticles and their application. Paper presented at the IOP conference series: Materials Science and Engineering
21. Eatemadi A, Daraee H, Karimkhanloo H, Kouhi M, Zarghami N, Akbarzadeh A, Abasi M, Hanifehpour Y, Joo SW (2014) Carbon nanotubes: properties, synthesis, purification, and medical applications. Nanoscale Res Lett 9(1):1–13
22. Gautam S, Hoque S, Gooding JJ (2021) Gold-coated magnetic nanoparticles as dispersible electrochemical biosensors for ultrasensitive biosensing. In: Biochemical sensors: nanomaterial-based biosensing and application in honor of the 90th birthday of Prof. Shaojun Dong. World Scientific, pp 59–83
23. Gloag L, Mehdipour M, Chen D, Tilley RD, Gooding JJ (2019) Advances in the application of magnetic nanoparticles for sensing. Adv Mater 31(48):1904385
24. Goh K, Karahan HE, Wei L, Bae T-H, Fane AG, Wang R, Chen Y (2016) Carbon nanomaterials for advancing separation membranes: a strategic perspective. Carbon 109:694–710
25. González-Merino AM, Hernández-Juárez A, Betancourt-Galindo R, Ochoa-Fuentes YM, Valdez-Aguilar LA, Limón-Corona ML (2021) Antifungal activity of zinc oxide nanoparticles in *Fusarium oxysporum-Solanum lycopersicum* pathosystem under controlled conditions. J Phytopathol 169(9):533–544
26. Guerra FD, Attia MF, Whitehead DC, Alexis F (2018) Nanotechnology for environmental remediation: materials and applications. Molecules 23(7):1760
27. Hashem A, Hossain MM, Al Mamun M, Simarani K, Johan MR (2021) Nanomaterials based electrochemical nucleic acid biosensors for environmental monitoring: a review. Appl Surf Sci Adv 4:100064
28. Hlongwane GN, Sekoai PT, Meyyappan M, Moothi K (2019) Simultaneous removal of pollutants from water using nanoparticles: a shift from single pollutant control to multiple pollutant control. Sci Total Environ 656:808–833
29. Hosseinpour-Mashkani SS, Sobhani-Nasab A (2017) Investigation the effect of temperature and polymeric capping agents on the size and photocatalytic properties of $NdVO_4$ nanoparticles. J Mater Sci: Mater Electron 28(21):16459–16466
30. Huang CB, Yao Y, Montes-García V, Stoeckel MA, Von Holst M, Ciesielski A, Samori P (2021) Highly sensitive strain sensors based on molecules-gold nanoparticles networks for high-resolution human pulse analysis. Small 17(8):2007593

31. Hussain CM (2020) Handbook of functionalized nanomaterials for industrial applications: Elsevier
32. Husen A, Siddiqi KS (2023) Advances in smart nanomaterials and their applications. Elsevier Inc., 50 Hampshire St., 5th Floor, Cambridge, MA 02139, USA
33. Islam MA (2020) Colloid and surface nature of birnessite (δ-MnO_2) and its application in wastewater treatment. Cell 1818:482533
34. Jamkhande PG, Ghule NW, Bamer AH, Kalaskar MG (2019) Metal nanoparticles synthesis: an overview on methods of preparation, advantages and disadvantages, and applications. J Drug Deliv Sci Technol 53:101174
35. Javed R, Zia M, Naz S, Aisida SO, Ain NU, Ao Q (2020) Role of capping agents in the application of nanoparticles in biomedicine and environmental remediation: recent trends and future prospects. J Nanobiotechnol 18:1–15
36. Jorio A, Saito R (2021) Raman spectroscopy for carbon nanotube applications. J Appl Phys 129(2):021102
37. Joshi N, Pandey DK, Mistry BG, Singh DK (2023) Metal oxide nanoparticles: synthesis, properties, characterization, and applications. Nanomaterials. Springer, pp 103–144
38. Karthik K, Nikolova MP, Phuruangrat A, Pushpa S, Revathi V, Subbulakshmi M (2020) Ultrasound-assisted synthesis of V_2O_5 nanoparticles for photocatalytic and antibacterial studies. Mater Res Innovations 24(4):229–234
39. Khan I, Saeed K, Khan I (2019) Nanoparticles: properties, applications and toxicities. Arab J Chem 12(7):908–931
40. Khanna K, Kohli SK, Handa N, Kaur H, Ohri P, Bhardwaj R, Yousaf B, Rinklebe J, Ahmad P (2021) Enthralling the impact of engineered nanoparticles on soil microbiome: A concentric approach towards environmental risks and cogitation. Ecotoxicol Environ Saf 222:112459
41. Kim M-K, Ki D-H, Na Y-G, Lee H-S, Baek J-S, Lee J-Y, Lee H-K, Cho C-W (2021) Optimization of mesoporous silica nanoparticles through statistical design of experiment and the application for the anticancer drug. Pharmaceutics 13(2):184
42. Krishna RN, Gayathri R, Priya V (2017) Nanoparticles and their applications-a review. J Pharm Sci Res 9(1):24
43. Kuswandi B, Hidayat M.A, Noviana E (2022) Based sensors for rapid important biomarkers detection. Biosensors Bioelectron 10:100246
44. Laycock A, Clark NJ, Clough R, Smith R, Handy RD (2022) Determination of metallic nanoparticles in biological samples by single particle ICP-MS: a systematic review from sample collection to analysis. Environ Sci Nano 9(2):420–453
45. Lee K-J, Park H-D (2016) Effect of transmembrane pressure, linear velocity, and temperature on permeate water flux of high-density vertically aligned carbon nanotube membranes. Desalin Water Treat 57(55):26706–26717
46. Li J, Wang X, Zhao G, Chen C, Chai Z, Alsaedi A, Hayat T, Wang X (2018) Metal–organic framework-based materials: superior adsorbents for the capture of toxic and radioactive metal ions. Chem Soc Rev 47(7):2322–2356
47. Liang Y, Song J, Dong H, Huo Z, Gao Y, Zhou Z, Tian Y, Li Y, Cao Y (2021) Fabrication of pH-responsive nanoparticles for high efficiency pyraclostrobin delivery and reducing environmental impact. Sci Total Environ 787:147422
48. Liu G, Lu M, Huang X, Li T, Xu D (2018) Application of gold-nanoparticle colorimetric sensing to rapid food safety screening. Sensors 18(12):4166
49. Liu P, Gao S, Wang Y, Zhou F, Huang Y, Huang W, Chang N (2020) Core-shell Ni@ C encapsulated by N-doped carbon derived from nickel-organic polymer coordination composites with enhanced microwave absorption. Carbon 170:503–516
50. Ma L, Dong X, Chen M, Zhu L, Wang C, Yang F, Dong Y (2017) Fabrication and water treatment application of carbon nanotubes (CNTs)-based composite membranes: a review. Membranes 7(1):16
51. Maddila S, Kerru N, Jonnalagadda SB (2022) Magnetic nanocatalysts for wastewater treatment. Indus Appl 97

52. Madhu A, Singh K (2017) Heavy metal removal from wastewater using various adsorbents: a review. J Water Reuse Desalin 7(4):387–419
53. Marouzi S, Sabouri Z, Darroudi M (2021) Greener synthesis and medical applications of metal oxide nanoparticles. Ceram Int 47(14):19632–19650
54. Mohamadiun M, Dahrazma B, Saghravani SF, Darban AK (2018) Removal of cadmium from contaminated soil using iron (III) oxide nanoparticles stabilized with polyacrylic acid. J Environ Eng Landsc Manag 26(2):98–106
55. Mollarasouli F, Zor E, Ozcelikay G, Ozkan SA (2021) Magnetic nanoparticles in developing electrochemical sensors for pharmaceutical and biomedical applications. Talanta 226:122108
56. Ognjanović M, Stanković DM, Ming Y, Zhang H, Jančar B, Dojčinović B, Prijović Ž, Antić B (2019) Bifunctional (Zn, Fe)$_3$O$_4$ nanoparticles: Tuning their efficiency for potential application in reagentless glucose biosensors and magnetic hyperthermia. J Alloy Compd 777:454–462
57. Potara M, Focsan M, Craciun A-M, Botiz I, Astilean S (2018) Polymer-coated plasmonic nanoparticles for environmental remediation: synthesis, functionalization, and properties. In: New polymer nanocomposites for environmental remediation. Elsevier, pp 361–387
58. Rafeeq H, Hussain A, Ambreen A, Waqas M, Bilal M, Iqbal H (2022) Functionalized nanoparticles and their environmental remediation potential: a review. J Nanostruct Chem 1–25
59. Rosas-Medellín D, Pérez-Salcedo K, Morales-Acosta D, Rodríguez-Varela F, Escobar B (2021) Green synthesis of Pt nanoparticles and their application in the oxygen reduction reaction. J Mater Res 36(20):4131–4140
60. Sánchez A, Recillas S, Font X, Casals E, González E, Puntes V (2011) Ecotoxicity of, and remediation with, engineered inorganic nanoparticles in the environment. TrAC, Trends Anal Chem 30(3):507–516
61. Shojaei S, Khammarnia S, Shojaei S, Sasani M (2017) Removal of reactive red 198 by nanoparticle zero valent iron in the presence of hydrogen peroxide. J Water Environ Nanotechnol 2(2):129–135
62. Siddiqi KS, Husen A (2018) Properties of zinc oxide nanoparticles and their activity against microbes. Nanoscale Res Lett 13(1):1–13
63. Slimani Y, Hannachi E (2020) Magnetic nanosensors and their potential applications. In: Nanosensors for smart cities. Elsevier, pp 143–155
64. Srinivas C, Kumar ER, Tirupanyam B, Meena SS, Bhatt P, Prajapat C, Rao TC, Sastry D (2020) Study of magnetic behavior in co-precipitated Ni–Zn ferrite nanoparticles and their potential use for gas sensor applications. J Magn Magn Mater 502:166534
65. Thobakgale L, Ombinda-Lemboumba S, Mthunzi-Kufa P (2022) Chemical sensor nanotechnology in pharmaceutical drug research. Nanomaterials 12(15):2688
66. Tyagi D, Wang H, Huang W, Hu L, Tang Y, Guo Z, Ouyang Z, Zhang H (2020) Recent advances in two-dimensional-material-based sensing technology toward health and environmental monitoring applications. Nanoscale 12(6):3535–3559
67. Wu Y, Zhao X, Shang Y, Chang S, Dai L, Cao A (2021) Application-driven carbon nanotube functional materials. ACS Nano 15(5):7946–7974
68. Yang X, Lan L, Li L, Liu X, Naumov P, Zhang H (2022) Remote and precise control over morphology and motion of organic crystals by using magnetic field. Nat Commun 13(1):2322
69. Yang Y, Lei Q, Li J, Hong C, Zhao Z, Xu H, Hu J (2022) Synthesis and enhanced electrochemical properties of AuNPs@ MoS2/rGO hybrid structures for highly sensitive nitrite detection. Microchem J 172:106904
70. Yildiz A, Celik F (2017) Atomic concentration effect on thermal properties during solidification of Pt-Rh alloy: a molecular dynamics simulation. J Cryst Growth 463:194–200
71. Zhang X, Zhu J, Wu Z, Wen W, Zhang X, Wang S (2023) Electrochemical sensor based on confined synthesis of gold nanoparticles@ covalent organic frameworks for the detection of bisphenol A. Anal Chim Acta 1239:340743

Current Trends and Future Applications of Silica Nanomaterials in Adsorption and Catalysis

Selvaraj Mohana Roopan, Mohamed Sulthan Hasan Fathima Afridha, and Gunabalan Madhumitha

Abstract Silica nanoparticles have recently been explored as one of the most significant types of materials in science and industry. Because of their unique qualities, including significant surface area, high porosity, cost-effectiveness, biocompatibility, as well as excellent physical, chemical and mechanical capabilities, research on silica nanoparticles has risen significantly. They have potential uses in separation, adsorption, catalysis and sensing. The ability to govern particle shape, porosity, size and crystallinity, as well as their presence in many forms such as mesoporous silica, hollow structures and dense silica, have led to their use in a variety of applications. Because of their excellent surface chemistry and large specific surface area, silica nanoparticles can operate as a possible adsorbent. Silica nanoparticles have created substantially active, selective and durable catalysts for various organic transformations. They also have small channels that serve as sturdy supports for active sites during catalysis. Nonetheless, due to recent improvements and future potential, mesoporous silica nanoparticles have caught the zeal of many researchers in the field of adsorption and catalysis. This book chapter focuses on the most recent breakthroughs of mesoporous silica nanoparticles in catalysis and adsorption.

Keywords Adsorbent · Adsorption · Catalysis · Catalyst · Mesoporous silica nanomaterials · Surface area

S. Mohana Roopan (✉) · M. S. H. Fathima Afridha · G. Madhumitha
Chemistry of Heterocycles and Natural Product Research Laboratory, Department of Chemistry, School of Advanced Sciences, Vellore Institute of Technology, Vellore, Tamil Nadu 632014, India
e-mail: mohanaroopan.s@vit.ac.in

R. K. Bachheti et al. (eds.), *Nanomaterials for Environmental and Agricultural Sectors*, Smart Nanomaterials Technology,
https://doi.org/10.1007/978-981-99-2874-3_7

1 Introduction

Nanomaterials made of silica are employed in various uses as extraordinary building blocks in materials science to create a range of useful materials [1–4]. Due to their distinctive qualities, such as their remarkable bioactivity, ease of surface modification, durability and economic viability, silicas are in high demand. The dimensions and topologies for commercial purposes are progressively increasing [5–7]. After Stöber et al. ground-breaking work on the fabrication of highly ordered silica, also recognized as Stöber SiO_2, researchers became interested in silica nanospheres [8]. Stöber silica's use is restricted in several disciplines due to its lack of pores and insufficient surface area. The creation of mesoporous silica later solved these problems, which are manufactured by creating pores and many surfaces using templates. Silica nanoparticles are categorized as either mesoporous or nanoporous, and the size of either type of nanoparticle can be altered by altering the composition of the surfactants used in their manufacture. Mesoporous silica nanomaterials have recently gained much attention because of their unique and adaptable physiochemical features. Mesoporous silica featuring clear-cut hexagonal porosity of 2–3 nm in size was first introduced by the Mobile Corporation in 1990 under the designation MCM-41. This marked the initial report of utilizing a cationic surfactant to make pores in silica. Dodecyl- or cetyltrimethylammonium (CTA) salts are typically used in numerous base-catalysed synthesis pathways; CTAC or CTAB compounds having chloride or bromide counterions were used to form the mesopores. Afterwards, numerous surfactant types, including cationic surfactants, anionic surfactant and neutral surfactants, were designed to make silica nanomaterials [9]. Tetraethyl orthosilicate is used as the source of silica in an acidic or basic environment. Various surfactants are used as substrates to hydrolyse the silica precursor to generate silica mesoporous of size 2–50 nm [10]. Diverse synthesis procedures have been documented in the quest to create silica nanomaterials with various topologies and porosities, viz., mesoporous and microporous. Stucky et al. claimed the development of Santa Barbara particles from a triblock copolymer in 1998 after the advent of MCM-41. The ability to precisely regulate the pore size discovered in SBA-15 is an important turning point in creating porous silica NMs.

Due to their appealing structural characteristics, low toxicity and excellent biocompatibility, silica NMs of various sorts have undergone substantial research for various applications [11–19]. Using various combinations of surfactant and inflating reagents, such as 1,3,5-triisopropyl benzene, this study led to numerous experiments on the production of pore-elongated SBA-15 and 16 with porosity sizes within the range of 5–30 nm [20–24]. Due to their superiority in this type of application, mesoporous silica viz., MCM-48, MCM-41, HMS and SBA-15 are believed to have vast surfaces, restricted pore widths and limited pore-size distributions. Furthermore, by adding functional groups, it is possible to boost its propensity for the target metal's surface [25–27]. Thus, functionalized mesoporous silica nanomaterials are the current booming catalyst in the field of catalysis and adsorption. This chapter

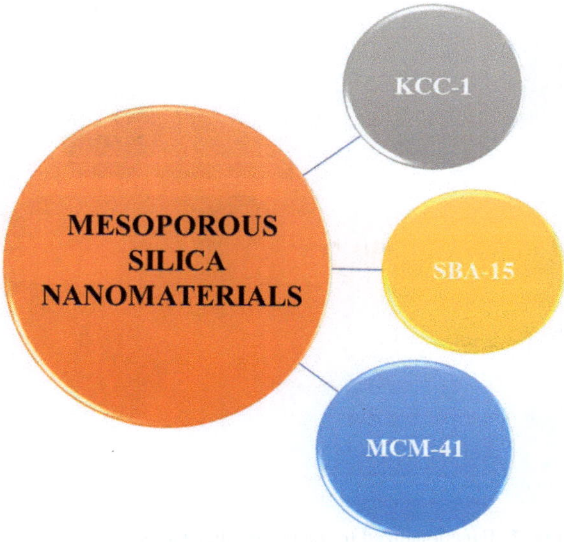

Fig. 1 Types of mesoporous silica nanomaterials

includes details about functionalized mesoporous silica nanomaterials in various catalysis and adsorption conditions (Fig. 1).

2 Functionalized Mesoporous Silica Nanomaterials in Catalysis

Common silica nanomaterials used in catalysis include KCC-1, SBA-15 and MCM-41, each of which are mesoporous materials (Fig. 2).

2.1 Activity of KCC-1

In order to accelerate the carbonylative SMC reaction, Pd nanoparticles implanted on KCC-1 fibrous nanosilica have been used as a catalyst (Scheme 1). Aryl iodide, aryl boronic acid and potassium carbonate in anisole were reacted in a stainless-steel autoclave along with a catalyst. The products were validated using ^{1}H and ^{13}C NMR and GC–MS methods. KCC-1-PEI/Pd showed excellent efficiency towards the carbonylative cross-coupling with less % loading of Pd. The catalyst also has led to the generation of high turnover numbers and turnover frequency [28].

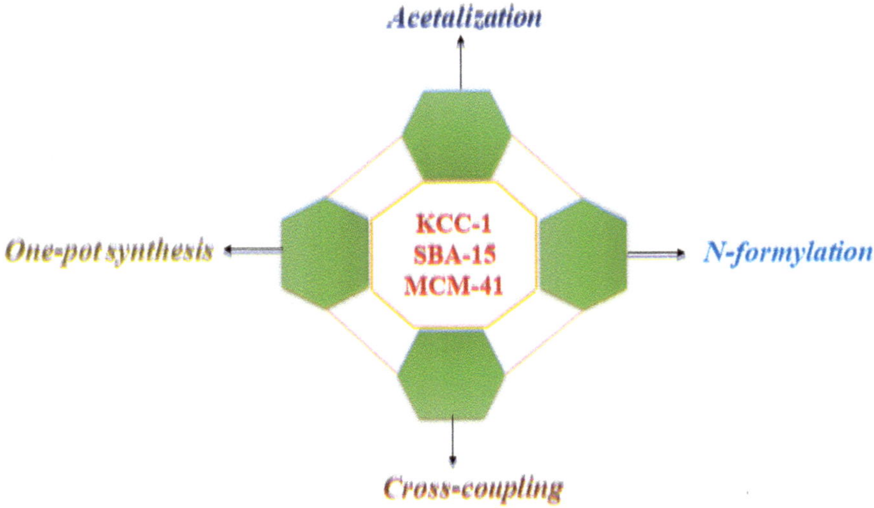

Fig. 2 Functionalized mesoporous silica nanomaterials in catalysis

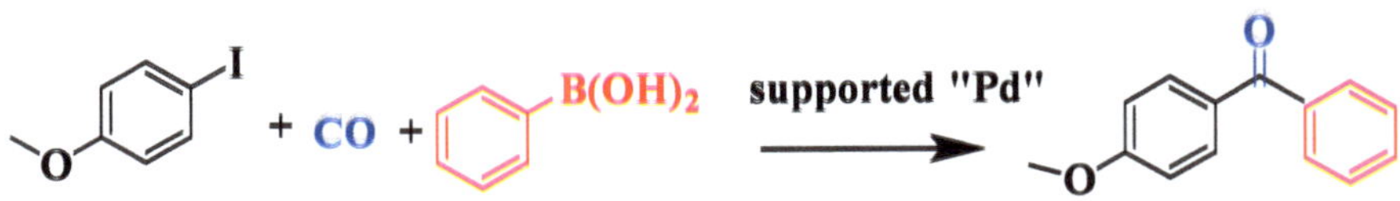

Scheme 1 Suzuki–Miyaura reaction (cross-coupling)

C–C and C–H cross-couplings have been facilitated by a Palladium–salen catalyst supported by Fe_3O_4/KCC-1/APTPOSS nanoparticles (Scheme 2). The catalyst was involved in the reaction of 2-acetyl-benzaldehyde to cyclopentadiene to produce pentafulvenes with different substrates. The catalyst exhibited high recoverability and reusability [29].

Gold nanoparticles of various sized were used as a support on KCC-1 fibrous nanospheres silica. The catalyst was employed for the oxidation of CO. The uniform and highly dispersed Au nanoparticles may well be settled within fibres of KCC-1-NH_2 by grafting $HAuCl_4$ followed by reduction. The catalyst was used for CO

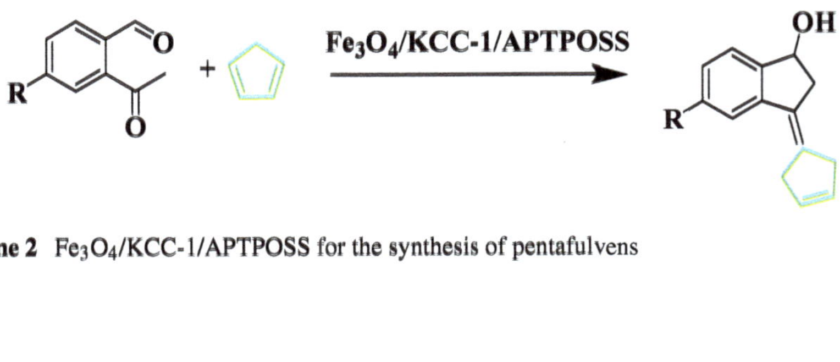

Scheme 2 Fe_3O_4/KCC-1/APTPOSS for the synthesis of pentafulvens

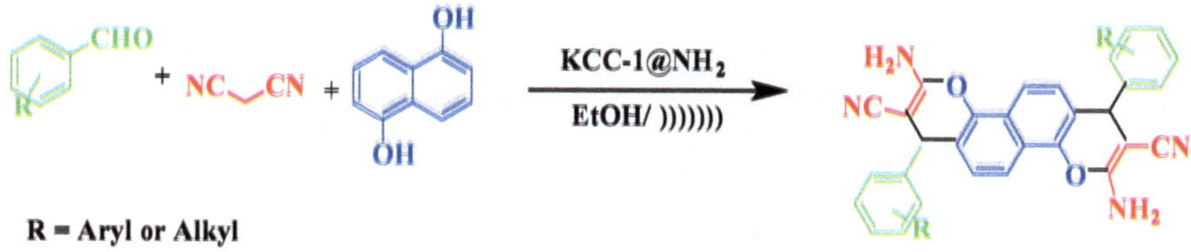

Scheme 3 Synthesis of chromenes

oxidation at temperatures between 100 and 300 °C since it showed no signs of aggregation. The larger size of Au nanoparticles failed to penetrate and showed less catalytic activity. This catalyst remains a benchmark in the oxidation of CO reactions [30].

Under ultrasonic irradiation, KCC-1 @NH_2 nanosilica has been deployed as a fundamental catalyst for the production of chromenes (Scheme 3). The catalyst has specific characteristics of high surface area, fibre surface morphology and great mechanical stability. The reaction involves malononitrile, aromatic aldehydes and 1,5-naphthalene diol as reactants and KCC-1@NH_2 as a catalyst in ethanol. The mixture was sonicated at 80 W power and 20 kHz for a suitable time. FT-IR, 1H-NMR and melting point analysis were used to evaluate the products. The catalyst potentially converted >92% of substrates with great efficiency [31].

The production of 2-oxazolidinones using CO_2 and propargylic amines has been effectively catalysed by the Ruthenium–Salen-bridged ionic network assisted by KCC-1 (Scheme 4). The products were produced with higher yields and shorter reaction times [32]. The catalyst exhibited high recyclability, which has been analysed via ICP-MS and confirmed the robustness of the catalyst.

The fibrous KCC-1@Pd/APTPOSS nanocatalyst has been used to produce pyrazolo[33, 34] cinnolines (Scheme 5). The catalyst's vast surface area and active sites aided the reaction's promotion. The catalyst exhibited high reusability ten times without diminishing the catalytic activity [35].

Spiroindenopyridazine-4H-pyran has been prepared with remarkable chemo selectivity using the KCC-1/SNS/Cr catalyst (Scheme 6). Ninhydrin, cyanoacetohydrazine, malononitrile and CH-acid, along with the catalyst, reacted in the presence of water. The catalyst has been confirmed as environmentally friendly for the synthesis. It showed potential characteristics of high yield and reusability [36].

Scheme 4 2-oxazolidinone synthesis with KCC-1/Salen/Ru(II) NPs

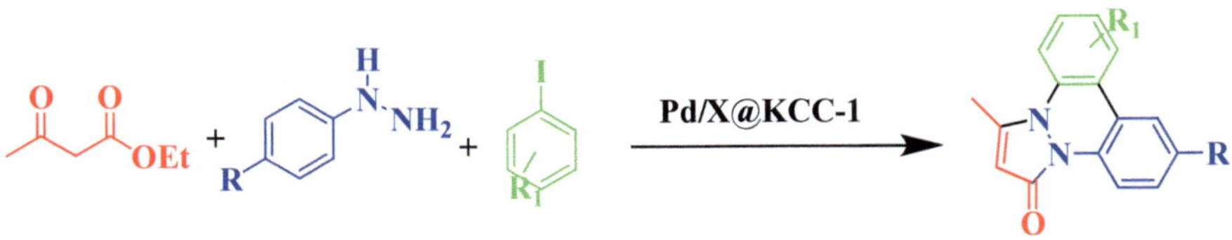

Scheme 5 KCC-1@Pd/APTPOSS for pyrazolo[33, 34] cinnolines synthesis

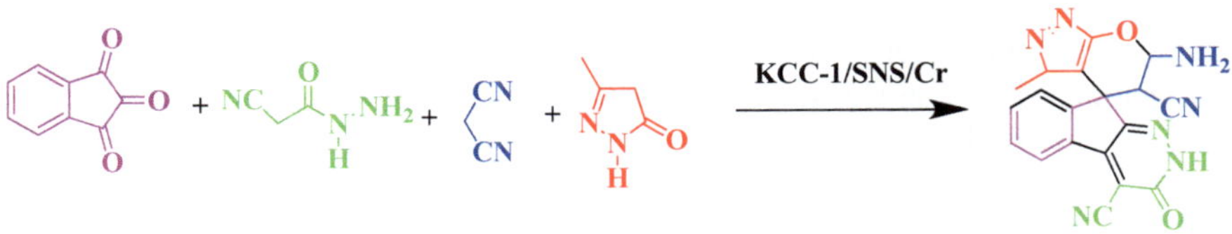

Scheme 6 Spiroindenopyridazine-4H-pyran synthesis

KCC-1 in spherical shape has been utilized as a template for preparing various nanomaterials, viz., C_3N_4 and porous carbon. The acidified KCC-1 has been employed as a template to produce C_3N_4. The material can be obtained by heating the silica cyanamide and template mixture at 550 °C. The yellow powder obtained as a product reacted with NH_4HF_2 solution to discard the template. The resulting carbon nitride manifested a similar shape to the template, exhibiting good photo-catalytic activity [37].

ZnO-modified KCC-1 has been employed for photodegradation of the palm oil mill effluent (POME) and textile wastewater containing persistent organic pollutants (POP). The catalyst has the potential to reduce 30.77% colour intensity at just 15 min. The BOD and COD test examination also confirmed the degradation of various chemical compounds. Thus, the catalyst can be utilized for various industrial wastewater treatments [38].

Paramagnetic mesoporous silica nanomaterials—Fe_3O_4@KCC-1 have been prepared by functionalization with 3-aminopropyltriethoxysilane (APTES). Candida rugosa lipase has been mounted on Fe_3O_4@KCC-1-NH_2 with glutaraldehyde as a linker. The prepared catalyst with an efficient lipase loading has manifested enzyme activity of nearly 630 U g^{-1} [39].

The in vivo one-pot hydrothermal approach has been used to synthesize a nickel catalyst mounted on fibrous KCC-1 silica nanoparticles. The catalyst has been applied to produce high H_2 and CO at a rate of 92 and 88%, respectively, from CH_4 and CO_2. The catalyst showed stability at 750 °C for 72 h [40].

Using the microemulsion process, discontinuous mordenite of lamellar silica—HM@KCC-1 impregnated with platinum has been developed. The catalyst's ability to catalyse the hydroisomerization of n-hexane has been investigated. The catalyst depicted enriched sites of Lewis's acid and modest Bronsted acid. The catalyst converted 75% n-hexane with about 98% isomer selectivity and 74% yield [41].

2.2 Activity of SBA-15

Pd was infused into a hybrid silica material SBA-15 with doubly charged DABCO and 3-chloropropyltrimethoxysilane for the catalysis of the SMC process (Scheme 7). The Pd@SBA-15/IL_{DABCO} synthesized biaryls using aryl halide, aryl boronic acid and potassium carbonate. NMR analysis of the ^{1}H and ^{13}C spectra was used to characterize the products. The catalyst paved the way for high yields in a short reaction time without the formation of by-products. The catalyst exhibited a TON of 1710 [42].

In order to provide a heterogeneous catalyst for the Knovenagal condensation, SBA-15 assisted in the production of imidazolium ionic liquids (SBA-15@IL-OAc) (Scheme 8). The catalyst exhibited greater performance at room temperature with yields of 90–98% [43].

In water medium, nitrobenzene is hydrogenated to aniline has been accomplished using SBA-15 functionalized with carboxylic acid and assisted by Pd catalyst. The hydrogenation process was effective in the absence of a solvent, and it considerably increased in rate when done in water. The catalyst exhibited a TOF of 3684.7 h^{-1}. The Pd/SBA-COOH has been recycled four times [44].

Ethylenediamine-SBA-15 functionalized with N^1-(3-(trimethoxysilyl) propyl) ethane-1,2-diamine (PrEn) has been sulfonated over 1° and 2° amine groups on the mesoporous surface. SBA-15@PrEn mounted with $NHSO_3H$ has been served as zwitterionic type-IL catalyst. This catalyst was employed for the N-formylation of amines. The catalyst based on sulfamic catalyses the processes as a Bronsted solid acid and has various advantages in terms of recoverability and recyclability (Scheme 9) [45].

Scheme 7 Synthesis of biaryl derivatives

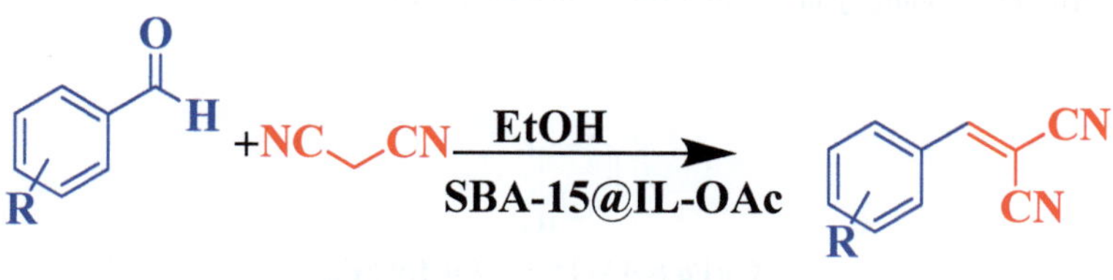

Scheme 8 SBA-15@IL-OAc catalysed Knoevenagel condensation

Scheme 9 N-formylation reaction

The telescoping preparation of 5-substituted tetrazole has been successfully performed using the catalyst SBA-15 incorporated with Glycine-M, with M corresponding to Cu and Ni (Scheme 10). The reaction involves nitrile and sodium azide as reactants with SBA-15@Glycine-M as a catalyst in PEG. High yields of the products were produced with minimal reaction time. The findings showed that SBA-15 incorporated with Glycine-Cu behaves as a more potent Lewis acid (LA) over SBA-15 incorporated with Glycine-Ni. Thus, the former exhibits better catalytic activity than the latter [46].

The mesoporous silica SBA-15 supported by imidazolium/PF_6 ionic liquid and ethylenediamine/CuI complex has been applied to synthesize triazoles by two methods (Schemes 11 and 12). For both methods A and B, the heterogeneous catalyst CuI@SBA-15/ ImPF6/PrEn is stable and recyclable. However, method B showed more efficiency having in mind the catalytic activity [47].

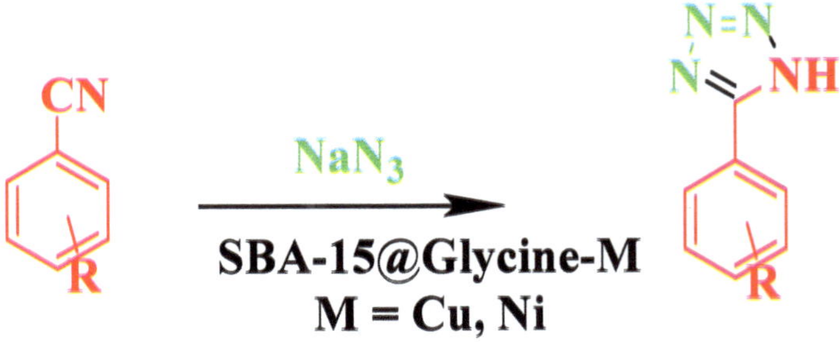

Scheme 10 Telescoping synthesis of 5-substituted tetrazoles

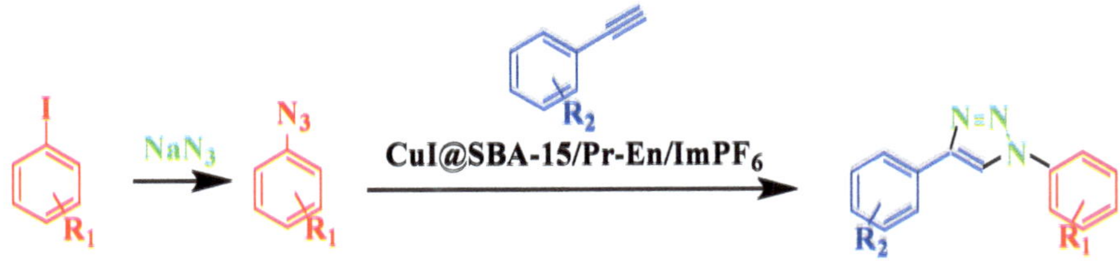

Scheme 11 Tandem synthesis of triazoles by method A

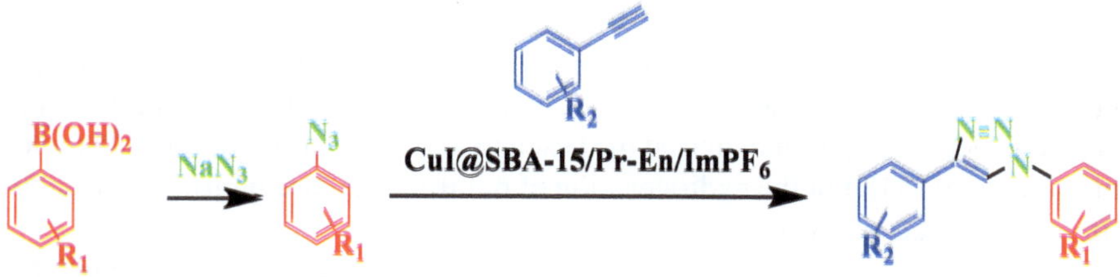

Scheme 12 The technique of tandem triazole synthesis

Scheme 13 Cycloaddition of benzyl halides

Cu-catalysed three-component azide-alkyne cycloaddition (CuAAC) reaction on SBA-15 assisted by Cu(II) complex has been investigated to synthesise 1,4-disubstituted 1,2,3-triazoles (Scheme 13). The reaction mixture included alkyne, organic halide, NaN_3 and Cu^{II}-Schiff base/SBA-15 as the catalyst. The process is eco-friendly and the yields of the products were fairly high. The catalyst also exhibited high recyclability [48].

From cytosine-chloropropyltriethoxysilane and tetraethoxysilane, cytosine functionalized with SBA-15 mesoporous nanomaterials has been produced (Scheme 14). This material can be used to condense aldehydes and ketones in the Knoevenagel approach and with malononitrile to form α, β unsaturated dicyanides in ethanol. The catalyst appeared more active and selective towards the end product and was easier to recover and reuse [49].

Cu-modified SBA-15 mesoporous material has been synthesized through a one-step method. The catalytic behaviour of Cu-SBA-15 has been tested for the deterioration of organic dyes. The studies depicted that the Cu(II) ions are infused uniformly onto the mesoporous silica thus, provides Lewis acidic sites leading to better catalytic activity [50].

Scheme 14 Knoevenagel condensation of aldehyde

SBA-15 silica supported by transition metal oxide $IrMO_x$ nanoparticles (M = Co, Fe and Ni) has been prepared by supporting IrM nanoparticles onto the SBA-15 mesoporous surface via the process of reduction and calcination. Amongst various metals, Ir-FeO_x/SBA-15 manifested the best application towards various substituted nitroaromatics [34]. For the hydrogenation of p-nitrophenyl acetonitrile, the catalyst offered improved catalytic efficiency and specificity.

The mesoporous carboxylic acid functionalized SBA-15 has been incorporated with silver nanoparticles. A significant growth in particle size, reaching up to 20 nm, has been observed for the Ag nanoparticles when they develop outside the mesopore. The catalyst has been applied for 4-nitrophenol to 4-aminophenol conversion. The catalyst demonstrated outstanding catalytic activity during the reduction process [51].

Mesoporous silica SBA-15 functionalized with several amines along with loading of copper nanoparticles has been employed for degradation applications. The mass and nature of the catalyst, Cu@SBA-3 have promoted the reduction reaction. The catalyst has been applied for the deterioration of Orange G dye and Methylene blue. The catalytic efficiency of Cu@SBA-3 has been compared for both dyes. Methylene blue has been reduced at a greater rate compared to Orange G, which implies the selectivity of the catalyst towards the former dye [33].

2.3 *Activity of MCM-41*

The cobalt(II) surface-supported complex of MCM-41 has been produced and presented as MCM-41@PDCA-Co. The catalyst was exploited to produce polyhydroquinolines (Scheme 15). The catalyst henceforth exhibited greater activity and high recyclability for six consecutive runs [52].

Pd Schiff base complex supported onto mesoporous MCM-41 was applied for reactions of Heck and Suzuki (Schemes 16 and 17). The former reaction involves a mixture of phenylboronic acid, aryl halide, sodium carbonate and catalyst (Pd-Py-MCM-41). The latter reaction involves aryl halide, butyl acrylate, potassium carbonate and the catalyst. It was discovered that the catalyst was outstanding in

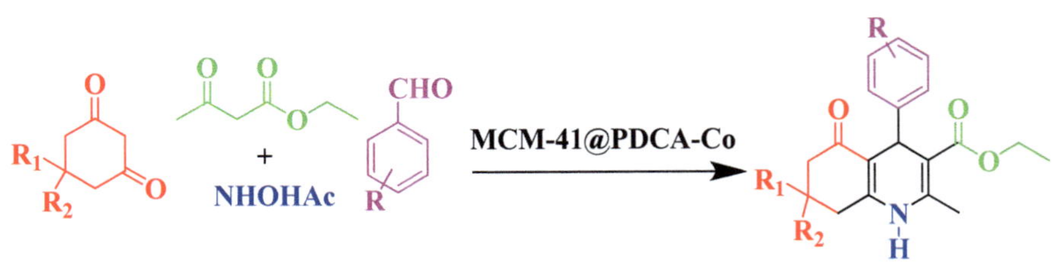

Scheme 15 Synthesis of different polyhydroquinolines using MCM-41@PDCA-Co

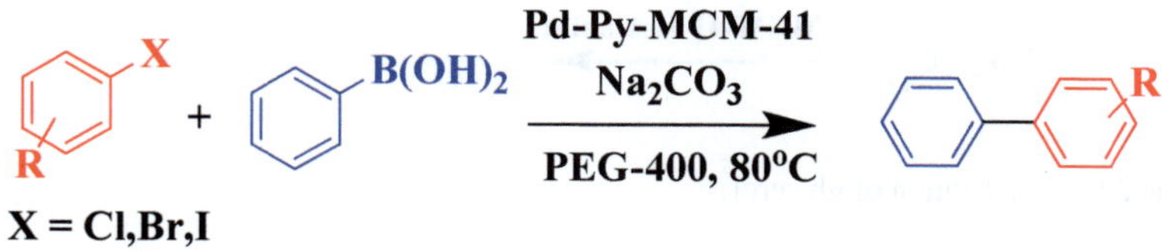

Scheme 16 Suzuki reaction

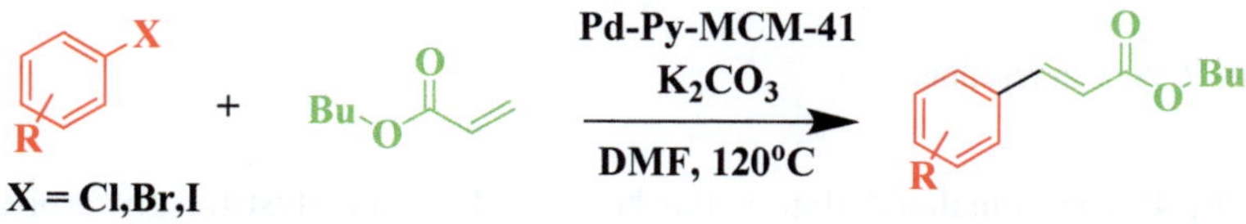

Scheme 17 Heck reaction

terms of non-toxicity, price, recoverability and stability. The catalyst depicted greater activity for Suzuki and Heck reactions [53].

Cytosine@MCM-41 modified by metals Cu and Ni has been utilized to produce pyranopyrazoles and tetrazoles (Schemes 18 and 19). 5 substituted 1H-tetrazole synthesis involves NaN_3, benzonitrile in the action of catalyst Ni/Cu-Cytosine@MCM41. Similarly, the synthesis of pyranopyrazoles involves ethyl acetoacetate, aldehyde, malononitrile and hydrazine hydrate along with a catalyst. The results confirmed the efficient catalytic activity of Ni/Cu-Cytosine@MCM41 for both reactions. The products were produced with significant yields in a minimal amount of time. The catalyst exhibited a high turnover frequency [54].

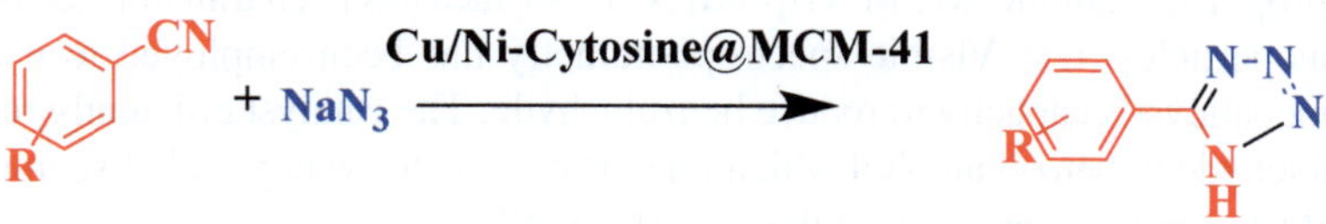

Scheme 18 5-Substituted 1H-tetrazole synthesis

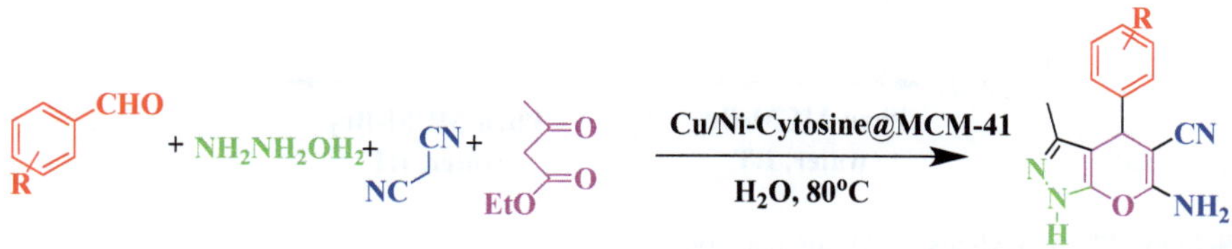

Scheme 19 Synthesis of pyranopyrazoles

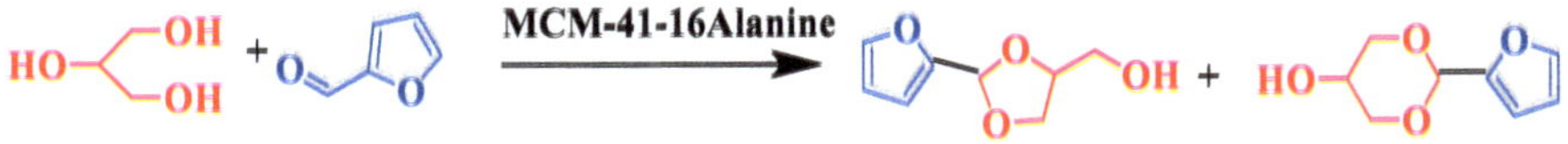

Scheme 20 Acetalization of glycerol

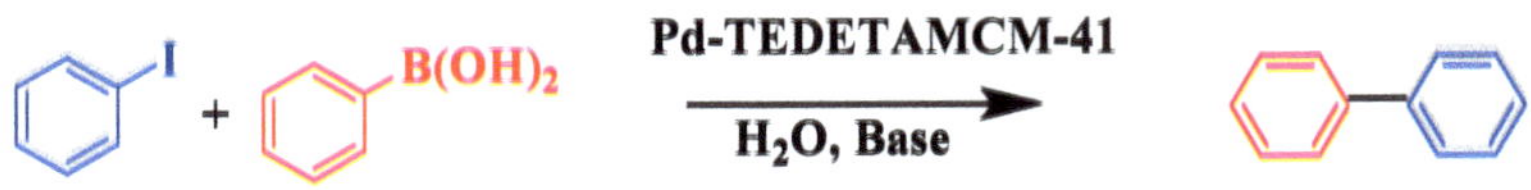

Scheme 21 Suzuki reaction

MCM-41-functionalized alanine has been used as a catalyst for glycerol acetalization to produce furfural. The reactants glycerol and furfural were involved in the reaction. The catalyst MCM-41-16Alanine manifested selectivity towards the desired product, 2-(furan-2-yl)1,3-dioxan-5-ol and (2-(furan-2-yl)-1,3-dioxolan-4-yl) methanol (Scheme 20). The catalyst achieved 90% conversion and 78% selectivity. Additionally, the catalyst is highly reusable and incredibly active [55].

Mesoporous MCM-41 supported by Pd-complex has been employed for C–C coupling reactions, i.e. Suzuki reaction (Scheme 21). The reaction involves aryl halide, sodium tetraphenylborate and Pd-TEDETA/MCM-41 as catalyst. The reaction has been optimized under various conditions. The result depicted better catalytic performance with reusability and stability [56].

In aqueous phase, for the Aza-Michael reaction, MCM-41 immobilized with phenanthrolinium dibromide has been used as a catalyst (Scheme 22). Aniline and acrylonitrile were used to conduct the aza-Michael reaction. By adding amines to β-nitrostyrenes, the reaction has been further expanded into manufacturing β-nitroamines along with the catalyst. The catalyst exhibited high stability and recoverability [57].

Free radical polymerization has been used to create MCM-41 complex with poly-N-isopropyl acrylamide. MCM-41/p-NIPAM's surface has been transformed by Ag–Pd nanoparticles. UV–Visible spectrophotometry has been employed to examine the nanocatalyst's capacity to reduce benzaldehyde. The catalyst efficiently reduced benzaldehyde to benzyl alcohol which may be due to the synergism between the Ag and Pd nanoparticles and the catalyst's surface [58].

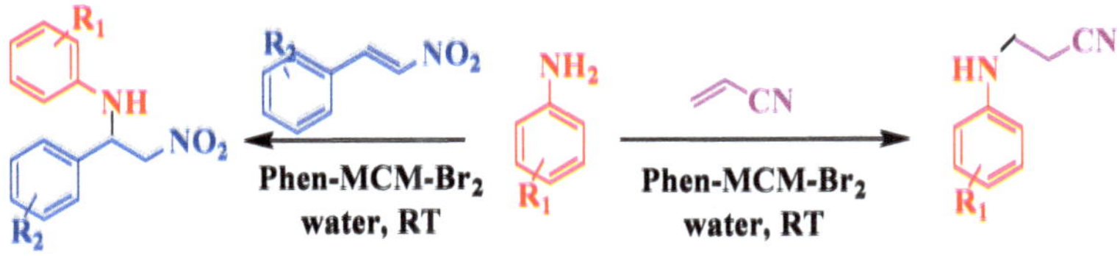

Scheme 22 Aza-Michael addition reaction

Ethane has been converted to ethylene via oxidative dehydrogenation using the Cr/MCM-41 catalyst. The catalyst was synthesized using a one-pot procedure with distinct addition sequences (Cr-TEOS and TEOS-Cr). As per the characterization results, adding Cr then TEOS showed better catalytic activity than the other one. Henceforth, Cr/MCM-41-Cr-TEOS (8%) catalyst efficiently converted ethane to ethylene [59].

Ni-incorporated MCM-41 carbon nanocomposites have been synthesized utilizing nickel nitrate as an additive via a one-pot synthetic approach. The addition of histidine has achieved the uniform distribution of nickel nanoparticles. The reduction of Ni^{2+} enriched the catalytic nature of nickel nanoparticles. The Ni-MCM-41 served as a powerful catalyst for nitrophenol reduction [60].

Fe_3O_4 nanoparticles coated with MCM-41 have been immobilized by L-ASNase utilizing 3-chloropropyltrimethoxysilane (CPTMS) acting as a surface modifying agent. Multiple characterization techniques have confirmed the nanocatalysts, viz., XRD, VSM, EDX, FTIR, SEM, EDX and zeta potential analysis. Due to its advantages, the Fe_3O_4 aided with MCM-41-Cl can be deployed as a substitute for immobilizing many types of enzymes [61].

The MCM-41 silica material has been prepared via the hydrothermal method and then modified using silver nitrate solution by chemical reduction method using $NaBH_4$. The obtained catalyst AgNPs@MCM-41 has been analysed for the deterioration of Methylene Blue with $NaBH_4$. The catalyst has also been tested for bacteria as an antibacterial agent. The catalyst exhibited excellent degradation of methylene blue within 7 min. The material also showed better antibacterial activity against the pathogens, which may be attributed by the silver and ammonium groups present in CTAB surfactant [62].

3 Functionalized Mesoporous Silica Nanomaterials in Adsorption (Fig. 3)

3.1 Heavy Metal Adsorption on Mesoporous Silica

Aminated and thiolated SBA-15 has been utilized as an adsorbent for the capture of hazardous metal ions collection. The aminated SBA-15 has been discovered to have a great propensity for metal ions, in contrast to the thiolated SBA-15, which has poorer interaction with copper ions [63].

Mesoporous silica MCM-41 synthesized via microwave method has been utilized as adsorbent materials to eliminate lead, copper and cadmium ions from varying concentrations. The catalyst has been found to possess an ordered hexagonal structure, high pore diameter and pore volume. The catalyst may have adsorbed Pb^{2+}, Cu^{2+} and Cd^{2+} with sorption capacities of 58.5, 36.3 and 32.3 mg/g at a pH range of 5–7. The adsorption process is homogeneous, confirmed through Langmuir isotherm [64].

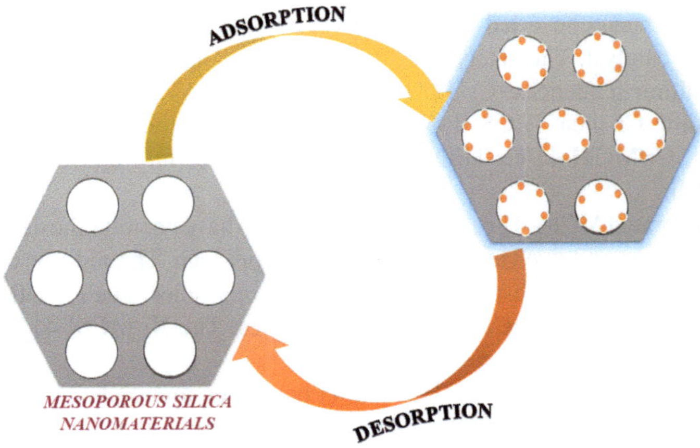

Fig. 3 Functionalized mesoporous silica nanomaterials in adsorption

Humongous mesoporous silica nanomaterials that have been functionalized with mercapto and amino groups have been created in various morphologies, viz., fibre-like, rod-like and platelet-like. The adsorption data has been revealed by Langmuir isotherms which presented that adsorption onto platelet adsorbent was quick compared to rod-like and fiber-like morphologies. Pb^{2+} has been readily adsorbed by all the morphologies of mercapto-functionalized silica adsorbents compared to amine-functionalized silica adsorbents. On the other hand, amine-functionalized silica adsorbent efficiently adsorbed Cu^{2+} ions compared to mercapto groups [65].

Mesoporous silica has reportedly been used for the adsorption of copper(II) ions due to its narrow distributions of pore diameters. The activity has been performed by batch adsorption studies which were correlated with Dubinin–Radushkevich, Freundlich and Langmuir theories. According to the findings, Cu^{2+} ion adsorption increased as the pore dimension dropped. The rapid Cu adsorption effect has been visualized in SBA-15 with 4 nm pores [66].

Mesoporous wormhole-framed silica functionalized by amine was synthesized by grafting 3-aminopropyltriethoxysilane (APTES). HMS-OH exhibited 90% adsorption of Rhodamine B dye and Methylene blue. HMS-NH_2 showed high toxic ions and CO_2 uptake [67].

It has been possible to create mesoporous silica SBA-15 complexed with N-propyl salicylaldimine and ethylenediamine propylenesalicyladimine. The catalyst has been effectively used as an adsorbent for Th(IV) and Eu(III) ions from aqueous system. The SBA/EnSA catalyst adsorbed Th(IV) ions at a higher rate compared to SBA/SA. As per the results, SBA/SA has been inferred by the Freundlich and Langmuir isotherms, whereas the SBA/EnSA catalyst is by Langmuir isotherms. The catalysts exhibited more excellent selectivity towards Th(IV) and Eu(III) ion adsorption [68].

3.2 *Mesoporous Silica for Drug Adsorption*

The drug can be adsorbed during the release process onto the silica substrates SBA-15 and Aerosil 200. The results revealed that drug molecules get more easily adsorbed onto the non-porous surface (Aersoil 200) than the mesoporous surface (SBA-15). This suggests the wet properties of the media and the adsorption of surfactant sodium dodecyl sulphate, onto the silica surface [69].

Mesoporous silica surface functionalized with 1-[3-(trimethoxysilyl)propyl] diethylenetriamine (DT) has been studied for drug delivery. The materials' ability to adsorb amino-modified has been found effective for the uptake of drugs. The result has also been attributed to the mesoporous size of all the materials. The material with high mesoporous size exhibited greater drug adsorption, possibly due to the space available on the surface for drug penetration [70].

The SBA-15, MCM-41 and amino-functionalized SBA-15 have been developed for both the release and adsorption of ampicillin. The outcomes highlighted that the surface had affected ampicillin adsorption and release compared to the pore size. Henceforth, SBA-15-NH_2 has been found to favour the release of ampicillin and also contributed to the adsorption process [71].

3-Aminopropyltriethoxysilane-modified SBA-15 and SBA-16 have been loaded onto antipyrine. There is a phosphate solution and stomach-like fluids with a pH of 7.2; the drug is effectively adsorbed on the mesoporous silica. The results concluded that the adsorption of antipyrine has been more efficient in pure silica than in amino-functionalized materials. Cubic-structured SBA-15 is more suitable for drug adsorption than hexagonal-structured SBA-15 [72].

3-Aminopropyl functionalized SBA-15 has been employed for the adsorption of cefazolin. The adsorption capacities have been studied by Langmuir, Freundlich and Dubbin–Radushkevich isotherms. The cefazolin has been adsorbed onto SBA-NH_2 as confirmed by Langmuir isotherms. The release of cefazolin drug has been prolonged for 7 days, showing the catalyst's potential activity [73].

4 Conclusions

Mesoporous silica has drawn much attention for both academic and industrial applications since Stöber silica was discovered in 1968. This is mostly because of its unique properties, notably uniform porosity, large surface-to-volume ratio, vast porosity and the capacity to synthesize various morphologies and forms. As discussed in this chapter, the various mesoporous silica nanomaterials have been widely utilized owing to their advantages. In the near future, diverse functionalized mesoporous silica nanomaterials could be applied in adsorption and catalysis.

References

1. Ciriminna R, Fidalgo A, Pandarus V, Beland F, Ilharco LM, Pagliaro M (2013) The sol–gel route to advanced silica-based materials and recent applications. Chem Rev 113(8):6592–6620
2. Kresge AC, Leonowicz ME, Roth WJ, Vartuli JC, Beck JS (1992) Ordered mesoporous molecular sieves synthesized by a liquid-crystal template mechanism. Nature 359(6397):710–712
3. Lin HP, Mou CY (2002) Structural and morphological control of cationic surfactant-templated mesoporous silica. Acc Chem Res 35(11):927–935
4. Yokoi T, Sakamoto Y, Terasaki O et al (2006) Periodic arrangement of silica nanospheres assisted by amino acids. J Am Chem Soc 128(42):13664–13665
5. Mbaraka IK, McGuire KJ, Shanks BH et al (2006) Acidic mesoporous silica for the catalytic conversion of fatty acids in beef tallow. Ind Eng Chem Res 45(9):3022–3028
6. Han Y, Lu Z, Teng Z, Liang J, Guo Z, Wang D, Han MY, Yang W (2017) Unraveling the growth mechanism of silica particles in the stober method: in situ seeded growth model. Langmuir 33(23):5879–5890
7. Tanabe K, Hölderich WF (1999) Industrial application of solid acid–base catalysts. Appl Catal A Gen 181(2):399–434
8. Stöber W, Fink A, Bohn E (1968) Controlled growth of monodisperse silica spheres in the micron size range. J Colloid Interface Sci 26(1):62–69
9. Lee G, Choi E, Yang S, Cho EB (2018) Tailoring pore size, structure, and morphology of hierarchical mesoporous silica using diblock and pentablock copolymer templates. J Phys Chem C 122(8):4507–4516
10. Sun Y, Ma K, Kao T, Spoth KA, Sai H, Zhang D, Kourkoutis LF, Elser V, Wiesner U (2017) Formation pathways of mesoporous silica nanoparticles with dodecagonal tiling. Nat Commun 8(1):1–10
11. Che S, Liu Z, Ohsuna T, Sakamoto K, Terasaki O, Tatsumi T (2004) Synthesis and characterization of chiral mesoporous silica. Nature 429(6989):281–284
12. Husen A (2023b) Smart nanomaterials from agricultural and horticultural products. Springer Nature Singapore Pte Ltd., 152 Beach Road, #21–01/04 Gateway East, Singapore 189721, Singapore
13. Husen A (2023a) Secondary metabolites based green synthesis of nanomaterials and their applications. Springer Nature Singapore Pte Ltd., 152 Beach Road, #21–01/04 Gateway East, Singapore 189721, Singapore
14. Husen A, Siddiqi KS (2023) Advances in smart nanomaterials and their applications. Elsevier Inc., 50 Hampshire St., 5th Floor, Cambridge, MA 02139, USA
15. Husen A (2022) Engineered nanomaterials for sustainable agricultural production, soil improvement and stress management. Elsevier Inc., 50 Hampshire St., 5th Floor, Cambridge, MA 02139, USA
16. Husen A, Iqbal M (2019) Nanomaterials and plant potential. Springer International Publishing AG, Gewerbestrasse 11, 6330 Cham, Switzerland. https://doi.org/10.1007/978-3-030-05569-1
17. Taghiyari HR, Jeffrey J Morrell, Husen A (2022) Emerging nanomaterials. In: Opportunities and challenges in forestry sectors. Springer Nature Switzerland AG, Gewerbestrasse 11, 6330 Cham, Switzerland. https://doi.org/10.1007/978-3-031-17378-3
18. Jin-Chul K, Madhusudhan A, Husen A (2021) Smart nanomaterials in biomedical applications. Springer Nature Switzerland AG, Gewerbestrasse 11, 6330 Cham, Switzerland
19. Wu SH, Mou CY, Lin HP (2013) Synthesis of mesoporous silica nanoparticles. Chem Soc Rev 42(9):3862–3875
20. Cao L, Man T, Kruk M (2009) Synthesis of ultra-large-pore SBA-15 silica with two-dimensional hexagonal structure using triisopropylbenzene as micelle expander. Chem Mater 21(6):1144–1153
21. Dacquin JP, Lee AF, Pirez C, Wilson K (2012) Pore-expanded SBA-15 sulfonic acid silicas for biodiesel synthesis. Chem Comm 48(2):212–214

22. Sanz R, Calleja G, Arencibia A, Sanz-Pérez ES (2013) CO_2 uptake and adsorption kinetics of pore-expanded SBA-15 double-functionalized with amino groups. Energy Fuels 27(12):7637–7644
23. Trofymluk O, Levchenko AA, Tolbert SH, Navrotsky A (2005) Energetics of mesoporous silica: investigation into pore size and symmetry. Chem Mater 17(14):3772–3783
24. Yi J, Kruk M (2015) Pluronic-P123-templated synthesis of silica with cubic Ia 3 d structure in the presence of micelle swelling agent. Langmuir 31(27):7623–7632
25. ALOthman ZA (2012) A review: fundamental aspects of silicate mesoporous materials. Materials 5(12):2874–2902
26. Kruk M, Jaroniec M, Sayari A (2000) New insights into pore-size expansion of mesoporous silicates using long-chain amines. Micropor Mesopor Mater 35:545–553
27. Yoshitake H (2005) Highly-controlled synthesis of organic layers on mesoporous silica: their structure and application to toxic ion adsorptions. New J Chem 29(9):1107–1117
28. Gautam P, Dhiman M, Polshettiwar V, Bhanage BM (2016) KCC-1 supported palladium nanoparticles as an efficient and sustainable nanocatalyst for carbonylative Suzuki-Miyaura cross-coupling. Green Chem 18(21):5890–5899
29. Hassankhani A, Sadeghzadeh SM, Zhiani R (2018) C-C and C–H coupling reactions by Fe_3O_4/KCC-1/APTPOSS supported palladium-salen-bridged ionic networks as a reusable catalyst. RSC Adv 8(16):8761–8769
30. Qureshi ZS, Sarawade PB, Hussain I, Zhu H, AlJohani H, Anjum DH, Hedhili MN, Maity N, D'Elia V, Basset JM (2016) Gold nanoparticles supported on fibrous silica nanospheres (KCC-1) as efficient heterogeneous catalysts for CO oxidation. Chem Cat Chem 8(9):1671–1678
31. Oboudatian HS, Safaei-Ghomi J (2022) Silica nanospheres KCC-1 as a good catalyst for the preparation of 2-amino-4H-chromenes by ultrasonic irradiation. Sci Rep 12(1):1–15
32. Saadati SM, Sadeghzadeh SM (2018) KCC-1 supported ruthenium-salen-bridged ionic networks as a reusable catalyst for the cycloaddition of propargylic amines and CO_2. Catal Lett 148(6):1692–1702
33. Abdulrasheed AA, Jalil AA, Hamid MY, Siang TJ, Fatah NA, Izan SM, Hassan NS (2020) Dry reforming of methane to hydrogen-rich syngas over robust fibrous KCC-1 stabilized nickel catalyst with high activity and coke resistance. Int J Hydrog Energy 45(36):18549–18561
34. Abidi MA, Hairom N, Jalil AA, Sidik DA, Madon R, Heirman N, Hamzah S, Jusoh N (2020) Photodegradation of different wastewater using modified fibrous silica (KCC-1) catalysts under different duration time. Int J 8(1.2)
35. Sadeghzadeh SM, Zhiani R, Emrani S (2017) Pd/APTPOSS@ KCC-1 as a new and efficient support catalyst for C-H activation. RSC Adv 7(40):24885–24894
36. Zhiani R, Sadeghzadeh SM, Emrani S (2018) Synthesis of spiroindenopyridazine-4 H-pyran derivatives using Cr-based catalyst complexes supported on KCC-1 in aqueous solution. RSC Adv 8(12):6259–6266
37. Rostamnia S, Doustkhah E, Zeynizadeh B (2016) Cationic modification of SBA-15 pore walls for Pd supporting: Pd@ SBA-15/ILDABCO as a catalyst for Suzuki coupling in water medium. Micropor Mesopor Mater 222:87–93
38. Rui Li J, Chen C, Lin HuY (2020) Novel and efficient knoevenagel condensation over mesoporous SBA-15 supported acetate-functionalized basic ionic liquid catalyst. Chem Sel 5(46):14578–14582
39. Ganji S, Bukya P, Liu ZW, Rao KS, Burri DR (2019) A carboxylic acid functionalized SBA-15 supported Pd nanocatalyst: an efficient catalyst for hydrogenation of nitrobenzene to aniline in water. New J Chem 43(30):11871–11875
40. Rostamnia S, Doustkhah E (2016) Covalently bonded zwitterionic sulfamic acid onto the SBA-15 (SBA-15/PrEn-$NHSO_3H$) reveals good bronsted acidity behavior and catalytic activity in N-formylation of amines. J Mol Catal A Chem 411:317–324
41. Tamoradi T, Ghorbani-Choghamarani A, Ghadermazi M, Veisi H (2019) SBA-15@ Glycine-M (M= Ni and Cu): two green, novel and efficient catalysts for the one-pot synthesis of 5-substituted tetrazole and polyhydroquinoline derivatives. Solid State Sci 91:96–107

42. Hosseini HG, Doustkhah E, Kirillova MV (2017) Combining ethylenediamine and ionic liquid functionalities within SBA-15: a promising catalytic pair for tandem Cu–AAC reaction. Appl Catal A Gen 548:96–102
43. Bagherzadeh M, Mahmoudi H, Amini M, Gautam S, Chae KH (2018) SBA-15-supported copper (II) complex: AN efficient heterogeneous catalyst for azide-alkyne cycloaddition in water. Sci Iran 25(3):1335–1343
44. Sharma S, Singh UP, Singh AP (2021) Synthesis of MCM-41 supported cobalt (II) complex for the formation of polyhydroquinoline derivatives. Polyhedron 199:115102
45. Nikoorazm M, Ghorbani-Choghamarani A, Jabbari A (2016) A facile preparation of palladium Schiff base complex supported into MCM-41 mesoporous and its catalytic application in Suzuki and Heck reactions. J Porous Mater 23(4):967–975
46. Nikoorazm M, Tahmasbi B, Gholami S, Moradi P (2020) Copper and nickel immobilized on cytosine@ MCM-41: as highly efficient, reusable and organic–inorganic hybrid nanocatalysts for the homoselective synthesis of tetrazoles and pyranopyrazoles. Appl Organomet Chem 34(11):e5919
47. Appaturi JN, Ramalingam RJ, AlLohedan HA, Khoerunnisa F, Ling TC, Ng EP (2021) Selective synthesis of dioxolane biofuel additive via acetalization of glycerol and furfural enhanced by MCM-41-alanine bifunctional catalyst. Fuel 288:119573
48. Hajjami M, Cheraghi M (2016) Synthesis of Pd-complex supported on MCM-41 and its catalytic activity for the C-C coupling reactions. Catal Lett 146(6):1099–1106
49. Hosseinzadeh R, Aghili N, Tajbakhsh M (2016) Synthesis, characterization and catalytic application of MCM 41 supported phenanthrolinium dibromide catalyst for aza-michael addition reaction in aqueous medium. Catal Lett 146(7):1194–1203
50. Zhang J, Zhang M, Yang C, Wang X (2014) Nanospherical carbon nitride frameworks with sharp edges accelerating charge collection and separation at a soft photocatalytic interface. Adv Mater 26(24):4121–4126
51. Ali Z, Tian L, Zhang B, Ali N, Zhang Q (2017) Synthesis of paramagnetic dendritic silica nanomaterials with fibrous pore structure (Fe_3O_4@ KCC-1) and their application in immobilization of lipase from Candida rugosa with enhanced catalytic activity and stability. New J Chem 41(16):8222–8231
52. Ibrahim M, Jalil AA, Khusnun NF, Fatah NA, Hamid MY, Gambo Y, Abdulrasheed AA, Hassan NS (2020) Enhanced n-hexane hydroisomerization over bicontinuous lamellar silica mordenite supported platinum (Pt/HM@ KCC-1) catalyst. Int J Hydrog Energy 45(36):18587–18599
53. Rajabi F, Fayyaz F, Luque R (2017) Cytosine-functionalized SBA-15 mesoporous nanomaterials: synthesis, characterization and catalytic applications. Micropor Mesopor Mater 253:64–70
54. Miao KK, Luo XL, Wang W, Guo SF, Cao FJ, Hu YQ, Chang PM, Feng GD (2019) One-step synthesis of Cu–SBA-15 under neutral condition and its oxidation catalytic performance. Micropor Mesopor Mater 289:109640
55. Yu H, Wu C, Wang S, Li T, Yin H (2021) transition metal oxide-modified Ir nanoparticles supported on SBA-15 silica for selective hydrogenation of substituted nitroaromatics. ACS Appl Nano Mater 4(7):7213–7220
56. Saikia D, Huang YY, Wu CE, Kao HM (2016) Size dependence of silver nanoparticles in carboxylic acid functionalized mesoporous silica SBA-15 for catalytic reduction of 4-nitrophenol. RSC Adv 6(42):35167–35176
57. Hachemaoui M, Boukoussa B, Ismail I, Mokhtar A, Taha I, Iqbal J, Hacini S, Bengueddach A, Hamacha R (2021) CuNPs-loaded amines-functionalized-SBA-15 as effective catalysts for catalytic reduction of cationic and anionic dyes. Colloids Surf A Physicochem Eng 623:126729
58. Rout L, Mohan A, Thomas AM, Ha CS (2020) Rational design of thermoresponsive functionalized MCM-41 and their decoration with bimetallic Ag–Pd nanoparticles for catalytic application. Micropor Mesopor Mater 291:109711
59. Al-Awadi AS, El-Toni AM, Alhoshan M, Khan A, Labis JP, Al-Fatesh A, Abasaeed AE, Al-Zahrani SM (2019) Impact of precursor sequence of addition for one-pot synthesis of Cr-MCM-41 catalyst nanoparticles to enhance ethane oxidative dehydrogenation with carbon dioxide. Ceram Int 45(1):1125–1134

60. Ghimire PP, Zhang L, Kinga UA, Guo Q, Jiang B, Jaroniec M (2019) Development of nickel-incorporated MCM-41–carbon composites and their application in nitrophenol reduction. J Mater Chem A 7(16):9618–9628
61. Ulu A, Noma SA, Koytepe S, Ates B (2019) Chloro-modified magnetic Fe_3O_4@ MCM-41 core–shell nanoparticles for L-asparaginase immobilization with improved catalytic activity, reusability, and storage stability. Appl Biochem Biotechnol 187(3):938–956
62. Asli B, Abdelkrim S, Zahraoui M, Mokhtar A, Hachemaoui M, Bennabi F, Ahmed AB, Sardi A, Boukoussa B (2022) Catalytic reduction and antibacterial activity of MCM-41 modified by silver nanoparticles. Silicon 18:1–12
63. Da'na E (2017) Adsorption of heavy metals on functionalized-mesoporous silica: a review. Micropor Mesopor Mater 247:145–157
64. Zhu W, Wang J, Wu D, Li X, Luo Y, Han C, Ma W, He S (2017) Investigating the heavy metal adsorption of mesoporous silica materials prepared by microwave synthesis. Nanoscale Res Lett 12(1):1–9
65. Lee JY, Chen CH, Cheng S, Li HY (2016) Adsorption of Pb (II) and Cu (II) metal ions on functionalized large-pore mesoporous silica. Int J Environ Sci Tech 13(1):65–76
66. Knight AW, Tigges AB, Ilgen AG (2018) Adsorption of copper (II) on mesoporous silica: the effect of nano-scale confinement. Geochem Trans 19(1):1–13
67. Ge S, He X, Zhao J, Duan L, Gu J, Zhang Q, Geng W (2017) Removal of cationic dyes, heavy metal ions, and CO_2 capture by adsorption on mesoporous silica HMS. Water Air Soil Pollut 228(12):1–10
68. Dolatyari L, Yaftian MR, Rostamnia S (2016) Adsorption characteristics of Eu (III) and Th (IV) ions onto modified mesoporous silica SBA-15 materials. J Taiwan Inst Chem Eng 60:174–184
69. McCarthy CA, Ahern RJ, Devine KJ et al (2018) Role of drug adsorption onto the silica surface in drug release from mesoporous silica systems. Mol Pharm 15(1):141–149
70. Martín A, Morales V, Ortiz-Bustos J et al (2018) Modelling the adsorption and controlled release of drugs from the pure and amino surface-functionalized mesoporous silica hosts. Micropor Mesopor Mater 262:23–34
71. Nairi V, Medda L, Monduzzi M, Salis A (2017) Adsorption and release of ampicillin antibiotic from ordered mesoporous silica. J Colloid Interface Sci 497:217–225
72. Goscianska J, Olejnik A, Nowak I (2017) APTES-functionalized mesoporous silica as a vehicle for antipyrine–adsorption and release studies. Colloids Surf A Physicochem Eng 533:187–196
73. Szewczyk A, Prokopowicz M, Sawicki W, Majda D, Walker G (2019) Aminopropyl-functionalized mesoporous silica SBA-15 as drug carrier for cefazolin: adsorption profiles, release studies, and mineralization potential. Micropor Mesopor Mater 274:113–126

Effect of Nanofertilizers on Plant Physiology, Metabolism and Associated Safety Issues

Bhupal Bhattacharya and Amit Kumar Mandal

Abstract Nanotechnology has been used in the development of nanofertilizers, which are designed to improve the efficiency of fertilizer delivery to crops. The use of nanotechnology in agriculture, specifically in the form of nanofertilizers, has a relatively short history, dating back to the early twenty-first century. The development of nanofertilizers aimed to improve the efficiency of fertilizer use, increase crop yields and reduce the environmental impact of traditional fertilizers. The advantages of nanofertilizers include increased nutrient uptake by plants, reduced leaching and run-off and improved soil health. Additionally, the use of nanofertilizers may lead to an increase in the cost of fertilizers and a potential for the development of pesticide resistance in pests. However, the use of nanofertilizers, which are fertilizers made up of nanoparticles, is a relatively new technology and the long-term effects of these fertilizers on both plant physiology and the environment are still not fully understood. Research is ongoing to better understand the benefits and risks associated with the use of nanofertilizers, including their impact on plant growth, soil fertility and the environment. It's important to carefully consider the potential benefits and risks of nanofertilizers, and to conduct further research to fully understand their impact on plant growth, soil fertility and the environment. In addition, regulations and standards for the production and use of nanofertilizers should be established and enforced to ensure that they are used safely and responsibly.

Keywords Nanotechnology and fertilizer · Law and ethics · Plant physiology and metabolism · Safety and ethical issues

B. Bhattacharya
Department of Law, Raignj University, Raiganj, West Bengal, India

A. K. Mandal (✉)
Department of Sericulture, Raignj University, Raiganj, West Bengal, India
e-mail: amitmandal08@gmail.com

R. K. Bachheti et al. (eds.), *Nanomaterials for Environmental and Agricultural Sectors*, Smart Nanomaterials Technology,
https://doi.org/10.1007/978-981-99-2874-3_8

1 Introduction

Plant metabolism is the chemical reactions and pathways that occur in plants to maintain life and carry out essential processes [27]. This includes photosynthesis, which converts light energy into chemical energy stored in glucose and other sugars and cellular respiration, which releases this stored energy to fuel plant growth and activity [13]. The study of plant physiology and metabolism helps us understand how plants function, adapt to their environment and interact with other species in their ecosystem. Nanofertilizers are fertilizer materials that have been processed at the nanoscale to enhance their efficacy. These fertilizers are tiny particles (typically less than 100 nm in size) that contain essential plant nutrients [24], such as nitrogen, phosphorus and potassium [22]. The small size of the particles allows for improved uptake and utilization of the nutrients by the plants [7], leading to increased growth and yield [22].

These materials have improved physical and chemical properties compared to their conventional counterparts, such as improved solubility, reactivity and stability. Fertilizers can deliver nutrients directly to the roots of plants, increasing the efficiency of nutrient uptake and reducing the amount of waste [1]. In terms of plant physiology, nanoparticles can have effects on the absorption and distribution of nutrients, water uptake and stress tolerance, leading to improved plant growth and health [2]. This can lead to improved plant growth and yield [31]. In plant physiology, the use of nanofertilizers has been shown to improve nutrient acceptance, enhance root development and increase plant stress tolerance. These benefits can result in improved plant growth and productivity [11].

In terms of plant physiology, the use of nanofertilizers can impact plant growth by altering the availability of essential nutrients and the plant's ability to absorb them [23]. The small size of the particles allows for more efficient absorption by plant roots, which can lead to improved growth, higher yield and better stress tolerance.

2 History and Background

The use of fertilizers to improve plant growth dates back to ancient civilizations. For example, the ancient Greeks and Romans used human and animal waste to fertilize their crops [3]. In the 1800s, scientists began to understand the role of specific nutrients in plant growth, and the commercial production of fertilizers began. In the early twentieth century, synthetic fertilizers were developed, allowing for more precise control over the nutrients available to plants [29]. The Green Revolution of the mid-twentieth century saw a massive increase in the use of synthetic fertilizers, leading to increased food production and improved crop yields in many parts of the world [4].

However, the use of nanofertilizers has a relatively short history compared to other forms of fertilizers. The concept of using nanotechnology in agriculture was

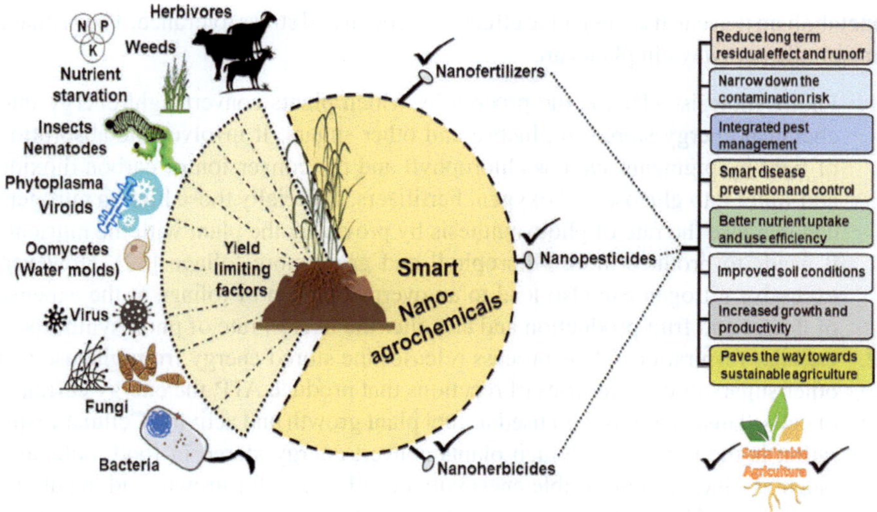

Fig. 1 Use of nanofertilizers and other nano-based products for improved plant growth and related activities (adopted from Kumar et al. [16])

first proposed in the late 1990s, and the first nanofertilizers were developed and commercialized in the early 2000s [10]. Today, fertilizers are widely used in agriculture and horticulture to improve plant growth and yields. They come in many forms, including inorganic fertilizers, organic fertilizers and slow-release fertilizers, and are used to supply essential nutrients such as nitrogen, phosphorus and potassium to plants.

The main advantage of nanofertilizers is that they can be designed to release nutrients in a controlled manner, which can lead to improved plant growth and reduced waste [16, 21] (Fig. 1). However, their use is still limited and further research is needed to fully understand their potential benefits and potential risks.

3 Fertilizer and Plant Metabolism

Metabolism in plants refers to the set of chemical reactions and processes that occur within the plant cells to sustain life and support growth. Plant metabolism refers to the set of chemical reactions and pathways that occur within plant cells to maintain life and carry out essential processes [28]. Fertilizers provide the plant with the essential nutrients it needs to support growth, development and overall health. However, the use of fertilizers can also lead to imbalanced nutrient ratios and changes in the plant's metabolism that can have negative effects on growth and stress tolerance. The primary metabolic pathways in plants are:

1. Photosynthesis—This is the process by which plants convert light energy into chemical energy stored in glucose and other sugars. It involves the absorption of light by pigments such as chlorophyll and the conversion of carbon dioxide and water into glucose and oxygen. Fertilizers, especially those high in nitrogen, can increase the rate of photosynthesis by providing the plant with the nutrients it needs to produce more chlorophyll and grow more foliage [12]. However, excessive nitrogen can also lead to an overproduction of foliage at the expense of flower and fruit production and can alter the overall rate of photosynthesis.
2. Cellular respiration—This process releases the stored energy from glucose and other sugars through a series of reactions that produce ATP, the energy currency of cells. This energy is then used to fuel plant growth and activity. Cellular respiration is the process by which plants convert energy stored in food molecules (such as glucose) into usable energy that can be used for growth and metabolic processes [25]. Fertilizers provide plants with the essential nutrients they need to support cellular respiration and other metabolic processes [30]. Without sufficient nutrients, cellular respiration can be impaired, reducing the plant's overall energy production and growth. However, over-fertilization can lead to imbalanced nutrient ratios, which can have negative effects on cellular respiration and other metabolic processes [17]. Additionally, excessive use of nitrogen-rich fertilizers can lead to an overproduction of foliage at the expense of flower and fruit production, reducing the overall productivity of the plant.
3. Biosynthesis—Biosynthesis refers to the production of organic compounds by living organisms, including plants and microorganisms, through metabolic processes. The organic compounds produced through biosynthesis are used for various purposes, including energy storage, structural support and as building blocks for more complex molecules. This refers to the production of cellular building blocks, such as lipids, amino acids and nucleotides, from simple precursors. These building blocks are then used to construct more complex structures such as cell membranes and proteins. Fertilizers can be synthetic, made from inorganic compounds, or organic, made from decomposing organic matter. The nutrients in fertilizers are used by crops to support growth and reproduction, and they can help to compensate for the natural nutrient loss that occurs due to soil depletion or other factors [20]. The choice of fertilizer to use will depend on the specific needs of the soil and the crops being grown, as well as other factors such as cost and environmental considerations.
4. Hormone metabolism—Hormone metabolism refers to the process by which hormones, chemical messengers developed by plants and animals, are produced, transported and regulated within an organism. Hormones play crucial roles in regulating various biological processes, including growth, development and response to stress and environmental changes [9]. Plants produce and respond to hormones, such as auxins, gibberellins and abscisic acid, that regulate growth and development [5]. These hormones play important roles in processes such as

germination, flowering and fruit development. The use of fertilizers can affect hormone metabolism in plants. Fertilizers provide essential nutrients for plant growth, and an adequate supply of these nutrients can lead to improved plant health and increased hormone production. However, the type and amount of fertilizer used can also have negative effects on hormone metabolism, such as disrupting the delicate balance of hormones in the plant and leading to stunted growth or other problems.

Plant metabolism is tightly regulated and balanced to maintain the plant's overall health and well-being. Abiotic factors, such as temperature, light and water availability, as well as biotic factors, such as herbivore damage, can impact plant metabolism and alter the rates and balances of these pathways [18]. Overall, plant metabolism is a complex and interrelated network of processes that work together to maintain the health and survival of the plant.

4 Issues with the Use of Fertilizers

There are several issues associated with the use of fertilizers:

1. Environmental Pollution: The use of fertilizers can have a significant impact on the environment, particularly if they are not used correctly. Overuse of fertilizers can lead to nitrogen and phosphorus pollution, which can harm aquatic ecosystems and contribute to the growth of harmful algae blooms. Fertilizers leach into groundwater and surface water, cause eutrophication and harm aquatic life. Further, it releases ammonia gas into the air, contributing to air pollution and potential harm to human health. It also contaminates soil and imbalance nutrient levels, reducing soil health and fertility.
2. Health Risks: The use of fertilizers can pose certain health risks if they are not handled and used properly. Some fertilizers contain heavy metals and other toxic substances that can be harmful to human health if ingested or inhaled. Inhalation of ammonia gas released during fertilizer application, can irritate the eyes, nose, throat and lungs and cause respiratory problems. Fertilizers irritates skin and eye from direct contact with dust or liquids. Ingestion of fertilizer that has contaminated food or water sources, leads to health problems such as gastrointestinal irritation, kidney damage and nerve damage. Chemical burns from fertilizer solutions can cause skin irritation and chemical damage.
3. Economic Costs: The use of chemical fertilizers can be expensive for farmers, especially in developing countries where the cost of inputs is a major constraint on agricultural production. The use of fertilizers can have both economic costs and benefits. On the cost side, the production and application of fertilizers can be expensive. The cost of raw materials, transportation and manufacturing can add up and be reflected in the price of the final product. Additionally, the application of fertilizers often requires specialized equipment, which can also add to the cost. On the benefit side, fertilizers can help increase crop yields and improve

the quality of crops, leading to higher profits for farmers. This can also lead to lower food prices for consumers. The use of fertilizers can also help increase the overall productivity of farmland, which can lead to economic growth in agricultural regions.

4. Soil Degradation: Soil degradation is a major concern associated with the overuse of chemical fertilizers. The excessive use of chemical fertilizers can cause several problems. Overuse of chemical fertilizers can lead to soil degradation, reducing its fertility and productivity over time. Chemical fertilizers can lower the pH of the soil, making it more acidic and less generous for plant growth. Chemical fertilizers typically contain just a few nutrients, whereas soil requires a complex mixture of nutrients to maintain its fertility. Overuse of chemical fertilizers can lead to imbalances that can harm plants and the soil. Chemical fertilizers can break down the soil structure, causing compaction and reducing its ability to retain water and air. Chemical fertilizers can leach into groundwater and contaminate drinking water sources. They can also contribute to water eutrophication, causing harm to aquatic ecosystems.
5. Greenhouse Gas Emissions: The production and transportation of chemical fertilizers are significant sources of greenhouse gas (GHG) emissions. The production of chemical fertilizers typically involves energy-intensive processes such as mining, manufacturing and transportation of raw materials, which can result in significant GHG emissions. The production of chemical fertilizers requires large amounts of energy and releases GHGs, including carbon dioxide (CO_2), nitrous oxide (N_2O) and methane (CH_4). The transportation of fertilizers from the production site to the fields further adds to GHG emissions, as it involves the use of fossil fuels and the emission of CO_2. The GHG emissions from the fertilizer industry contribute to global warming and climate change, which have far-reaching impacts on agriculture, ecosystems and human populations.

5 Issues with Fertilizers and Plant Physiology

The use of fertilizers can lead to a number of issues related to plant physiology. Some of the most common problems include:

1. Chemical imbalances: When plants do not receive the necessary nutrients in the right amounts, it can lead to physiological changes and metabolism disruptions. However, it's important to use fertilizers correctly as over-fertilization can lead to negative effects on plant health and the environment. Over-fertilization or use of the wrong type of fertilizer can lead to an imbalance of essential nutrients in the soil, which can cause stunted growth, yellowing of leaves and other symptoms of nutrient deficiency.
2. Soil pH changes: The pH of soil can greatly impact the availability of nutrients to plants and therefore, their growth and health. Soil that is too acidic or alkaline can make it difficult for plants to absorb nutrients, even if the soil contains sufficient amounts. Using fertilizers can also alter the soil pH, leading to further imbalances.

Additionally, some fertilizers, if not used correctly, can lead to excess buildup of nutrients in the soil, leading to toxicity issues for plants. Fertilizers can alter the pH of soil, making it too acidic or too alkaline for certain plant species.

3. Toxicity: Some fertilizers, especially those that are high in nitrogen, can be toxic to plants if applied in excessive amounts [26]. Over-fertilization can result in the buildup of certain nutrients, such as nitrogen, in the soil, which can lead to toxicity in plants. This can cause a range of symptoms, including stunted growth, yellowing of leaves and even death in severe cases. In addition to nutrient toxicity, some fertilizers contain toxic elements or chemicals, such as heavy metals, that can harm plants and the environment.
4. Nitrate pollution: Excessive use of nitrogen-rich fertilizers can lead to nitrate pollution of groundwater and surface waters [6], which can be harmful to aquatic life and humans. Nitrates are highly soluble and can leach into water sources, potentially contaminating drinking water and causing harm to aquatic life. When plants absorb excess nitrates, it can lead to a range of physiological problems, such as stunted growth, reduced crop yields and increased susceptibility to disease and pests. Excess nitrates in the soil can also alter the balance of other nutrients and beneficial microorganisms in the soil, leading to further problems for plant growth and health.
5. Soil compaction: Overuse of heavy equipment and machinery to apply fertilizers can lead to soil compaction, which can reduce the amount of air and water that reaches plant roots, resulting in poor growth and health [19]. Compacted soil has a denser structure, making it difficult for roots to penetrate and access these essential resources. Using heavy equipment or applying fertilizer in excess can lead to soil compaction, exacerbating the problem. This can cause stunted growth, reduced crop yields and increased susceptibility to disease and pests.

The use of fertilizers can also lead to a number of issues related to plant metabolism. Some of the most common problems include:

1. Imbalanced nutrient ratios: Fertilizers can provide excess amounts of certain nutrients, which can disrupt the balance of nutrients required for optimal plant metabolism. Imbalanced nutrient ratios in soil can affect plant growth and health. Different plants have different nutritional requirements, and when certain nutrients are lacking or present in excess, it can lead to imbalances that can impact plant growth. Using fertilizers that have an imbalanced nutrient ratio can also lead to problems, as it can further disrupt the soil's natural nutrient balance. Excessive application of certain nutrients, such as nitrogen, can lead to the buildup of these nutrients in the soil, affecting the uptake of other important nutrients by the plants.
2. Altered photosynthesis: Photosynthesis is a crucial process in plant growth and health, as it allows plants to convert light energy into chemical energy, which is used for growth and reproduction. Excessive nitrogen in the soil can lead to an overproduction of foliage at the expense of flower and fruit production, and can also affect the rate of photosynthesis.

3. Stunted root growth: Stunted root growth can negatively impact plant growth and health by reducing the plant's ability to absorb water and nutrients from the soil [8]. Over-fertilization can lead to stunted root growth, which can reduce the plant's ability to absorb water and nutrients from the soil. High levels of nitrogen can encourage foliage advancement. Soil compaction, imbalanced nutrient ratios and soil toxicity can also contribute to stunted root growth.
4. Decreased stress tolerance: Plants that are exposed to imbalanced soil nutrients, soil toxicity or excess fertilizer can have decreased stress tolerance, making them more susceptible to environmental stressors, such as drought, disease and pests [14]. Fertilizers can alter the plant's metabolism and reduce its ability to cope with environmental stress factors, such as drought or high temperatures. Excessive use of fertilizer, particularly nitrogen-based fertilizers, can cause rapid plant growth, which can increase the plant's demand for water and nutrients, leading to a higher risk of stress. Soil compaction, altered root growth and imbalanced nutrient ratios can also contribute to decreased stress tolerance in plants.
5. Increased disease susceptibility: Excessive use of fertilizer, particularly nitrogen-based fertilizers, can encourage rapid plant growth, which can increase the plant's susceptibility to disease [15]. This is because an overabundance of lush foliage can create a favorable environment for disease-causing pathogens to thrive. Excessive fertilization can lead to rapid growth and an abundance of foliage, which can create an environment conducive to the spread of diseases and pests. Imbalanced soil nutrients, soil toxicity and altered root growth can also contribute to increased disease susceptibility in plants.

6 Ethical and Legal Issues in Use of Fertilizer

The use of fertilizers for plant growth is regulated by various laws and regulations that vary depending on the location and jurisdiction. Some of the key legal issues related to the use of fertilizers for plant growth include:

1. Environmental regulations: The use of fertilizers can impact the environment by polluting water sources, creating air pollution and contributing to soil degradation. To address these impacts, many countries have enacted environmental regulations to limit the use and disposal of fertilizers.
2. Labeling requirements: Fertilizers must be labeled correctly and accurately, including information about their composition, application rates and potential hazards.
3. Health and safety regulations: The handling, storage and use of fertilizers can pose health and safety risks, particularly in concentrated forms. Regulations are in place to protect workers and the general public from these risks.
4. Waste disposal regulations: The disposal of fertilizer waste, including bags and packaging, must comply with local and national regulations, which may include restrictions on landfilling or incineration.

5. Economic regulations: Fertilizer prices are often regulated to protect farmers and consumers. Governments may set price controls, regulate the distribution and sales of fertilizers or provide subsidies to support the use of certain fertilizers.
6. Intellectual property rights: Fertilizer products can be protected by patents, trademarks and other intellectual property rights, which can limit the use and distribution of the product.
7. Use of genetically modified crops: The use of genetically modified crops that are designed to be used with specific fertilizers may be subject to additional regulations.

7 Reasons for Failure of Laws in Controlling and Regulating Unethical Use of Fertilizers

There can be several reasons for the failure of laws in controlling and regulating the use of fertilizers for plant growth and metabolism, including:

1. Lack of enforcement resources: Governments may not have the resources or personnel to effectively enforce regulations, particularly in rural areas or in developing countries.
2. Corruption: Corruption can undermine the enforcement of regulations, particularly in areas where there are limited resources for enforcement.
3. Lack of public education: The general public may be unaware of the risks associated with the use of fertilizers, and may not be able to understand the importance of following regulations.
4. Complex regulations: Regulations can be complex, difficult to understand and confusing for both regulators and users, making it difficult for regulators to enforce the laws and for users to comply with them.
5. Economic factors: The use of fertilizers is often motivated by economic considerations, and regulations that are seen as too restrictive or costly may be ignored by users.
6. Resistance from industry: The fertilizer industry may resist regulations that they believe will harm their business, and may lobby against regulations or challenge them in court.
7. Inadequate regulations: Regulations may be outdated or inadequate to address the complex and evolving nature of the fertilizer industry, and may not address unethical practices such as the use of substandard or counterfeit fertilizers.

Addressing these factors is crucial for the effective enforcement of laws and regulations related to the use of fertilizers for plant growth and metabolism. This can include strengthening enforcement resources, improving public education, simplifying regulations and addressing corruption and industry resistance.

8 Conclusion

In conclusion, proper fertilizer management is essential for maintaining the balance of nutrients required for optimal plant metabolism and avoiding these issues. Overall, proper fertilizer management and application are crucial for ensuring healthy plant growth and avoiding these potential physiological issues.

It is important to use fertilizers properly and in moderation to ensure optimal plant health and productivity. The smaller size of nanofertilizers allows for better absorption by plant roots, leading to more efficient use of the fertilizer. However, there is still limited research on the long-term effects and potential risks of using nanofertilizers, so their use remains a topic of debate in the scientific and agricultural communities.

Using fertilizers in a balanced and appropriate manner, taking into account the specific needs of the soil and the crops being grown is highly needed. Overuse of fertilizers can lead to soil and water pollution, while underuse can result in poor plant growth and reduced crop yields. Proper fertilizer management is crucial for ensuring that plants have the necessary nutrients to support optimal cellular respiration and inclusive growth.

Overall, while fertilizers are necessary for boosting crop yields and feeding the growing global population, it's important to use them responsibly and find sustainable alternatives to minimize their negative impacts.

9 Suggestions

Fertilizers are essential for increasing crop yields and feeding a growing population. However, improper use and storage of fertilizers can lead to environmental pollution. To minimize the environmental impact of fertilizers, it is important to follow best management practices such as using the right amount of fertilizer, applying it at the right time and in the right place and properly storing and disposing of any unused fertilizer. Additionally, using alternative methods of soil fertility management, such as cover crops and composting, can help reduce the need for chemical fertilizers and reduce their associated environmental risks.

To prevent soil degradation, it is important to use fertilizers in moderation and to regularly incorporate organic matter into the soil to maintain its fertility and health. In addition, using a diverse mix of fertilizers, including organic fertilizers, can help maintain a healthy soil environment.

Reducing the GHG emissions associated with the production and transportation of chemical fertilizers requires a multi-faceted approach, including improving energy efficiency in production processes, reducing transportation distances and promoting the use of low-carbon energy sources. Incorporating practices such as cover cropping, crop rotation and the use of organic fertilizers can help reduce the overall dependence on chemical fertilizers, ultimately reducing their associated GHG emissions.

To mitigate the GHG emissions from the fertilizer industry, it is important to promote sustainable practices in fertilizer production and transportation, such as the use of renewable energy sources and more energy-efficient processes. Reducing the use of chemical fertilizers through the adoption of sustainable agricultural practices, such as soil conservation, crop rotation and the use of cover crops, can help reduce GHG emissions.

It's important to regularly test soil pH and adjust as needed to ensure that the soil provides optimal conditions for plant growth. Also, using fertilizers appropriately and in the right amounts can help to avoid soil and plant health issues.

To avoid soil compaction and its negative effects on plant physiology, it's important to use proper tillage techniques and minimize heavy equipment traffic on the soil. Using cover crops, mulch and other soil-building practices can also help to improve soil structure and reduce compaction. Applying fertilizer at the recommended rates and in the proper way can help to maintain soil health and support healthy plant growth.

It's important to use fertilizers correctly and in the right amounts to avoid the negative effects on photosynthesis and plant growth. A balanced supply of all essential nutrients, including macronutrients (nitrogen, phosphorus and potassium) and micronutrients (iron, zinc and magnesium), can support optimal photosynthesis and healthy plant growth. Regular soil testing and monitoring of plant health can also help to ensure that soil conditions are optimal for plant growth and development. To avoid imbalanced nutrient ratios and their negative effects on plant growth, it's important to regularly test soil and adjust fertilization practices as needed to maintain optimal nutrient levels. Using fertilizers with a balanced nutrient ratio and applying them correctly, following recommended rates and guidelines, can also help to support healthy plant growth.

To avoid stunted root growth and its negative effects on plant growth, it's important to use fertilizers correctly, following recommended application rates and guidelines. Using cover crops, mulch and other soil-building practices can also help to improve soil structure, reduce compaction and support healthy root growth. Regular soil testing and monitoring of plant health can also help to identify and address any soil or fertilizer-related issues that may be affecting root growth.

To avoid decreased stress tolerance and its negative effects on plant growth, it's important to use fertilizers correctly. Regular soil testing and monitoring of plant health can also help to ensure that soil conditions are optimal for plant growth and development and to address any soil or fertilizer-related issues that may be affecting plant stress tolerance. Using conservation practices, such as crop rotation, cover cropping and reduced tillage, can help to improve soil health and reduce stress on plants.

A balanced supply of all essential nutrients, including macronutrients (nitrogen, phosphorus and potassium) and micronutrients (iron, zinc and magnesium), can support optimal plant growth and health. Regular monitoring of plant health can also help to identify and address any soil or fertilizer-related issues that may be affecting disease susceptibility. Using disease-resistant plant varieties and following good agricultural practices, such as crop rotation and sanitation, can help to reduce the risk of disease in crops.

It's important for fertilizer manufacturers, distributors and users to be aware of and comply with relevant legal requirements to ensure the safe and responsible use of fertilizers for plant growth.

Addressing the crucial factors which assist in ignoring the effective enforcement of regulations related to the unethical use of fertilizers. This can include strengthening enforcement resources, improving public education, simplifying regulations and addressing corruption and industry resistance. It's important to regularly review and update regulations to ensure they are adequate to address current challenges and practices in the fertilizer industry.

References

1. Adisa IO, Reddy Pullagurala VL, Peralta-Videa JR, Dimkpa CO, Elmer WH, Gardea-Torresdey JL, White JC (2019) Recent advances in nano-enabled fertilizers and pesticides: a critical review of mechanisms of action. Environ Sci Nano 6(7):2002–2030
2. Ali S, Mehmood A, Khan N (2021) Uptake, translocation, and consequences of nanomaterials on plant growth and stress adaptation. J Nanomater 2021:1–17
3. Antonkiewicz J, Łabętowicz J (2016) Chemical innovation in plant nutrition in a historical continuum from ancient Greece and Rome until modern times. Chem-Didactics-Ecol-Metrol 21(1–2):29–43
4. Ashley K, Cordell D, Mavinic D (2011) A brief history of phosphorus: from the philosopher's stone to nutrient recovery and reuse. Chemosphere 84(6):737–746
5. Bari R, Jones JDG (2009) Role of plant hormones in plant defence responses. Plant Mol Biol 69:473–488
6. Chand S, Ashif M, Zargar MY, Ayub BM (2011) Nitrate pollution: a menace to human, soil, water and plant. Univers J Environ Res Technol 1(1)
7. Dimkpa CO, Fugice J, Singh U, Lewis TD (2020) Development of fertilizers for enhanced nitrogen use efficiency–Trends and perspectives. Sci Total Environ 731:139113
8. Fageria NK, Moreira A (2011) The role of mineral nutrition on root growth of crop plants. Adv Agron 110:251–331
9. Fahad S, Nie L, Chen Y, Chao W, Xiong D, Saud S, Hongyan L, Cui K, Huang J (2015) Crop plant hormones and environmental stress. Sustain Agric Rev 15:371–400
10. Fellet G, Pilotto L, Marchiol L, Braidot E (2021) Tools for nano-enabled agriculture: fertilizers based on calcium phosphate, silicon, and chitosan nanostructures. Agronomy 11(6):1239
11. Fincheira P, Tortella G, Seabra AB, Quiroz A, Diez MC, Rubilar O (2021) Nanotechnology advances for sustainable agriculture: current knowledge and prospects in plant growth modulation and nutrition. Planta 254:1–25
12. Hidayati N, Anas I (2016) Photosynthesis and transpiration rates of rice cultivated under the system of rice intensification and the effects on growth and yield. HAYATI J Biosci 23(2):67–72
13. Holding DR, Streich AM (2013) Plant growth processes: transpiration, photosynthesis, and respiration. In: The board of reagents of the University of Nebraska, pp 1–10
14. Kovács S, Kutasy E, Csajbók J (2022) The multiple role of silicon nutrition in alleviating environmental stresses in sustainable crop production. Plants 11(9):1223
15. Kumar R, Kumar R, Prakash O (2019) Chapter-5 the impact of chemical fertilizers on our environment and ecosystem. Chief Ed 35:69
16. Kumar A, Choudhary A, Kaur H, Mehta S, Husen A (2021) Smart nanomaterial and nanocomposite with advanced agrochemical activities. Nanoscale Res Lett 16(156):1–26

17. Lai CH, Settinayake ARH, Yeo WS, Lau SW, Jong TK (2019) Crop nutrients review and the impact of fertilizer on the plantation in malaysia: a mini review. Commun Soil Sci Plant Anal 50(17):2089–2105
18. Ogbe AA, Finnie JF, Van Staden J (2020) The role of endophytes in secondary metabolites accumulation in medicinal plants under abiotic stress. S Afr J Bot 134:126–134
19. Pahalvi HN, Rafiya L, Rashid S, Nisar B, Kamili AN (2021) Chemical fertilizers and their impact on soil health. In: Microbiota and biofertilizers, vol 2: ecofriendly tools for reclamation of degraded soil environs, pp 1–20
20. Palm CA, Myers RJK, Nandwa SM (1997) Combined use of organic and inorganic nutrient sources for soil fertility maintenance and replenishment. Replenishing Soil Fertil Afr 51:193–217
21. Prasad R, Kumar V, Prasad KS (2014) Nanotechnology in sustainable agriculture: present concerns and future aspects. Afr J Biotechnol 13(6):705–713
22. Qureshi A, Singh DK, Dwivedi S (2018) Nano-fertilizers: a novel way for enhancing nutrient use efficiency and crop productivity. Int J Curr Microbiol App Sci 7(2):3325–3335
23. Rahman SU, Wang X, Shahzad M, Bashir O, Li Y, Cheng H (2022) A review of the influence of nanoparticles on the physiological and biochemical attributes of plants with a focus on the absorption and translocation of toxic trace elements. Environ Pollut 119916
24. Raliya R, Saharan V, Dimkpa C, Biswas P (2017) Nanofertilizer for precision and sustainable agriculture: current state and future perspectives. J Agric Food Chem 66(26):6487–6503
25. Saltveit ME (2004) Respiratory metabolism. In: The commercial storage of fruits, vegetables, and florist and nursery stocks, p 68
26. Savci S (2012) Investigation of effect of chemical fertilizers on environment. APCBEE Proc 1:287–292
27. Seigler DS (1998) Plant secondary metabolism. Springer Science & Business Media
28. Shao H-B, Chu LY, Lu Z-H, Kang C-M (2008) Primary antioxidant free radical scavenging and redox signaling pathways in higher plant cells. Int J Biol Sci 4:1–8
29. Smil V (2004) Enriching the earth: Fritz Haber, Carl Bosch, and the transformation of world food production. MIT press
30. Soetan KO, Olaiya CO, Oyewole OE (2010) The importance of mineral elements for humans, domestic animals and plants: a review. Afr J Food Sci 4(5):200–222
31. Zulfiqar F, Navarro M, Ashraf M, Akram NA, Munné-Bosch S (2019) Nanofertilizer use for sustainable agriculture: advantages and limitations. Plant Sci 289:110270

Benefits, Future Prospective, and Problem Associated with the Use of Nanopesticides

Afshan Muneer, Sana Zia, Tean Zaheer, Rao Zahid Abbas, Mahreen Fatima, Attia Nawaz, Amjad Islam Aqib, Tauseef ur Rehman, and Muhammad Imran

Abstract Pests of livestock and agriculture present plenty of challenges when it comes to overall productivity. Owing to the non-judicious application of pesticides, the bugs of concern are developing resistance against therapeutic doses. The broader part of the picture is even more important, where the non-target organisms (including humans) are adversely affected. One of the smart alternatives to conventionally utilized pesticides is the use of nanopesticides. The nanopesticide application could be extended to the field based on laboratory-scale success. There is a need to evaluate the exact dosage and the forms of nanoparticle composite or coated or loaded pesticides. Also, the withdrawal times of these nanoformulations are deemed necessary to be calculated. The chapter overviews the practical benefits and pitfalls related to

A. Muneer
Department of Zoology, Cholistan University of Veterinary and Animal Sciences, Bahawalpur 63100, Pakistan

S. Zia
Government Sadiq College Women University, Bahawalpur 63100, Pakistan

T. Zaheer (✉) · R. Z. Abbas · M. Imran
Department of Parasitology, Faculty of Veterinary Science, University of Agriculture, Faisalabad, Pakistan
e-mail: teanzaheer942@gmail.com

M. Fatima
Faculty of Biosciences, Cholistan University of Veterinary and Animal Sciences, Bahawalpur 63100, Pakistan

A. Nawaz
Department of Microbiology, Cholistan University of Veterinary and Animal Sciences, Bahawalpur 63100, Pakistan

A. I. Aqib
Department of Medicine, Cholistan University of Veterinary and Animal Sciences, Bahawalpur 63100, Pakistan

T. Rehman
Department of Parasitology, Faculty of Veterinary and Animal Sciences, Islamia University Bahawalpur, Bahawalpur 63100, Pakistan

R. K. Bachheti et al. (eds.), *Nanomaterials for Environmental and Agricultural Sectors*, Smart Nanomaterials Technology,
https://doi.org/10.1007/978-981-99-2874-3_9

the application of nanopesticides in plants and animal sciences. Most importantly, evidence of safety within the environment connected to using nanoparticles on crops/ animals is required. These research directions, if taken seriously, can help convince the end consumer to adopt nanopesticides at the commercial scale.

Keywords Nano-pesticides · Nano Agrochemicals · Pest Resistance · Antibiotic Resistance · Nano particles

1 Introduction

The organic chemical components of pesticides are responsible for their heavy use in agricultural lands and urban environments despite alternative methods of pest management. The majority of registered pesticides infiltrate the neural systems of insects and animal husbandry via the primary mode of action. They are neurotoxic and are thought to exacerbate certain neurodegenerative disorders such as Parkinson's disease [1]. Various unique products, such as pest and insect-killing chemicals, have been launched in the market recently to reduce neurotoxic substances. As pesticides have been widely used to fight against pests, they have also had a devastating effect on humans and other living organisms, with a rise in human poisoning. Globally, crop diseases are also caused by fungi, in addition to pests [2]. Pesticides are generally viewed as having a limited range of applications because of a number of problems associated with them. Pesticides applied for pest control are typically either lost in the environment or are unable to reach their target areas [3]. Overuse of chemical fertilizers and pesticides during the green revolution of the late 20th century in India resulted in a great loss of soil biodiversity and soil health as a result of expanding production to feed the growing population. Pesticide overuse has also led to the development of pesticide resistance, a major agricultural problem. It is possible to solve such problems through nanotechnology [4]. A recent trend in nanoparticulate materials has been the use of biopolymers due to their eco-friendliness and ability to deliver persistent releases of pesticide active ingredients. As an alternative source of raw materials, natural polymers and gums are biocompatible, sustainable, biodegradable, economical, and non-toxic [5]. Nanopesticides are a potential alternative because they can be used as 'smart delivery systems', releasing pesticides in a controlled manner over time. In this way, pollution and the hazards it poses to the environment would be reduced [6]. The ability of silica-based nanoparticles advised in agriculture to kill pests is attributed to their ability to boost pathogen resistance and deal with biological stresses. Because of their nanoscale size and intrinsic antibacterial action, silica nanoparticles appear to inhibit fungal growth and decrease pest resistance; yet, high doses of these nanoparticles cause toxicity and damage to plants and livestock [7]. Moreover, the WHO has found that low doses of unstructured silica are non-pesticides that pose no danger to humans. Fertilizers containing nutrients measuring 30–40 nm in size are called nanofertilizers. Because of their huge specific area, they can keep adequate nutrient ions and slowly or gradually release them in the needed concentration as the crop requires [1]. Nanofertilizers have a higher surface area, tiny size, and reactivity than bulk fertilizers, which can increase diffusion and solubility, nutrient availability to plants, and boost agricultural production. Using fertilizer carriers to build smart fertilizers, nanofertilizers demonstrated the feasibility of investigating nanostructured materials as a new facility to improve nutrient utilization while minimizing environmental degradation.

2 Resistance to Pests and Public Health Concerns

2.1 Resistance Status of Pesticides in Agriculture

Pesticide use in agricultural production has triggered a major experiment in macroevolution worldwide, creating ecosystems that can be good subjects for evolutionary research [8] and eco-evolutionary research [9]. The development of pesticide resistance (the development of genetic, metabolic, physiological, and behavioral adaptations in arthropods to withstand plant allelochemicals and pesticides, in this paper, pesticide resistance poses a major challenge to global agriculture and is an excellent example of how human activity has led to rapid evolution [10]. According to the definition, this sort of resistance occurs when a population's susceptibility to a toxin is decreased genetically through exposure to the toxin in the field [11]. Because pesticide treatment has a strong directional selection effect on pests, it typically occurs immediately after introducing a new synthetic component [12]. Recent research has suggested that an individual species' evolutionary history may influence its likelihood of resisting, making certain species more likely to resist [13, 14, 15]. We can define cross-resistance as the result of an earlier exposure to a different, unrelated toxin during the evolutionary history of a species. There is a greater likelihood that resistance to cross-resistance will evolve in generalist arthropods [13]. Their detoxification or digestion of plant defense compounds is more extensive than that of specialized species [16]. Insects that eat herbaceous plants are particularly sensitive to pesticides due to evolutionary interactions between plants and arthropods [14]. Some insect species seem more prone to developing pesticide resistance [17]. However, the amount of pesticides used may obscure this trend [18, 19] (Fig. 1).

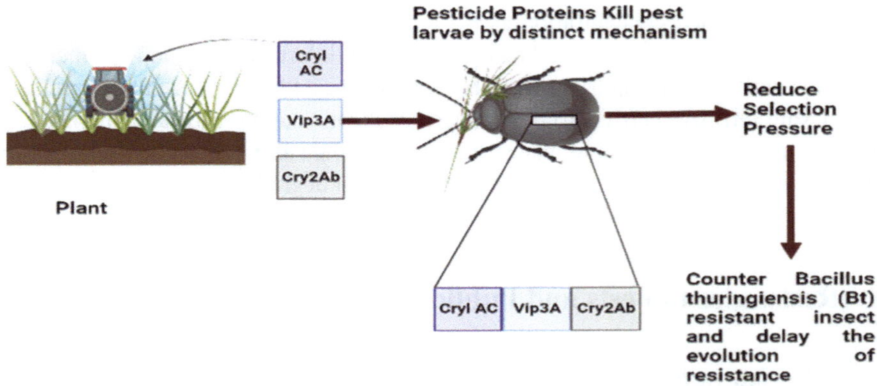

Fig. 1 Schematic representation of utilization of Bacillus thuringiensis toxin (Bt) Pyramids to overcome crop pest resistance. Field trial of Bt toxin pyramids showcasing enhanced control against resistant crops pests. The stacked combination of multiple Bt toxins in the pyramid design provides a sustainable solution to combat the evolving pest resistance, ensuring improved crop protection and yield

2.2 *Resistance Status of Pests in Livestock*

Globally, parasitic diseases pose a significant obstacle to animal health and product performance. They may result from endoparasites living inside the body or ectoparasites like ticks, mites, flies, fleas, midges, etc. that live on the surface of the body. A very harmful external parasite of mammals, birds and reptiles throughout the world, ticks are an important ecto-parasite [20]. Farmers make their own pest control practices, as they rely primarily on the application of synthetic chemical insecticides to take immediate action [21]. Many countries with widespread tick populations have shown resistance to various acaricide classes [22]. Different continents have different patterns of antibiotic consumption and use, influenced by animal species, regional production patterns, types of production systems, intensive or extensive farming, farming purposes (commercial, industrial, or domestic), the lack of an antibiotic-use policy or legislative framework, population size, and socioeconomic status [23, 24]. In developing countries, the use of non-essential antibiotics in livestock feed to boost growth is usually unregulated [25]. Several factors contribute to the continued use of these non-essential antibiotics in livestock farming, including the expansion and concentration of farmland, inadequate government policies and controls over antibiotic use and sales, a reduction in infection control measures, and farmers' unwillingness to adopt changes to their farm practices that have been delegated to them [26]. It is generally believed that the amount and rate of antibiotic consumption in developing nations' farming sectors can be influenced by how farmers obtain these antibiotics (over the counter and use them (multidrug practices and existing conditions. Infection prevalence and level are high, state management and development strategies need to be more robust, livestock zone planning needs to be improved, and livestock husbandry practices are inadequate despite being part of an integrated

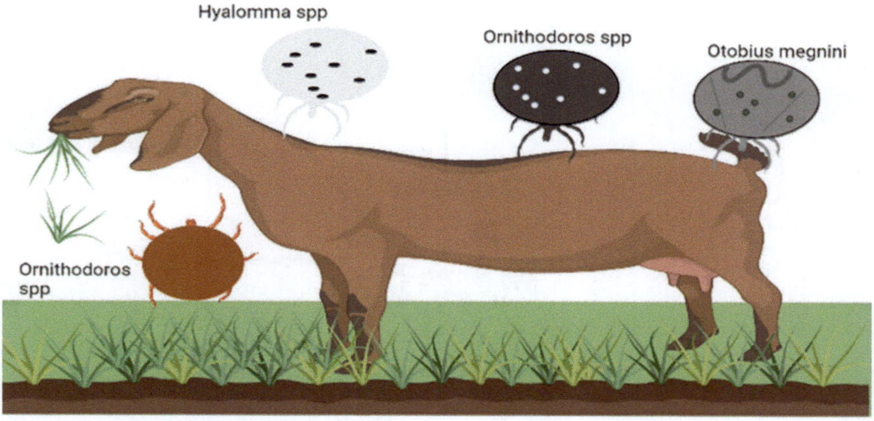

Fig. 2 Infestation of goats by pests while on grazing

agricultural system [27, 28]. In livestock production, antibiotics are routinely used to prevent disease and enhance growth. This promotes antibiotic resistance among commensal and pathogenic bacteria. The environmental impacts of antibiotic resistance are caused by the non-metabolization of antibiotics and waste products released into the environment due to antibiotic resistance. There is still a possibility that these waste products can influence the bacteria population, resulting in antibiotic resistance, and frequent use of antibiotics will no longer be effective against these pathogens [18] (Fig. 2).

2.3 *Public Health Concerns About Pesticides or Pest Resistance*

However, because the agricultural land available is scarce, sufficient production and enhancing the productivity of existing agricultural land are critical components of solving the food security of organisms around the globe [29]. In agriculture, pesticides are chemical substances used to prevent, eradicate, and neutralize all types of pests, including animals, weeds, insects, and microorganisms [30]. Inhalation, ingestion, contact with the skin, or passage through the placenta are all possible routes for pesticides to enter the human body. Yet, accidental pesticide exposure to humans can cause adverse effects on the human health (Fig. 3).

Some pesticides have biomarkers in blood serum, sperm, ovarian follicular fluid, umbilical cord blood, urine, breast milk, and, meconium [31]. Even though pesticides have a wide range of chemical structures and biological mechanisms of action, significant progress is being made in identifying their relevant mechanisms of action. Lindane, a pesticide identified by the International Agency for Research on Cancer as a carcinogen or potential carcinogen in humans, is among the pesticides identified

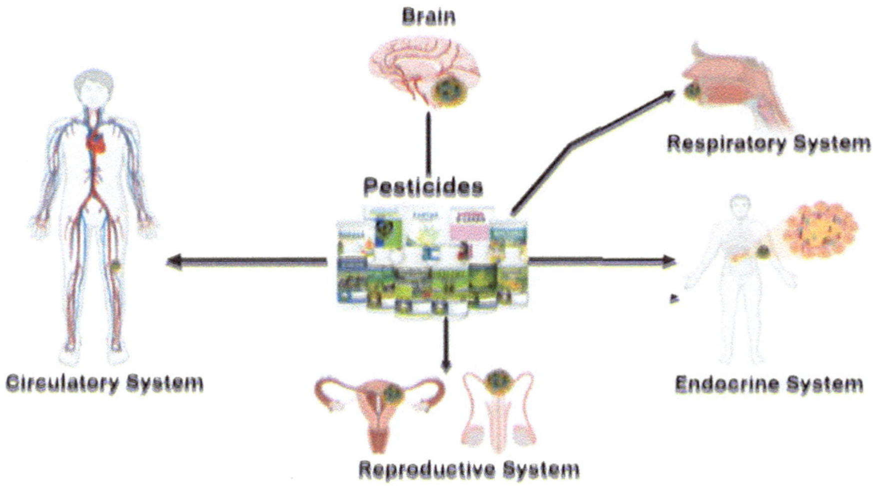

Fig. 3 Salient consequences of exposure of pesticides on organs of human body. Damage to the organs is caused by the persistence and effective degradation of pesticide-based contaminants in the environment

as carcinogens or potential carcinogens. Agricultural, public health, and residential settings use malathion to control insects. It is still manufactured in Resistance status of pests in Livestock in large quantities worldwide. The malathion molecule can be radially absorbed and disseminated before being consumed to produce bioactive malaxon, which causes chromosome damage and DNA damage in humans [32]. Diazinon is commonly used to control insects in agriculture and the home and garden. As a result of diazinon's effects on chromosomal structure in human cells [33], chromosomal damage occurs. Among all herbicides, glyphosate is the most commonly used broad-spectrum herbicide. There have been reports of increased micronuclei, a biomarker of damage to chromosomes, in several populations following glyphosate spraying [34]. Following xenobiotic exposure in humans, many metabolic byproducts are produced, which can lead to a cascade of additional biological effects. As a result of this relatively early molecular epidemiological finding, more data can be used to evaluate the weight of evidence, and some biomarkers can be detected long before clinical illness occurs. However, not all these biomarkers contribute to carcinogenesis [35].

Furthermore, the frequent use of pesticides caused pest resistance in livestock, human, and agriculture. In addition to the nature of the species and the environment as a whole, resistance models often focus on the factors that influence the selection of alleles that confer resistance. It is difficult to capture the complexity needed to make meaningful estimates, particularly when it comes to specific pests [36]. Climate change can affect resistance evolution by causing multivoltine pests to mature at different ages [37]. In response to shorter generation times, resistance evolution is expected to accelerate. Climate change can impact selection pressures on resistance alleles by affecting the rate at which chemicals break down in the environment [38]. When organisms are subjected to extreme environmental conditions, their susceptibility to toxins increases [39]. Most pesticide resistance research focuses on the proximal biochemical mechanism contributing to the reduced sensitivity to the pesticide in question, such as mutations in target sites or overexpression of target sites, metabolic breakdown, or pesticide efflux. On the other hand, resistance generation and dissemination are evolutionary mechanisms that can improve resistance risk assessment and management strategies [40, 41]. In this way, resistance can be seen as an example of evolutionary rescue [42], addressing issues such as adaptability to changing settings, variation sources, and the emergence of novel characteristics.

3 Benefits of Nanopesticides

A nanopesticides is a nanomaterial engineered to protect plants, minimize application losses, cover leaves more effectively, increase stability, and reduce formulation ingredients. An active ingredient in nanopesticide formulations is an encapsulating system such as nanoemulsions, polymeric nanoparticles, and lipid nanoparticles consisting of self-organizing systems such as liposomes, dendrimers, and metallic nanoparticles. It is a pesticide formulated with nanomaterials for agricultural applications. It can be fixed on hybrid substrates, encapsulated in matrixes, or equipped with functionalized nanocarriers that are stimulated externally or trigger enzymes. To explore pesticide activity in nanocarriers, several materials can interact with nano-sized particles based on their shape and properties, including silica, lipids, polymer, copolymer, ceramics, metals, carbon, and others [43].

By increasing the solubility, availability, and durability of agrochemicals, nanopesticide formulations can tremendously enhance the control of pathogens, weeds, and insects in plants [44]. While nanomaterials are borderline cytotoxic and genotoxic, their features are also borderline.

By controlling and targeting the release of biocide, nanoformulations are different from conventional pesticide formulations because they increase the soluble active ingredients' solubility [45]. Applying a smaller amount of active ingredient per area may be necessary to achieve sustained delivery of active ingredients that remain effective for longer periods. Controlled-release formulations reduce production costs, nontarget effects, and phytotoxicity by reducing production costs, nontarget effects, and phytotoxicity. Various physical and chemical methods are discussed for preparing nanocapsules, nanospheres, nanogels, and micelles, the most commonly synthesized controlled-release formulations. Hydrophobic active ingredients can be dispersed using polymer matrixes in aqueous solutions with high selectivity and without affecting their biocidal properties [46]. For active ingredients to be released correctly, the chemical properties of a polymeric matrix, bond strength, and biocide molecule size must be considered. The active ingredient polymer diffuses or disassembles upon contact with water and receiving the appropriate stimuli [47].

Nanoformulations encapsulating polymers are better than conventional formulations because they are controlled-release formulations [28]. It is possible to encapsulate hydrophobic or hydrophilic bioactive compounds in controlled-release formulations. This technology is widely used in agricultural applications. The formation of liposomes from lecithin and micelles can encapsulate hydrophobic and hydrophilic organic compounds [19, 48]. Since Chitosan is biodegradable, non-toxic, and adsorbent, it can serve as a valuable carrier for controlled delivery in addition to reducing pesticides, enhancing the stability of unstable core materials, suppressing sharp odors caused by the released chemicals, and securing biocompatibility with carrier systems. Chitosan matrixes protect active ingredients encapsulated in them from the environment, controlling their release and protecting them from degradation [38]. The loss of active ingredients with a short half-life can be reduced [49], and their activity can be extended [50]. Since nanoformulations possess substantially different properties from conventional bulk pesticides, there are still concerns about increased health hazards from inhalation to skin penetration [51] (Fig. 4).

As a general rule, agrochemical technology is used to prevent crop areas from being invaded by pests (pathogens, parasitic weeds, harmful insects), as well as to diagnose soil fertility, control livestock, and enhance fishery productivity. Although indiscriminate use of pesticides has adversely affected production, they also resulted in resistance to pathogens and insects, increased demand for agrochemicals, and an imbalance in the environment [52].

Nanoparticulates are efficient and unpredictable due to their large surface areas, making them appealing to crop pests. Excessive and unsystematic use of agrochemicals intensifies pathogen resistance, reduces nitrogen fixation and biodiversity, and increases pesticide bioaccumulation. Humans, the ecosystem, and sustainable agriculture are adversely affected as a result [53].

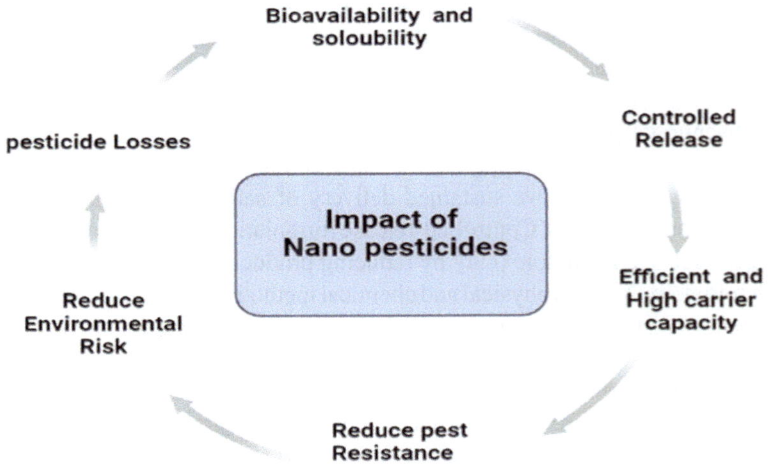

Fig. 4 Benefits of nanopesticides

3.1 *Nanopesticides as Antimicrobials*

Metal and metal oxide nanoparticles obtained applications and medical, environmental protection, forestry, and agriculture sectors [54, 55, 56, 57, 58, 59]. Compared to conventional formulations and microparticles, these nanoparticles have higher surface-to-volume ratios, larger pores, and more flexible pores [60, 61]. These formulations may reduce conventional formulations' toxic and harmful effects, including poor solubility, unwanted side effects, and toxic environmental effects [62, 63]. It has been proposed that metal and metal oxide nanoparticles act biocidal via three modes: (1) they kill microorganisms via photocatalysis by releasing superoxide radicals that destroy their molecular structures, (2) they rupture cell membranes as a result of the accumulation of metal nanoparticles, and (3) they disrupt DNA replication by taking metallic ions [64]. In an environmentally friendly manner, nanoparticles can be formulated together with conventional pesticides as active or active ingredients alone. Nanoformulations have excellent electronic properties, high adsorption capacities, and enhanced ion exchange capabilities [65].

Many papers have been published on using silver nanoparticles in pest control and the early detection of plant pathogens [8, 66, 67]. As a result of their bactericidal, antifungal, antiviral, and antibacterial properties, these compounds have proven particularly effective against phytopathogens [68, 69]. The antimicrobial properties of silver make it more effective than synthetic fungicides against microorganisms [70]. The active ingredients in AgNP-based pesticide formulations are more readily deposited on the target species than in conventional formulations (Fig. 5).

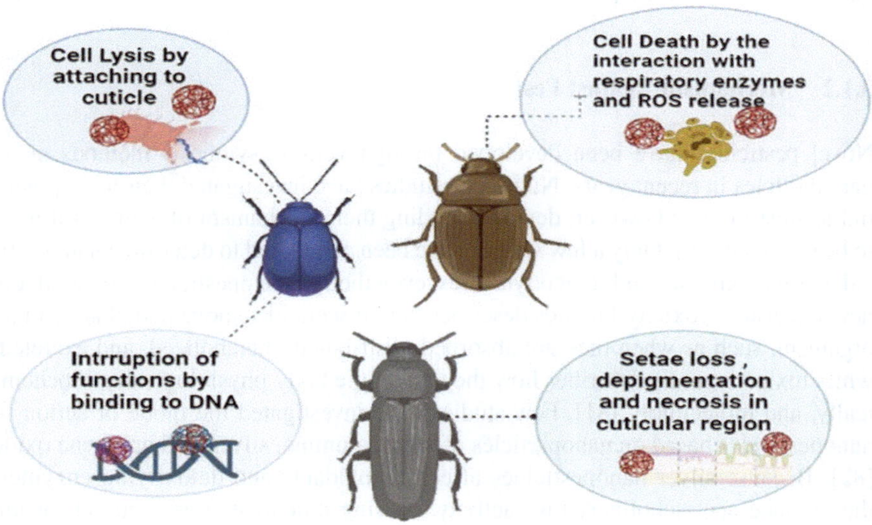

Fig. 5 Mechanism action of Nanopesticides against ticks. After the administration nanoparticles produce reactive oxygen species disturbs the redox balance in the body and cause apoptosis, cell lysis, and setae loss

3.1.1 Livestock Ticks

More than 80% of the world's cattle are infected by ticks, the most important vector of disease [71, 72]. Anaplasmosis, babesiosis, borreliosis, and ehrlichiosis are some diseases transmitted by ticks to humans and animals [73]. These insects transmit infectious agents in greater numbers than any other group of blood-feeding arthropods, affecting livestock, wildlife, and pets worldwide [74]. In spite of this, chemical acaricides currently on the market have been found to have several disadvantages, including: increasing resistance among ticks, pollution of the environment, contamination of meat and milk from livestock and a high cost [74]. Acaricide resistance can be mitigated by developing new acaricidal and repellent products [75]. Consequently, it is crucial to explore alternative methods of controlling tick vectors [76].

Humans are unintentional hosts for several tick species. Many risk factors contribute to tick infestations in developing countries, which are higher than in developed countries. To protect against tick bites, synthetic insecticides and biopesticides are mostly used [77]. The world's largest cattle population is affected by 80% of tick-borne diseases (TBDs), which pose a major threat to the livestock industry. The threat of tick-borne disease has been assessed primarily based on environmental factors (precipitation, temperature, and humidity), organic factors, and human factors (land use and agriculture). Tick load in animals has been reported to be affected by certain precipitation conditions, including humidity and temperature. Observations have shown that moderate rain and high humidity provide favorable microclimatic conditions for tick proliferation, leading to their invasion of animals [78]. Anthropod pests and vectors of economic importance are highly susceptible to metal, metal oxide, and carbon nanoparticles, especially those derived from eco-sustainable sources [79].

3.1.2 Mechanism Against Pest

Novel pesticides have been developed through various synthesis methods using nanoparticles in recent years. Numerous studies have investigated their toxic potential against insects; however, details regarding their mechanism of action still need to be provided [80]. Only a few studies have been conducted to determine nanopesticides' toxicokinetics and toxicodynamics, even though nanopesticides are relatively new materials. Toxicity kinetics describes how insecticides move and change in an organism, such as when they are absorbed, distributed, metabolized, and excreted, while toxicodynamic describes how they affect the body physiologically, biochemically, and molecularly [81]. Few studies have investigated the mode of action of nanopesticides based on nanoparticles of silica, alumina, silver, and graphene oxide [82]. Because silver nanopesticides affect antioxidants and detoxifying enzymes, they reduce acetylcholinesterase activity, causing oxidative stress and cell death. The insect genes are either up-regulated or down-regulated, resulting in developmental damage and reproductive failure by either reducing protein synthesis and gonadotropin release [83]. Metal nanoparticles can denaturant organelles and enzymes, resulting in cell death, since they bind S and P on proteins and nucleic

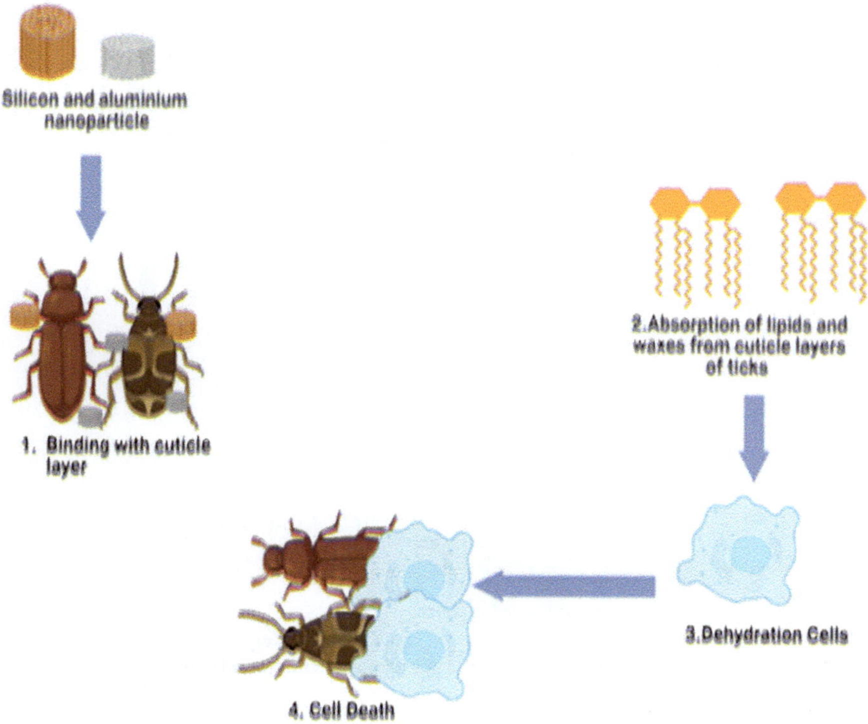

Fig. 6 Mechanism of Action of Silicon and Aluminum nanopesticides on ticks (1) Nanoparticles attached with the cuticle layer of the tick due to their smaller size (2) Nanoparticles absorb lipids and waxes from the cutical layer of ticks (3) Rapid absorption of lipids cause dehydration in cells (cell shrinkage) (4) Disturbance in the cell homeostasis cause cell death

acids, reducing membrane permeability. It has been shown that gold nanoparticles can also interfere with the activity of trypsin, affecting how cells develop and reproduce [84, 85]. Insects are dehydrated by physico-sorption of waxes and lipids by nanopesticides such as aluminum oxide and silicon dioxide [86, 87] (Figs. 6 and 7).

3.2 *Comparison of Nanopesticides Over Commonly Used Pesticides*

3.2.1 Environmental Issues with Conventional Pesticides

The use of pesticides increases crop yields and improves food safety, reduces energy consumption, and improves quality of life. Food production has increased dramatically due to conventional pesticides and agrochemicals used to feed the rapidly

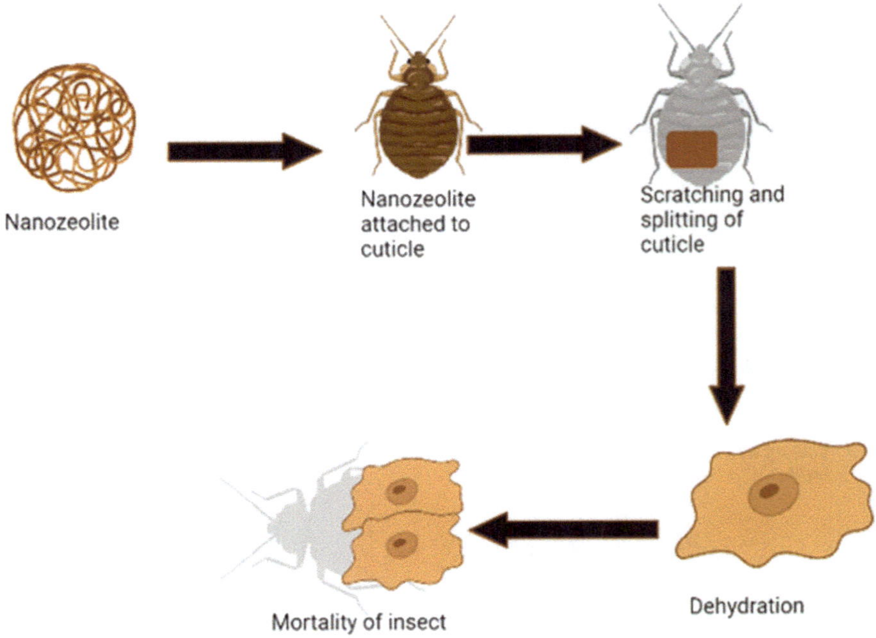

Fig. 7 Nanozeolite are crystalline aluminosilicates damaged cuticle and causes dehydration

expanding human population [88]. It is difficult for agriculture to maintain its trend of increasing yields in the twenty-first century due to several challenges. Due to AIs' poor water solubility, non-selective behavior, and uncontrollable release into the environment, pesticide formulations have several problems [89]. The contamination of ecosystems caused by agrochemical residues, poisoning crops and foods, and affecting their nutritional value has recently been the topic of many studies [90]. A depiction of possible pathways for environmental pollution from pesticide applications is provided. However, 99.9% of pesticides leaked into the environment around them and reached their targets only 0.1% of the time. Thus, inefficient delivery contributed to pollution of water, soil contamination, pest and pathogen resistance, biodiversity loss, and species extinction (e.g., bees) [66]. Another issue to be addressed is the complex interactions between pesticide mixtures, which can result in unpredictable hazard effects.

In the current market, pesticides come in several forms, but the majority are emulsifiable concentrates (ECs), oil-in-water emulsions (O/Ws), or alternatives to these formulations [91]. It typically comprises flammable, toxic, and expensive organic solvents or surfactant emulsifier blends to ensure spontaneous emulsion with water. However, emulsification requires much energy, so they are unsuitable [92, 93]. O/W emulsions do not suffer from these shortcomings because the solvent is removed, and nonionic surfactants, block polymers, and polymeric surfactants are used instead.

Table 1 Mode of action of different nanoparticles

Nanoparticles	Mode of action	References
Silica NPs	Insects die due to physisorbed lipids disrupting the protective barrier	[100]
Zinc	Resulting in the targeted oxidation of cells and their death	[101]
Au	Interruption of production and development	[85]
Polystyrene nanoparticle	CYP450 enzyme activity was inhibited	[102]
Ag nanoparticle	Enhanced antioxidant enzyme levels in moth larval guts following nano-induced oxidative stress	[103]

A new formulation based on micro- and nanoemulsions was developed that provides NPs sizes ranging from 20 to 100 nm to address these disadvantages [94]. Different types of plants require different microemulsion formulations to control broad-spectrum diseases, including plant growth regulators and systemic fungicides [95]. A microemulsion, by contrast, is more stable, or is too expensive for handlers to prepare on-site. Commercial agrochemical production can be maximized by stirring with a high velocity, homogenizing with high pressure, or using ultrasound generators instead of nanoemulsions, which are energy-intensive. Several advantages can be found in microemulsions over other conventional formulations. Among the benefits are improved tank mix compatibility, improved stability, reduced low flammability, reduced handler toxicity (because the formulation contains little solvent), as well as improved efficacy because surfactants have a high solubilizing power, which allows for greater penetration and uptake [90]. Although these substrates have many advantages, they also have some disadvantages, including a low amount of active ingredients (30%), a high amount of surfactants (20%), and a limited number of surfactant systems [96]. Using nanodispersion or nanosuspension as a formulation concept can partially overcome these limitations by combining active ingredients (prepared by a specific procedure) with nanocrystals or nanoparticles of 50 nm [97] to form nanodispersions that have similar properties to solutions). Even though triclosan and novaluron have been reported, this approach has not been widely utilized [98, 99] (Table 1).

3.2.2 Efficacy

It is also possible to improve pesticides' performance and efficiency and their adverse environmental effects by developing sustainable release systems using nanoparticles. Due to their ability to penetrate plant cells easily, nanoparticles are often referred to as 'nanocarriers.' Their ability to deliver products accurately is because they are customized to transfer specific biomolecules to specific cells, tissues, or organisms according to the conditions [104]. It is possible to engineer metals, metal oxides, silica, carbon, and semiconductor nanomaterials for tracking and delivery, with physically and chemically unique properties [105]. In addition to enhancing seed vigor and plant growth, nanodelivery vehicles can also protect crops from diseases and pests, and even modify their DNA [106]. A nanoparticle's ability to be absorbed by plants includes interacting with a carrier protein, creating new pores, and attaching to ion channels [107]. In some studies, the phytotoxicity of nanoparticles was examined, as well as their ability to penetrate plant cell walls [108] and their effects on plant growth [109]. NPs have been used successfully to deliver biomolecules to plants since Torney et al. [110] reported the concept. In recent years, other researchers have developed this idea for use in a variety of fields [111, 112]. Small nanoparticles of very small sizes are absorbed by plant cells [103, 113].

A plant's internal systems can be adversely affected by NPs because they can easily reach them, causing major changes to the plant's growth, metabolism, and growth efficiency. Several phytotoxicity studies conducted by Khodakovskaya et al. [114] indicate that root elongation is not necessarily a reliable indicator of the toxicity of NPs to plants. A large number of phytotoxicity studies conducted by Khodakovskaya et al. [114] indicate that root elongation is not necessarily a reliable indicator of the toxicity of NPs to plants. The positive effects of carbon nanotubes on tomato plants can be attributed to their ability to penetrate seed coats and increase water absorption (2009). It has been demonstrated that application of nano-ZNO particles at certain concentrations accelerates the growth of mung bean seedlings, *Vigna radiata* (L.) R. Wilczek, and gram, *Cicer arietinum* L. [115]. There was no significant difference between castor seeds exposed to gold nanoparticles and those exposed to silver nanoparticles [113]. A transmission electron microscope (TEM) analysis of plant and insect tissues showed that nanoparticles penetrated into cell organelles and were confined to mitochondria and nuclei, indicating that pesticides and fertilizers could be applied directly to plants [113]. It was confirmed that plants synthesized NPs through UV-visualization spectrophotometry.

SEM, TEM, and XRD were also used to confirm NP synthesis. These results were complemented by energy-dispersive X-ray spectroscopy and Fourier transform infrared spectroscopy [116].

4 Methods of Application

4.1 Nanoparticles as Nanocarriers for Pesticide Delivery

In the industrial arena, a concept similar to pest control was developed, called a "pesticide delivery system," utilizing concepts from the medical field, where Nurse Practitioners have delivered therapeutics successfully (PDS) [117], based on the successes of NPs in medical treatment. For the active ingredients to be effective at their fullest biological efficacy and reduce harmful effects, their concentrations and durations must be specified [3]. It is crucial to release pesticides in a controlled manner to maximize their biological efficiency and minimize their adverse effects [117]. NPs have many advantages as nanocarriers, including a large surface area, the ability to attach both single and multiple pesticide molecules, and the ability to transfer mass efficiently to the target. The chances of pesticides being released gradually over time are higher when encapsulated, reducing the need to apply highly concentrated and perhaps toxic pesticides at first and repeatedly after that. Moreover, NPS delay degradation-related losses of efficacy.

Active pesticide molecules can be loaded on nanoparticles using various methods, including adsorption, covalent attachment with ligands, encapsulation, and entrapment. Polymers (e.g., nanocarriers) can release active molecules slowly and control by controlling their degradation properties, bonding between ingredients, and environmental conditions. Polymers (soft nanoparticles), synthetic silica, titania, alumina, silver, copper, and natural minerals/clays (inorganic nanoparticles) can all be used to deliver pesticides [118, 119], nanotechnology is effective in neem oil and essential oils, as well as in garlic essential oil [120, 121]. The essential oils of *Artemisia arborescens* L (Asterales: Asteraceae) and *Lippia sidoides* L (Lamiales: Verbenaceae) are among the oils of *Lippia sidoides* L (Lamiales: Verbenaceae) and *Catharanthus roseus* (L.) G. Don [122] and juniper oil [123].

4.2 Nanoencapsulation for the Handling of Pesticides

The discovery of carbon nanotubes in 1991 made it possible to use them as a safer method for handling pesticides and reducing their environmental impacts via nanoencapsulation. Nanometers weigh only a sixth of a gram and are a sixth of the thickness of steel atoms, but they conduct electricity better than copper and are 100 times stronger than steel. The example of nanomaterials being beneficial in this situation is exceptional [124]. The potential of silica-based NPs as agrochemical delivery systems has attracted considerable interest among the various NPs available. Because NPs are structurally flexible, they can form a wide range of shapes and sizes and form pores for loading biomolecules [123, 125, 126]. Silica nanoparticles can be solid or mesoporous: solid silica nanoparticles (SSN) are those that are solid or mesoporous (MSN) are those that are mesoporous [127, 128]. These MSNs consist of well-ordered pores that allow molecules, such as proteins, to be loaded efficiently [129]. Additionally, MSN surfaces can be modified, allowing them to be customized for particular experiments [130]. In addition, MSNs were used to fertilize land and water through the slow release of urea [128]. When gold-plated MSN surfaces were bombarded, NP density and penetration of plant cell walls were enhanced [112], resulting in phytotoxicity in rice plants [131]. Silica nanoparticles calcined from non-porous silica could be transported into roots of *Arabidopsis thaliana* (L.) Heynh. It does not cause phytotoxicity (Brassicales: Brassicaceae) [127].

4.3 Clay Based Nanopesticide Formulations

Nanoclay refers to thin sheets of silicate materials found in volcanic ash, such as montmorillonite clay. The thickness of these particles is 1 nm and the width is 70–150 nm [132, 133]. A neutral and hydrophobic active ingredient can be adsorbent on these surfaces and delivered in a controlled manner. A neutral and hydrophobic active ingredient can be adsorbent on these surfaces and delivered in a controlled manner. In agriculture, they are considered to be potential nanocarriers due to their biocompatibility and economic viability. Agricultural applications consider them potential nanocarriers due to their biocompatibility and economic viability. A clay-based material has proven to be an innovative approach for delivering active ingredients in an environmentally friendly manner. An investigation of the controlled release of clay materials using carboxymethylcellulose and three clays, bentonite, kaolinite, and Fuller's earth, was performed by (Choudhary et al. 2006). Silica nanocapsules that release active ingredients in a controlled manner have been extremely challenging to develop [10]. In addition to improving pesticide loading, silica nanoparticles possess a charged nature. As a result of the modification of clay with organic cations, the clay was more able to adsorb hydrophobic active ingredients [134]. The dimensional stability and thermochemical stability of clay-based nanocarriers also make them suitable for controlling the release of active ingredients [135]. Research has been conducted on developing biodegradable clay nanocarriers made from plants in recent years [136]. In spite of these advancements, clay nanocarriers still contain crystalline impurities, which prevent their use in surface modification techniques or synthesis procedures.

4.4 Ease of Application

In the majority of non-organic commercial arable farms (59% of 946), productivity and profitability will not be adversely affected by the production situation. A number of research projects have attempted to reduce pesticide use (e.g., by 42%) on most arable farms (e.g., by 42%) (Lechenet et al. 2012). It is crucial to control pesticides and fertilizers sustainably under diverse production conditions in order to improve efficacy while reducing active ingredients (AIs). To find the ultimate solution, nanotechnology can be used to manipulate materials at the nanoscale to achieve many unique physicochemical properties. The field of nanotechnology deals with nanometer-scale materials and is one of the most promising research areas. Several shortcomings inherent in existing products can be overcome by using nanomaterials, including cost, fabrication strategies, functionality, and overall performance (by converting their bulk forms to nanomaterials). In addition to its numerous potential applications in medicine, food processing, agriculture, pharmaceuticals, materials science, electronics, and energy, this technology's emerging nature has made it a valuable tool for many fields [67, 137, 138]. In agriculture, nanotechnology can, for example, help balance crop nutrients against deficits and surpluses, monitor water quality, treat seeds, control pests, germinate seeds, distribute pesticides and fertilizers, detect toxic agrochemicals, and reduce their toxic effects. Several companies manufacture plant protection products. The agri-food sector also offers applications, including packaging for food, animal husbandry (involving both detoxification and nanomedicine), and environmental applications (such as purifying water, retaining water, and remediating pollutants). A nano-based artificial intelligence has been developed for pesticides to achieve higher efficacy while reducing pesticide volumes and increasing yields [139].

Nanoformulations of pesticides or nanopesticides must offer various benefits (such as enhanced effectiveness and durability, high dispersion and wettability, biodegradability in soil and environment, low toxicity, photogenetic properties). It would be better to reduce AIs and make them more effective pesticides so that they can be used to protect crops from insects and diseases [140]. As nanopesticide research has developed rapidly over the past decade, researchers seek to develop nanopesticides that are less damaging to the environment and more target-specific without sacrificing efficacy. [141]. It should therefore be possible to reduce non-target plant damage and decrease environmental pollution with target-specific nanopesticides.

5 Challenges to Use Nanopesticides

5.1 Toxicities

Nanopesticides are used in the formulation of registered agrochemical ingredients (AcI) to improve performance compared to non-technical AcI and counteract the drawbacks of existing agrochemicals. Understanding the response of plants to exposure to engineering nanopesticides and accurately predicting toxicity depends on thorough research that addresses some challenges and aspects that still need to be in use [142].

Using nanocarriers is one of the most frequently used strategies to solve major problems in modern agriculture, particularly balancing increased yields with environmental impacts and reducing pesticide use rates [92, 29]. However, achieving all production agriculture goals without a full understanding of the difficulties raises risks at a time when nanoparticles and ACI are more dangerous to human health and ecosystems [143]. According to previous studies, water tables, rivers and lakes are contaminated by half of the pesticides that are applied, either by leaching or wind runoff (Fig. 8).

Nanopesticides must be targeted in a controlled environment on a stimulus-based basis to effectively combat pests, reduce environmental pollution, prevent bioaccumulation effects, and threaten ecosystems and human health. Over the past few years, nanotechnology has been based on understanding nanoparticles' toxicity and genotoxicity in humans, as well as surfactants, solvents, wetting agents and stimulants

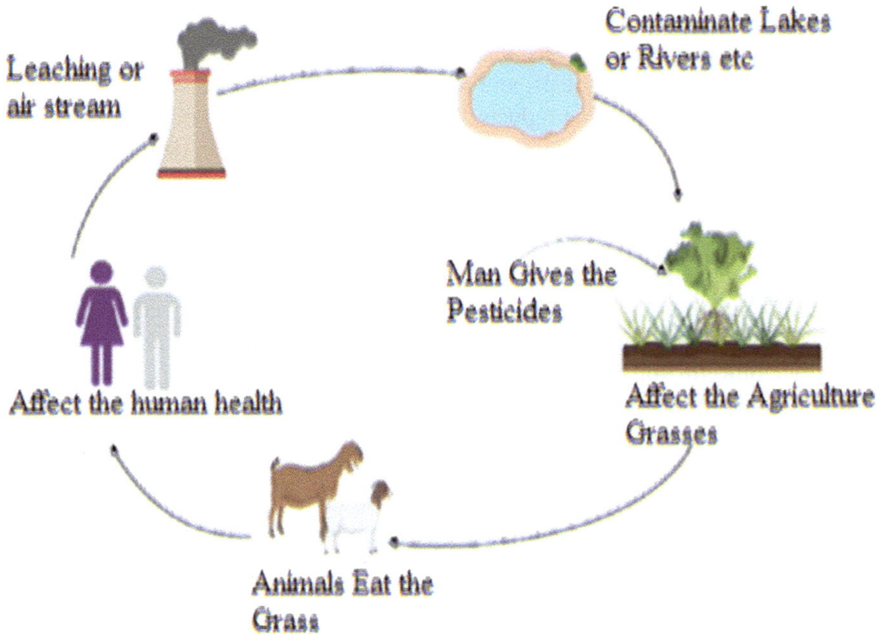

Fig. 8 Circulation of Pesticides impacting One Health

and their associated risks and toxic effects. However, when ACI and nanotechnology are linked, it seems that the toxicity risk of reaching all targets is increased [44, 131].

When sprayed, pesticides such as the herbicide clomazone pose harmful toxic risks to animals and people as they wash away crops from nearby areas. Reproductive problems are usually brought about by placental cytotoxicity [144]. Acute exposure to nanoparticles and clomazone was assessed using tadpoles (*L. catesbeianus*). In experimental groups exposed to clomazone, macrophage clustering (melano-macrophage centers), eosinophil accumulation, and liver function lipidosis increased. In this research, the mixture of nanoparticles polymeric had no significant effect on the liver's ability to protect tadpoles.

5.2 *Issues Related to Preparations*

Due to the fertilizers smart controlled release profiles, agrochemicals essential for increasing crop productivity, using nanotechnology-based delivery systems has introduced significant opportunities for revamping agricultural practices [44]. These systems are essential to increase the effectiveness of agrochemicals and fertilizers in agriculture [141]. However, the future of agriculture with nanopesticides appears bright. Still, man-hazardous agrochemicals that can cross biological barriers are a significant concern because they can permanently damage vital organs, such as the blood vessel barrier, the blood-retinal barrier, and the blood–brain barrier. Researchers are increasingly concerned about the dangers of exposure to toxic and genotoxic materials in bulk materials and nanochemical properties such as dimensions and charge. Researchers are increasingly concerned about the dangers of exposure to toxic and genotoxic materials in bulk materials and nanochemical properties such as dimensions and charge [145]. Innovative nanomaterials include nanopesticides designed for plant protection, reduced uses, increased foliar coverage, improved stability, and lower ingredient concentrations.

A nanopesticide is a pesticide encapsulated in a matrix or functionalized as a nanocarrier for environmental stimuli or enzyme-mediated stimuli. Nanosized particles are thought to probe pesticide activity for nanocarriers made from various materials, including silica, lipid, polymer, copolymer, ceramic, metal, carbon, and others [53].

Nanoformulations are known for contributing to a reduction in the degradation of active ingredients, an increase in water solubility, and increased bioavailability. Specifically, to reduce the quality and quantity of agricultural products and foods to prevent the spread of native pests and damage to plants and plants' harvests [146].

Formulation's effects on a species' behavior nanopesticides in environmental factors, the ecosystem, agricultural workers, and sector involved in agriculture still need to be fully understood [54].

It has been noted that nanoparticles can have toxic effects due to their biomimetic properties, high capacity for distribution, and potential for bioaccumulation in food and water environments as well as in all animals, especially mammals [52]. As a result of soil leaching phenomena, agricultural and industrial wastewater runoff during rainfall can enter water supplies, affecting water quality, and increasing human exposure time and ecosystems. For humans, multiple adverse effects associated with individual sensitivity and time of exposure to nanoparticles cause acute and chronic pathological manifestations in various respiratory, cardiovascular, lymphatic, autoimmune, neurologic, and cancer systems. Due to bioaccumulation and special nanoparticle properties, these side effects may appear immediately after exposure or years later [147].

5.3 Economics of Preparations

Although nanotechnology offers many benefits for agricultural industries, few products are available based on it [148]. A major reason for this low level of commercialization is that most studies are carried out in universities and research institutes or by small businesses (start-ups and spin-offs) to address specific agricultural needs. In fact, large companies are acquiring more patents annually [149]. Using nanotechnology-based pesticides is associated with many complex and poorly understood risks. The transport, bioaccumulation, and degradation rates of nano-based pesticides differ significantly from those of conventional pesticides, but more study is needed to understand them fully. Furthermore, there need to be more practical methods to understand the mechanism of action of nano-based pesticides, which hinders their development. To better understand how nano-based pesticides interact with both terrestrial and aquatic organisms, it is crucial to assess how long they persist in the environment [150].

Furthermore, the mechanism of action of nano-based pesticides is not well understood, which hinders their development. Determining how long nano-based pesticides persist in the environment is critical to better understanding how they interact with both terrestrial and aquatic organisms. The initial costs of developing a nanopesticide are high, and the current lack of widespread crop protection makes it impossible for such products to generate positive financial returns. Furthermore, a major barrier to the application of nanotechnology in agriculture is the need for more regulation. The high cost of registering a new active compound is another barrier to commercializing nano-based pesticides [104, 151].

The following is a list of the biggest obstacles that must be removed to boost the commercialization of pesticides based on nanotechnology:

1. Standard techniques for accurate risk–benefit analysis.
2. Improving our understanding of how nanomaterials interact with pathogens or plants.
3. Establish methods to monitor nanopesticides in the environment and assess how they affect human health and food safety.
4. Lack of a consensus definition for nanomaterials
5. Incorporating stimuli-responsive nanopesticides into international legislation to ensure their safety and effectiveness
6. Organizations working on nano-based products lack a global network for effective communication [45].

5.4 Farm Profitability

Growing evidence shows that pesticide use in agriculture causes environmental problems and health, especially for those directly affected [152]. In temperate climates, intensive farming techniques with production of extremely specialized crops and heavy reliance on mineral fertilizers and pesticides dominate the agricultural sector [153]. Over the past few years, pesticides have been used more and more. In 2013, 7% of the population received water exceeding the maximum permitted pesticide level. The French government is simultaneously promoting agricultural principles, emphasizing integrated pest management to reduce the need for pesticides [154].

Farming techniques to reduce using pesticides on arable farms while maintaining crop productivity and profitability. The effectiveness of pesticide reduction strategies on arable farms is still debated, as is whether less pesticide use increases agriculture's profitability and productivity [155]. According to some studies, pesticides are important for reducing pest infestations and ensuring high food levels, and reducing pesticide use can result in significant yield and financial losses [58]. Several studies have indicated that pesticides threaten the viability of agriculture, but that significant reductions in pesticide use are often accompanied by higher levels of agricultural performance, such as increased crop productivity [156].

For the farm, biophysical and socioeconomic contexts were described using Twenty- two different variables. Because these variables may directly or indirectly affect productivity and profitability, they were included in regression models in which the treatment frequency index (TFI) is correlated. In order to improve the prediction accuracy, regression models were fitted using the Lasso (minimum absolute shrinkage and selection operator) method, which employs only a subset of the variables instead of all of them. The fitted models calculated marginal TFI effects on productivity and profitability, corresponding to changes in productivity or profitability caused by a one-unit increase in TFI. TFI effects were modified to account for socioeconomic and biophysical contexts using estimated interaction effects. The biophysical context into account in order to estimate the degree of uncertainty in the estimated marginal TFI effects, bootstrapping was used to calculate confidence intervals [156].

6 Future Perspective

In the future, nanoparticles will play an important role in fertilization and pesticides. It is still in its infancy, but nanotechnology research in agriculture is rapidly developing. Due to their novel physicochemical properties, nanostructured agrochemicals are being investigated for plant growth and protection [157]. This nanoparticle is resistant to plant pests and has a sustained release due to its nanobiointeractions, transport and fate in plants and the environment [77]. With nanomaterials, the agricultural revolution will provide new opportunities for delivering nutrients, pesticides, and genetic material [158]. Despite this, nanoparticles pose several health risks due to their properties. Dry nanoproducts, for example, may be prone to inhalation due to their ease of suspension when suspended in the air.

Alternatively, nanoparticle suspensions in water can cause ionization or aggregation, resulting in submicrometer or micrometer-sized particles [77]. By conjugating metal/metal oxide nanoparticles with biodegradable polymers, nanocomposites can be developed that are easy to translocate within plant tissues and can exert antifungal activity in plants. Due to their biocompatible and eco-friendly properties, future research should encourage biodegradable polymers [159]. Applications based on nanoparticles have progressed significantly. Due to this, nanomaterials can be applied to agriculture to enhance quality and provide various benefits. There is, however, a major obstacle to the use of nanoparticles because of their toxicity. As a result, a variety of rational strategies are being developed to control toxic effects. An environmentally friendly method of synthesis of nanoparticles involves using biological entities or their products to prepare nanoparticles. A further promising field is the bioconjugation of nanoparticles with bioactive molecules and their encapsulation with bioactive molecules. For biomimetic nanomaterials to be successfully applied and commercialized in agriculture, different experts should collaborate to design biomimetic nanomaterials for evaluation [149]. For novel nanocomposites to succeed, size and stability must be considered. Antifungal management would be significantly improved by preparing a size-controlled nanocomposite. Maintaining stability throughout the period [160]. As a tool for agriculture and pest management, nanotechnology has immense potential [161] as it can be used to apply nanoparticles directly to soil, seeds, or plants to prevent them from attacking pathogens.

The use of silver nanoparticles in hydroponics systems and plant soil is now widely accepted as an effective method of eliminating microorganisms. There are advantages to using silver nanoparticles over common chemicals that cause multidrug resistance [162]. Nanopesticides reduce splash damage typically caused by conventional pesticides and EOs spraying, as the droplet size decreases due to nanoformulations [163]. The use of silver is also excellent for stimulating plant growth [164].

The use of silver metallic nanoparticles as insecticides is promising since silver metallic nanoparticles possess bactericidal, fungicidal, virucidal, pneumatical, and insecticidal properties [165]. These nanoparticles are also free from toxic chemicals, so they are safe for use and compatible with biological entities. Additionally, AgNP provides the advantage of managing multiple plant pathogens, as complex diseases are often encountered in the field. Multiple cellular mechanisms may be involved in these effects, but the mechanism of action may be non-specific. Thus, AgNP is a broad-spectrum antimicrobial agent that is effective against fungi and bacteria associated with plant diseases [166, 167]. Several root-associated fungal pathogens (e.g., Gaeumannomyces graminis and Rhizoctonia solani) may be susceptible to AgNP's antifungal properties. As a result of protection from additional stress from other pathogens, pathogens treated with AgNP may be more tolerant of root-knot nematode damage [168].

6.1 Combination Compounds

Any pesticide formulation that contains either as a whole or as a component of the engineered structure, engineered nanomaterials with biocidal properties are used as active ingredients is referred to as a nanopesticide [92]. A nanopesticide's primary purpose is to reduce the severity of fungal, bacterial, and oomycete diseases in plants. Due to the nanoscale properties of these chemicals, it is anticipated that these changes will increase their effectiveness, require smaller application doses, and maintain, if not boost, their productivity. In fact, using Engineered nanomaterials as pesticides to defend plants resistance a of pathogens has recently attracted more attention. Notably, using lower rates of ENM pesticides than their conventional counterparts would prevent overapplication, runoff of the environment of the active ingredients, and consequent environmental contamination. Along with these advantages, less energy and water would also be used to produce the materials. Together, these things will reduce the financial burden of farmers' pesticide inputs [169].

Studies have shown that ENMs have the potential to function as superior substitutes for traditional pesticides, which has sparked even more concern for the development of antimicrobials that contain NMs, either as polymers. The development of nano-enabled antimicrobials has been achieved using various techniques, including inorganic and organic polymers with various morphologies [170]. With varying levels of biodegradability, polymer-based nanoformulations include nanospheres, nanocapsules, nanogels, and nanofibers. Nanospheres contain evenly dispersed active ingredients throughout the polymeric matrix, whereas nanocapsules contain concentrated active ingredients around the center [171]. However, nanogels are cross-linked networks of biopolymers with pores filled with the active ingredient. Standards for organic farming have been proposed to be met by nanogels with active ingredients such as copper, essential oils, or pheromones. As previously mentioned, pesticides can be nanoencapsulated to release the active ingredient gradually or slowly. Alternately, using nanoemulsions of water can also improve the solubility and effectiveness of antimicrobials [172]. Some substances have shown the activity of antimicrobial. Notably, some of these components are nutrients that plants need. Plant diseases were controlled, yields were boosted, and nutrient use efficiency was improved using nutrient-element-based ENMs. The Ag, Cu, and Zn metals and metallic oxides have received the most research attention. However, research has also been done on other materials like Mn, Ti, and Ce, as well as biopolymers like chitosan and -D-glycan nanoparticles [173].

6.2 *Use of Stabilizing Agents*

Whether conventional and nanoforms can enhance the biological and chemical effects of pesticides, which is estimating the level of relationship with the environment, has not yet been studied [60]. It is important to consider factors such as molecular geometry, free rotation of the atoms, chains, branches, and rings of bonded atoms, as well as the composition of the element in the region where the pesticide is applied [174]. Different functional groups result in different pesticide solubility and reactivity.

A pesticide's ability to dissolve in water is important to its effectiveness. From a chemistry standpoint, pesticides that are highly water soluble will be less volatile. Although it is clear that the surface area of a pesticide in the nanoform is greater than in the bulk form, it will be important to discuss how dilution can aid in manipulating solubility. According to their surface properties, nanoparticles are usually in a suitable solvent dispersed (hydrophilic or hydrophobic). Instead of "colloidal solutions", nanotechnologists should refer to them as "colloidal dispersions". A common ground of preference is a solvent/solvent system that allows the best. A nanoformulation of a pesticide can be formulated as either (a) carrier plus payload or (b) direct application. In the initial scenario, solubility property of carrier molecule will also affect how the payload (pesticide) disperses in water. The size limit of the payload influences its diffusion from the carrier matrix, and colloidal principles govern its diffusion. Regarding pesticides, solubility follows the chemistry principle "like dissolves like." After coming into contact with water, a pesticide in nanoform (let's say glyphosate in nanoform without a carrier). Choose pesticide nanoforms over macroforms or traditional forms based on two distinct factors: dispersion and solubility. So far, most AC nanoformulations have been based on carrier + payload modules. This only makes sense in terms of penetration and bioavailability. As a result, pesticide water solubility is not affected by physical application methods (nano or bulk).

The volatility or air solubility of a pesticide expressed as the log of vapor pressure affects its active performance (VP). A pesticide with a higher VP will outgas more quickly and escape before reaching its target [175]. Using Twenty- two different variables, the biophysical and socioeconomic context of the farm was described. These variables may have direct or indirect effects on productivity and profitability compared to the treatment frequency index (TFI), so they were included in regression models that correlated these variables with TFI. Low volatility or good solubility of nanomaterials, preferably in water, is one of the important aspects of nanosafety in use. VP of pesticides should be low for long-term protection. Volatility changes over time based on seasonal factors. The transition from the liquid pesticide to the gas phase occurs more rapidly at higher temperatures (in tropical climates). Additionally, the site of application (such as vegetation, soil, etc.) affects the pesticide.

Nanosafety in applications requires nanomaterials to be less volatile or soluble in water. It is recommended that long-term pesticides have lower VPs. Volatility changes with time based on climatic conditions. The transition from a liquid pesticide to a gas phase happens more quickly at higher temperatures (in tropical climates). Additionally, the site of application (such as foliage, soil, etc.) affects the pesticide's volatilization. Nanopesticides might evaporate more quickly than conventional forms because of their greater surface area. Nanocomposite ACs have demonstrated less volatilization than a control [176]. However, for matrix-free nanoformulations, volatility control also needs to be considered. The method of application also affects the volatility rate. Release from soil particles is required before the nanopesticide can move along the soil–air interface and volatilize (applied in incorporated mode) [177].

7 Conclusion

The agrochemical industry is focused on laying down crops regardless of their safety or conventional pest management. To meet the agricultural demand for productivity while preserving human health and ecosystems, a great deal of research and financial support must be provided. A current perspective highlights concerns about the development of pathogens resistant to antibiotics and fungicides, and suggests practical solutions that can be applied to this problem through the development of pesticides derived from nanoparticles with different antimicrobial mechanisms of action and the development of target gene systems for metallic and bimetallic nanoparticles. Using a variety of materials and techniques, experimental research has developed technology-based pesticides to resolve inconveniences. Agricultural production can be remodeled while ensuring the preservation of ecosystems and food safety. Nanocarriers are coated with stimuli-responsive compounds that can control their release systems, allowing them to be released by the modified release systems. It is important to pay attention to the possible adverse effects of nanoparticles on the environment and nontarget organisms as well as the development of environmentally safer nanopesticides.

References

1. Ascherio A, Chen H, Weisskopf MG, O'Reilly E, McCullough ML, Calle EE, Thun MJ (2006) Pesticide exposure and risk for Parkinson's disease. Ann Neurol: Off J Am Neurol Assoc Child Neurol Soc 60(2):197–203
2. Fisher MC, Henk DA, Briggs CJ, Brownstein JS, Madoff LC, McCraw SL, Gurr SJ (2012) Emerging fungal threats to animal, plant and ecosystem health. Nature 484(7393):186–194
3. Ghormade V, Deshpande MV, Paknikar KM (2011) Perspectives for nano-biotechnology enabled protection and nutrition of plants. Biotechnol Adv 29(6):792–803
4. Duhan JS, Kumar R, Kumar N, Kaur P, Nehra K, Duhan S (2017) Nanotechnology: The new perspective in precision agriculture. Biotechnol R 15:11–23
5. Kumar R, Kumar N, Rajput VD, Mandzhieva S, Minkina T, Saharan BS, Duhan JS (2022) Advances in biopolymeric nanopesticides: a new eco-friendly/eco-protective perspective in precision agriculture. Nanomaterials 12(22):3964
6. Pradhan S, Mailapalli DR (2020) Nanopesticides for pest control. Sustain Agricul Rev 40:43–74
7. Barik TK, Sahu B, Swain V (2008) "Nanosilica—from medicine to pest control." Parasitol Res 103:253–258
8. Karlsson Green K, Stenberg JA, Lankinen Å (2020) Making sense of integrated Pest management (IPM) in the light of evolution. Evol Appl 13:1791–1805
9. Hendry AP, Gotanda KM, Svensson EI (2017) Human influences on evolution, and the ecological and societal consequences. Philos Trans R Soc B 372:20160028
10. Chen YH, Schoville SD (2018) Editorial overview: ecology: ecological adaptation in agroecosystems: novel opportunities to integrate evolutionary biology and agricultural entomology. Curr Opinion Insect Sci 26:iv– viii
11. Tabashnik BE, Brévault T, Carrière Y (2013) Insect resistance to Bt crops: lessons from the first billion acres. Nat Biotechnol 31:510–521
12. Palumbi SR (2001) Humans as the World's greatest evolutionary force. Science 293:1786–1790
13. Dermauw W, Pym A, Bass C, Van Leeuwen T, Feyereisen R (2018) Does host plant adaptation lead to pesticide resistance in generalist herbivores? Curr Opinion Insect Sci 26:25–33
14. Hardy NB, Peterson DA, Ross L, Rosenheim JA (2018) Does a plant-eating insect's diet govern the evolution of insecticide resistance? Comparative tests of the pre-adaptation hypothesis. Evol Appl 11:739–747
15. Walsh TK, Heckel DG, Wu Y, Downes S, Gordon KHJ, Oakeshott JG (2022) Determinants of insecticide resistance evolution: comparative analysis among heliothines. Annu Rev Entomol 67:387–406
16. Cheng T, Wu J, Wu Y, Chilukuri RV, Huang L, Yamamoto K et al (2017) Genomic adaptation to polyphagy and insecticides in a major east Asian noctuid pest. Nat Ecol Evol 1:1747–1756
17. Crossley MS, Snyder WE, Hardy NB (2021) Insect-plant relationships predict the speed of insecticide adaptation. Evol Appl 14:290–296
18. Cogliani C, Goosens H, Greko C (2011) Restricting antimicrobial use in food animals: lessons from Europe. Microbe 6:274–279
19. Demetzos, C. (2015). Advanced drug delivery nanosystems: Perspectives and regulatory issues. In: GeNeDis 2014. Springer, Cham, pp 195–198
20. Rajput ZI, Hu SH, Chen WJ, Arijo AG, Xiao CW (2006) Importance of ticks and their chemical and immunological control in livestock. J Zhejiang Univ. Science B 7(11):912

21. Gadhave KR, Hourston JE, Gange AC (2016) Developing soil microbial inoculants for pest management: can one have too much of a good thing? J Chem Ecol 42:348–356
22. Yessinou RE, Akpo Y, Adoligbe C, Adinci J, Assogba MN, Koutinhouin B, Farougou S (2016) Resistance of tick Rhipicephalus microplus to acaricides and control strategies. J Entomol Zool Stud 4(6):408–414
23. Van Boeckel TP, Brower C, Gilbert M, Grenfell BT, Levin SA, Robinson TP, Teillant A, Laxminarayan R (2015) Global trends in antimicrobial use in food animals. Proc Natl Acad Sci USA 112:5649–5654
24. Bester LA, Essack SY (2012) Observational study of the prevalence and antibiotic resistance of Campylobacter spp. From different poultry production systems in KwaZulu-Natal, South Africa. J Food Prot 75:154–159
25. Maron DF, Smith TJ, Nachman KE (2013) Restrictions on antimicrobial use in food animal production: an international regulatory and economic survey. Glob Health 9:48
26. Kaplan J, Bowman M, Cordova C, Kar A (2014) Pharming chickens: it's time for the US poultry industry to demonstrate antibiotic stewardship; Natural Resources Defence Council: New York, NY, USA
27. Bashahun DGM, Odoch TA (2015) Assessment of antibiotic usage in intensive poultry farms in Wakiso District, Uganda. Livest Res Rural Dev 27:247
28. Rudzinski WE, Dave AM, Vaishnav UH, Kumbar SG, Kulkarni AR, Aminabhavi TM (2002) Hydrogels as controlled release devices in agriculture. Des Monomers Polym 5(1):39–65
29. Popp J, Pető K, Nagy J (2013) Pesticide productivity and food security. A review. Agron Sustain Develop 33(1):243–255
30. Grube A, Donaldson D, Kiely T, Wu L (2011) Pesticide industry sales and usage. 2006 and 2007 market estimates. D.O.o.P.P. Biological and economic analysis, office of chemical safety and pollution prevention, United States Environmental Protection Agency, Editor, Washington, DC
31. Colborn T (2006) A case for revisiting the safety of pesticides: a closer look at neurodevelopment. Environ Health Perspect 114(1):10–17
32. International Agency for Research on Cancer (2017) Some organophosphate insecticides and herbicides in IARC Monograph on the evaluation of carcinogenic risk to humans. International Agency for Research on Cancer, Lyon, France
33. Hatjian BA, Mutch E, Williams FM, Blain PG, Edwards JW (2000) Cytogenetic response without changes in peripheral cholinesterase enzymes following exposure to a sheep dip containing diazinon in vivo and in vitro. Mutat Res 472:85–92
34. Bester LA, Essack SY (2012) Observational study of the prevalence and antibiotic resistance of Campylobacter spp. From different poultry production systems in KwaZulu-Natal, South Africa. J Food Prot 75:154–159
35. Alavanja MC, Ross MK, Bonner MR (2013) Increased cancer burden among pesticide applicators and others due to pesticide exposure. CA Cancer J Clin 63:120–142
36. Stratonovitch P, Elias J, Denholm I, Slater R, Semenov MA, Guedes RNC (2014) An individual-based model of the evolution of pesticide resistance in heterogeneous environments: control of Meligethes aeneus population in oilseed rape crops. PLoS ONE 9:e115631
37. Tobin PC, Nagarkatti S, Loeb G, Saunders MC (2008) Historical and projected interactions between climate change and insect voltinism in a multivoltine species. Glob Change Biol 14:951–957
38. Kashyap PL, Xiang X, Heiden P (2015) Chitosan nanoparticle based delivery systems for sustainable agriculture. Int J Biol Macromol 77:36–51
39. Polson KA, Brogdon WG, Rawlins SC, Chadee DD (2012) Impact of environmental temperatures on resistance to organophosphate insecticides in Aedes aegypti from Trinidad. Rev Panam Salud Publica 32:1–8
40. MacLean RC, Hall AR, Perron GG, Buckling A (2010) The population genetics of antibiotic resistance: integrating molecular mechanisms and treatment contexts. Nature Rev Genet 11(6):405–414

41. Neve P, Busi R, Renton M, Vila-Aiub MM (2014) Expanding the eco-evolutionary context of herbicide resistance research. Pest Manage Sci 70(9):1385–1393
42. Alexander HK, Martin G, Martin OY, Bonhoeffer S (2014) Evolutionary rescue: linking theory for conservation and medicine. Evol Appl 7(10):1161–1179
43. Agostini A, Mondragón L, Coll C, Aznar E, Marcos MD, Martínez-Máñez R, Sancenón F, Soto J, Pérez-Payá E, Amorós P (2012) Dual enzyme-triggered controlled release on capped nanometric silica mesoporous supports. ChemistryOpen 1(1):17–20
44. Yadav RK, Singh NB, Singh A, Yadav V, Bano C, Khare S, Niharika (2020) Expanding the horizons of nanotechnology in agriculture: recent advances, challenges and future perspectives. Vegetos 33(2):203–221. https://doi.org/10.1007/s42535-019-00090-9
45. Margulis-Goshen K, Magdassi S (2013) Nanotechnology: an advanced approach to the development of potent insecticides. In: Advanced technologies for managing insect pests. Springer, Dordrecht, pp 295–314
46. Peteu SF, Oancea F, Sicuia OA, Constantinescu F, Dinu S (2010) Responsive polymers for crop protection. Polymers 2(3):229–251
47. Xin X, He Z, Hill MR, Niedz RP, Jiang X, Sumerlin BS (2018) Efficiency of biodegradable and pH-responsive polysuccinimide nanoparticles (PSI-NPs) as smart nanodelivery systems in grapefruit: in vitro cellular investigation. Macromol Biosci 18(7):1800159
48. Yang Z, Kang SG, Zhou R (2014) Nanomedicine: de novo design of nanodrugs. Nanoscale 6(2):663–677
49. Adak T, Kumar J, Shakil NA, Walia S (2012) Development of controlled release formulations of imidacloprid employing novel nano-ranged amphiphilic polymers. J Environ Sci Health B 47(3):217–225
50. Bajpai SK, Thomas V, Mohan YM, Sreedhar B (2007) A versatile strategy to fabricate hydrogel-silver nanocomposites and investigation of their antimicrobial activity. J Colloid Interface Sci 315:389–395
51. Cao L, Zhang H, Cao C, Zhang J, Li F, Huang Q (2016) Quaternized chitosan-capped mesoporous silica nanoparticles as nanocarriers for controlled pesticide release. Nanomaterials 6(7):126
52. Bombo AB, Pereira AES, Lusa MG, de Medeiros Oliveira E, de Oliveira JL, Campos EVR, de Jesus MB, Oliveira HC, Fraceto LF, Mayer JLS (2019) A mechanistic view of interactions of a nanoherbicide with target organism. J Agric Food Chem 67(16):4453–4462. https://doi.org/10.1021/acs.jafc.9b00806
53. Vigneshwaran N, Kathe AA, Varadarajan PV, Nachane RP, Balasubramanya RH (2006) Biomimetics of silver nanoparticles by white rot fungus, *Phaenerochaete chrysosporium*. Colloids Surf B, Biointerfaces 53(1):55–59. https://doi.org/10.1016/j.colsurfb.2006.07.014
54. Huang B, Chen F, Shen Y, Qian K, Wang Y, Sun C, Zhao X, Cui B, Gao F, Zeng Z, Cui H (2018) Advances in targeted pesticides with environmentally responsive controlled release by nanotechnology. Nanomaterials 8(2). https://doi.org/10.3390/nano8020102
55. Husen A, Siddiqi KS (2023) Advances in smart nanomaterials and their applications. Elsevier Inc., 50 Hampshire St., 5th Floor, Cambridge, MA 02139, USA
56. Husen A (2022) Engineered nanomaterials for sustainable agricultural production, soil improvement and stress management. Elsevier Inc., 50 Hampshire St., 5th Floor, Cambridge, MA 02139, USA
57. Husen A, Iqbal M (2019) Nanomaterials and plant potential. Springer International Publishing AG, Gewerbestrasse 11, 6330 Cham, Switzerland. https://doi.org/10.1007/978-3-030-05569-1
58. Jacquet F, Butault J-P, Guichard L (2011) An economic analysis of the possibility of reducing pesticides in French field crops. Ecol Econ 70:1638–1648. https://doi.org/10.1016/j.ecolecon.2011.04.003
59. REX Consortium (2013) Heterogeneity of selection and the evolution of resistance. Trends Ecol Evol 28:110–118
60. Minguela JV, Cunha JPAAR (2011) Manual of application of phytosanitary products. Aprenda Facil Editora, Vicosa, Brazil, p 588

61. Vellingiri K, Philip L, Kim KH (2017) Metal–organic frameworks as media for the catalytic degradation of chemical warfare agents. Coord Chem Rev 353:159–179
62. He L, Liu Y, Mustapha A, Lin M (2011) Antifungal activity of zinc oxide nanoparticles against *Botrytis cinerea* and *Penicillium expansum*. Microbiol Res 166:207–215
63. Pinto RV, Antunes F, Pires J, Graça V, Brandão P, Pinto ML (2017) Vitamin B3 metal organic frameworks as potential delivery vehicles for therapeutic nitric oxide. Acta Biomater 51:66–74
64. Chatterjee AK, Chakraborty R, Basu T (2014) Mechanism of antibacterial activity of copper nanoparticles. Nanotechnology 25(13):135101
65. Masoomi MY, Bagheri M, Morsali A (2016) High adsorption capacity of two Zn-based metal-organic frameworks by ultrasound assisted synthesis. Ultrason Sonochem 33:54–60
66. Goswami L, Kim KH, Deep A, Das P, Bhattacharya SS, Kumar S, Adelodun AA (2017) Engineered nano particles: nature, behavior, and effect on the environment. J Environ Manage 196:297–315
67. Kilfoyle BE, Sheihet L, Zhang Z, Laohoo M, Kohn J, Michniak-Kohn BB (2012) Development of paclitaxel-TyroSpheres for topical skin treatment. J Control Release 163(1):18–24
68. Chen X, Schluesener HJ (2008) Nanosilver: a nanoproduct in medical application. Toxicol Lett 176(1):1–12
69. Gonfa YH, Gelagle AA, Hailegnaw B, Kabeto SA, Workeneh GA, Tessema FB, Tadesse MG, Wabaidur SM, Dahlous KA, Abou Fayssal S, Kumar P, Adelodun B, Bachheti A, Bachheti RK (2023) Optimization, characterization, and biological applications of silver nanoparticles synthesized using essential oil of aerial part of *Laggera tomentosa*. Sustainability 15(1):797
70. Aziz N, Pandey R, Barman I, Prasad R (2016) Leveraging the attributes of *Mucor hiemalis*-derived silver nanoparticles for a synergistic broad-spectrum antimicrobial platform. Front Microbiol 7:1984
71. FAO (1984) Tick-borne diseases control: a practical field manual 1. Rome, Italy.https://doi.org/10.1017/S0031182003004682
72. El-Tokhy A, Hamdon HAK, Kassab AY, Amer MES (2019) Novel acaricides by nanotechnology against cattle ticks and their effect on physiological and productive performance. J Sustain Agricul Sci 45(3):81–96
73. Guglielmone AA, Robbins RG, Apanaskevich DA, Petney TN, Estrada-Peña A, Horak IG (2014) The hard ticks of the world (Acari: Ixodida: Ixodidae) Springer. Dordrecht, The Netherlands
74. Sonenshine DE, Lane RS, Nicholson WL (2002) Ticks (Ixodida). In: Mullen G, Durden L (eds) Medical and veterinary entomology. Academic, San Diego, pp 517–558
75. Abdel-Ghaffar F, Al-Quraishy S, Mehlhorn H (2015) Length of tick repellency depends on formulation of the repellent compound (icaridin= Saltidin®): tests on Ixodes persulcatus and Ixodes ricinus placed on hands and clothes. Parasitology Res 114:3041–3045
76. Benelli G (2016) Tools to fight ticks: a never-ending story? News from the front of green acaricides and photosensitizers. Asian Pac J Trop Dis 6(8):656–659
77. Cuenca JB, Tirado N, Vikström M, Lindh CH, Stenius U, Leander K, Berglund M, Dreij K (2019) Pesticide exposure among Bolivian farmers: associations between worker protection and exposure biomarkers. J Expo Sci Environ Epidemiol 30:730–742
78. Chepkwony R, van Bommel S, van Langevelde F (2021) Interactive effects of biological, human and environmental factors on tick loads in Boran cattle in tropical drylands. Parasit Vectors 14(1):1–8
79. Athanassiou CG, Kavallieratos NG, Benelli G, Losic D, Usha Rani P, Desneux N (2018) Nanoparticles for pest control: current status and future perspectives. J Pest Sci 91:1–15
80. Rai AB, Halder J, Kodandaram MH (2014) Emerging insect pest problems in vegetable crops and their management in India: an appraisal. Pest Manage Horticul Ecosyst 20(2):113–122
81. Alzogaray RA, Zerba EN (2017) Rhodnius prolixus intoxicated. J Insect Physiol 97:93–113
82. Benelli G (2018) Mode of action of nanoparticles against insects. Environ Sci Pollut Res 25(13):12329–12341
83. Nair PMG, Choi J (2011) Identification, characterization and expression profiles of *Chironomus riparius* glutathione S-transferase (GST) genes in response to cadmium and silver nanoparticles exposure. Aquat Toxicol 101(3–4):550–560

84. Patil CD, Borase HP, Suryawanshi RK, Patil SV (2016) Trypsin inactivation by latex fabricated gold nanoparticles: a new strategy towards insect control. Enzyme Microb Technol 92:18–25
85. Small T, Ochoa-Zapater MA, Gallello G, Ribera A, Romero FM, Torreblanca A, Garcerá MD (2016) Gold-nanoparticles ingestion disrupts reproduction and development in the German cockroach. Sci Total Environ 565:882–888
86. Arumugam G, Velayutham V, Shanmugavel S, Sundaram J (2016) Efficacy of nanostructured silica as a stored pulse protector against the infestation of bruchid beetle, *Callosobruchus maculatus* (Coleoptera: Bruchidae). Appl Nanosci 6(3):445–450
87. Debnath N, Das S, Seth D, Chandra R, Bhattacharya SC, Goswami A (2011) Entomotoxic effect of silica nanoparticles against *Sitophilus oryzae* (L.). J Pest Sci 84(1):99–105
88. Gianessi LP (2013) The increasing importance of herbicides in worldwide crop production. Pest Manag Sci 69(10):1099–1105
89. Beggel S, Werner I, Connon RE, Geist JP (2010) Sublethal toxicity of commercial insecticide formulations and their active ingredients to larval fathead minnow (*Pimephales promelas*). Sci Total Environ 408(16):3169–3175
90. Green JM (2014) Current state of herbicides in herbicide-resistant crops. Pest Manag Sci 70(9):1351–1357
91. Knowles A (2009) Global trends in pesticide formulation technology: the development of safer formulations in China. Outlooks Pest Manage 20(4):165–170
92. Kah M, Hofmann T (2014) Nanopesticide research: current trends and future priorities. Environ Int 63:224–235
93. Kah M, Beulke S, Tiede K, Hofmann T (2013) Nanopesticides: state of knowledge, environmental fate, and exposure modeling. Crit Rev Environ Sci Technol
94. Song S, Liu X, Jiang J, Qian Y, Zhang N, Wu Q (2009) Stability of triazophos in self-nanoemulsifying pesticide delivery system. Colloids Surf, A 350(1–3):57–62
95. Robinson DKR, Salejova-Zadrazilova G (2010) Observatory NANO (2010) Nanotechnologies for nutrient and biocide delivery in agricultural production. Working Paper Version
96. Lawrence MJ, Warisnoicharoen W (2006) Recent advances in microemulsions as drug delivery vehicles. Nanoparticulates as Drug Carriers 125–171
97. Müller RH, Junghanns JU (2006) Drug nanocrystals/nanosuspensions for the delivery of poorly soluble drugs. Nanopartícul Drug Carriers 1:307–328
98. Zhang H, Wang D, Butler R, Campbell NL, Long J, Tan B, Duncalf DJ, Foster AJ, Hopkinson A, Taylor D, Angus D, Rannard SP (2008) Formation and enhanced biocidal activity of water-dispersable organic nanoparticles. Nat Nanotechnol 3(8):506–511
99. Elek N, Hoffman R, Raviv U, Resh R, Ishaaya I, Magdassi S (2010) Novaluron nanoparticles: formation and potential use in controlling agricultural insect pests. Colloids Surf A: Physicochem Eng Aspects 372(1–3):66–72
100. Yadav J, Jasrotia P, Kashyap PL, Bhardwaj AK, Kumar S, Singh M, Singh GP (2021) Nanopesticides: current status and scope for their application in agriculture. Plant Prot Sci 58(1):1–17
101. Zaheer T, Ali MM, Abbas RZ, Atta K, Amjad I, Suleman A, Khalid Z, Aqib AI (2022) Insights into nanopesticides for ticks: the superbugs of livestock. Oxidative Med Cellular Longevity
102. Fröhlich E, Kueznik T, Samberger C, Roblegg E, Wrighton C, Pieber TR (2010) Size-dependent effects of nanoparticles on the activity of cytochrome P450 isoenzymes. Toxicol Appl Pharmacol 242(3):326–332
103. Yasur J, Rani PU (2015) Lepidopteran insect susceptibility to silver nanoparticles and measurement of changes in their growth, development and physiology. Chemosphere 124:92–102
104. Deshpande MV (2019) Nanobiopesticide perspectives for protection and nutrition of plants. Nano-Biopesticides Today Future Perspect. https://doi.org/10.1016/B978-0-12-815829-6.00003-6
105. Kunzmann A, Andersson B, Vogt C, Feliu N, Ye F, Gabrielsson S, Toprak MS, Buerki-Thurnherr T, Laurent S, Vahter M, Krug H, Fadeel B (2011) Efficient internalization of silica-coated iron oxide nanoparticles of different sizes by primary human macrophages and dendritic cells. Toxicol Appl Pharmacol 253(2):81–93

106. Kole C, Kole P, Randunu KM, Choudhary P, Podila R, Ke PC, Rao AM, Marcus RK (2013) Nanobiotechnology can boost crop production and quality: first evidence from increased plant biomass, fruit yield and phytomedicine content in bitter melon (*Momordica charantia*). BMC Biotechnol 13(1):1–10
107. Rico CM, Majumdar S, Duarte-Gardea M, Peralta-Videa JR, Gardea-Torresdey JL (2011) Interaction of nanoparticles with edible plants and their possible implications in the food chain. J Agric Food Chem 59(8):3485–3498
108. Lee CW, Mahendra S, Zodrow K, Li D, Tsai YC, Braam J, Alvarez PJ (2010) Developmental phytotoxicity of metal oxide nanoparticles to *Arabidopsis thaliana.* Environ Toxicol Chem Int J 29(3):669–675
109. Balalakshmi C, Gopinath K, Govindarajan M, Lokesh R, Arumugam A, Alharbi NS, Kadaikunnan S, Khaled JM, Benelli G (2017) Green synthesis of gold nanoparticles using a cheap *Sphaeranthus indicus* extract: impact on plant cells and the aquatic crustacean *Artemia nauplii.* J Photochem Photobiol, B 173:598–605
110. Torney F, Trewyn BG, Lin VSY, Wang K (2007) Mesoporous silica nanoparticles deliver DNA and chemicals into plants. Nat nanotechnol 2(5):295–300
111. Martin-Ortigosa S, Valenstein JS, Lin VSY, Trewyn BG, Wang K (2012a) Gold functionalized mesoporous silica nanoparticle mediated protein and DNA codelivery to plant cells via the biolistic method. Adv Function Mater 22(17):3576–3582
112. Martin-Ortigosa S, Valenstein JS, Sun W, Moeller L, Fang N, Trewyn BG, Wang K (2012b) Parameters affecting the efficient delivery of mesoporous silica nanoparticle materials and gold nanorods into plant tissues by the biolistic method. Small 8(3):413–422
113. Yasur J, Rani PU (2013) Environmental effects of nanosilver: impact on castor seed germination, seedling growth, and plant physiology. Environ Sci Pollut Res 20:8636–8648
114. Khodakovskaya MV, de Silva K, Nedosekin DA, Dervishi E, Biris AS, Shashkov EV, Zharov VP (2011) Complex genetic, photothermal, and photoacoustic analysis of nanoparticle-plant interactions. Proc Nat Acad Sci 108(3):1028–1033
115. Mahajan P, Dhoke SK, Khanna AS (2011) Effect of nano-ZnO particle suspension on growth of mung (Vigna radiata) and gram (Cicer arietinum) seedlings using plant agar method. J Nanotechnol
116. Rajan R, Chandran K, Harper SL, Yun SI, Kalaichelvan PT (2015) Plant extract synthesized silver nanoparticles: an ongoing source of novel biocompatible materials. Ind Crops and Prod 70:356–373
117. Tsuji K (2001) Microencapsulation of pesticides and their improved handling safety. J Microencapsul 18(2):137–147
118. Anjali CH, Khan SS, Margulis-Goshen K, Magdassi S, Mukherjee A, Chandrasekaran N (2010) Formulation of water-dispersible nanopermethrin for larvicidal applications. Ecotoxicol Environ Safe 73(8):1932–1936
119. Jerobin J, Sureshkumar RS, Anjali CH, Mukherjee A, Chandrasekaran N (2012) Biodegradable polymer based encapsulation of neem oil nanoemulsion for controlled release of Aza-A. Carbohydr Polym 90(4):1750–1756
120. Yang JJ, Miao F, Pickett MD, Ohlberg DA, Stewart DR, Lau CN, Williams RS (2009) The mechanism of electroforming of metal oxide memristive switches. Nanotechnology 20(21):215201
121. Lai F, Wissing SA, Müller RH, Fadda AM (2006) Artemisia arborescens L essential oil-loaded solid lipid nanoparticles for potential agricultural application: preparation and characterization. Aaps Pharmscitech 7:E10–E18
122. Pavunraj M, Baskar K, Duraipandiyan V, Al-Dhabi NA, Rajendran V, Benelli G (2017) Toxicity of Ag nanoparticles synthesized using stearic acid from Catharanthus roseus leaf extract against Earias vittella and mosquito vectors (Culex quinquefasciatus and Aedes aegypti). J Clust Sci 28:2477–2492
123. Athanassiou CG, Kavallieratos NG, Evergetis E, Katsoula AM, Haroutounian SA (2013) Insecticidal efficacy of silica gel with Juniperus oxycedrus ssp. oxycedrus (Pinales: 17. Cupressaceae) essential oil against Sitophilus oryzae (Coleoptera: Curculionidae) and Tribolium confusum (Coleoptera: Tenebrionidae). J Econ Entomol 106(4):1902–1910

124. Lok C (2010) Nanotechnology: small wonders. Nature 467 (7311):18–21
125. Campbell JL, Arora J, Cowell SF, Garg A, Eu P, Bhargava SK, Bansal V (2011) Quasi-cubic magnetite/silica core-shell nanoparticles as enhanced MRI contrast agents for cancer imaging. PloS one 6(7):e21857
126. Jang HR, Oh HJ, Kim JH, Jung KY (2013) Synthesis of mesoporous spherical silica via spray pyrolysis: Pore size control and evaluation of performance in paclitaxel pre-purification. Microporous Mesoporous Mater 165:219–227
127. Slomberg DL, Schoenfisch MH (2012) Silica nanoparticle phytotoxicity to Arabidopsis thaliana. Environ Sci Technol 46(18):10247–10254
128. Wanyika H, Gatebe E, Kioni P, Tang Z, Gao Y (2012) Mesoporous silica nanoparticles carrier for urea: potential applications in agrochemical delivery systems. J Nanosci Nanotechnol 12(3):2221–2228
129. Popat A, Hartono SB, Stahr F, Liu J, Qiao SZ, Lu GQM (2011) Mesoporous silica nanoparticles for bioadsorption, enzyme immobilisation, and delivery carriers. Nanoscale 3(7):2801–2818
130. Trewyn BG, Slowing II, Giri S, Chen HT, Lin VSY (2007) Synthesis and functionalization of a mesoporous silica nanoparticle based on the sol–gel process and applications in controlled release. Acc Chem Res 40(9):846–853
131. Mustafa IF, Hussein MZ (2020) Synthesis and technology of nanoemulsion-based pesticide formulation. Nanomaterials (Basel, Switzerland) 10(8). https://doi.org/10.3390/nano10081608
132. Hakamy A, Shaikh FUA, Low IM (2015) Characteristics of nanoclay and calcined nanoclay-cement nanocomposites. Compos Part B: Eng 78:174–184
133. Saba N, Jawaid M, Alothman OY, Paridah MT, Hassan A (2016) Recent advances in epoxy resin, natural fiber-reinforced epoxy composites and their applications. J Reinf Plast Compos 35(6):447–470
134. Rodrigues M, Passos P (2013) Patterns of interpersonal coordination in rugby union: analysis of collective behaviours in a match situation. Adv Phys Educ 3(04):209
135. Rani A, Kumar A, Lal A, Pant M (2014) Cellular mechanisms of cadmium-induced toxicity: a review. Int J Environ Health Res 24(4):378–399
136. Mattos EV, Machado LA, Williams ER, Goodman SJ, Blakeslee RJ, Bailey JC (2017) Electrification life cycle of incipient thunderstorms. J Geophys Res: Atmos 122(8):4670–4697
137. Ma J, Chen D, Li Y, Chen Y, Liu Q, Zhou X, Qian K, Li Z, Ruan H, Hou Z, Zhu X (2018) Zinc phthalocyanine-soybean phospholipid complex based drug carrier for switchable photoacoustic/fluorescence image, multiphase photothermal/photodynamic treatment and synergetic therapy. J Control Release 284:1–14
138. Mohammadi M, Shaegh SAM, Alibolandi M, Ebrahimzadeh MH, Tamayol A, Jaafari MR, Ramezani M (2018) Micro and nanotechnologies for bone regeneration: recent advances and emerging designs. J Control Release 274:35–55
139. Kah M, Weniger AK, Hofmann T (2016) Impacts of (nano) formulations on the fate of an insecticide in soil and consequences for environmental exposure assessment. Environ Sci Technol 50(20):10960–10967
140. Chauhan N, Dilbaghi N, Gopal M, Kumar R, Kim KH, Kumar S (2017) Development of chitosan nanocapsules for the controlled release of hexaconazole. Int J Biol Macromol 97:616–624
141. Prasad R, Bhattacharyya A, Nguyen QD (2017) Nanotechnology in sustainable agriculture: recent developments, challenges, and perspectives. Front Microbiol 8:1014. https://doi.org/10.3389/fmicb.2017.01014
142. Sanzari I, Leone A, Ambrosone A (2019) Nanotechnology in plant science: to make a long story short. Front Bioeng Biotechnol 7:120. https://doi.org/10.3389/fbioe.2019.00120
143. Pérez-Iglesias JM, Franco-Belussi L, Natale GS, de Oliveira C (2019) Biomarkers at different levels of organisation after atrazine formulation (SIPTRAN 500SC(®)) exposure in Rhinella schineideri (Anura: Bufonidae) Neotropical tadpoles. Environ Pollut (Barking, Essex: 1987) 244:733–746. https://doi.org/10.1016/j.envpol.2018.10.073

144. de Oliveira CR, Fraceto LF, Rizzi GM, Salla RF, Abdalla FC, Costa MJ, Silva-Zacarin ECM (2016) Hepatic effects of the clomazone herbicide in both its free form and associated with chitosan-alginate nanoparticles in bullfrog tadpoles. Chemosphere 149:304–313. https://doi.org/10.1016/j.chemosphere.2016.01.076
145. Zielińska A, Costa B, Ferreira MV, Miguéis D, Louros JMS, Durazzo A, Lucarini M, Eder P, Chaud MV, Morsink M, Willemen N, Severino P, Santini A, Souto EB (2020) Nanotoxicology and nanosafety: safety-by-design and testing at a glance. Int J Environ Res Public Health 17(13). https://doi.org/10.3390/ijerph17134657
146. Syafrudin M, Kristanti RA, Yuniarto A, Hadibarata T, Rhee J, Al-Onazi WA, Algarni TS, Almarri AH, Al-Mohaimeed AM (2021) Pesticides in drinking water-a review. Int J Environ Res Public Health 18(2). https://doi.org/10.3390/ijerph18020468
147. Ferreira AJ, Cemlyn-Jones J, Robalo Cordeiro C (2013) Nanoparticles, nanotechnology and pulmonary nanotoxicology. Rev Port Pneumol 19(1):28–37. https://doi.org/10.1016/j.rppneu.2012.09.003
148. Kah M, Kookana RS, Gogos A, Bucheli TD (2018) A critical evaluation of nanopesticides and nanofertilizers against their conventional analogues. Nat Nanotechnol 13(8):677–684. https://doi.org/10.1038/s41565-018-0131-1
149. Baker S, Volova T, Prudnikova SV, Satish S, Prasad MN (2017) Nanoagroparticles emerging trends and future prospect in modern agriculture system. Environ Toxicol Pharmacol 53:10–17. https://doi.org/10.1016/j.etap.2017.04.012
150. Villaverde JJ, Sevilla-Morán B, López-Goti C, Alonso-Prados JL, Sandín-España P (2018) Considerations of nano-QSAR/QSPR models for nanopesticide risk assessment within the European legislative framework. Sci Total Environ 634:1530–1539. https://doi.org/10.1016/j.scitotenv.2018.04.033
151. Lade BD, Gogle DP, Lade DB, Moon GM, Nandeshwar SB, Kumbhare SD (2019) Nanobiopesticide formulations: application strategies today and future perspectives. In: Nanobiopesticides today and future perspectives. https://doi.org/10.1016/B978-0-12-815829-6.00007-3
152. Tisdell C, Wilson C, Tisdell C (2001) Why farmers continue to use pesticides despite environmental, health and sustainability costs. Ecol Econ 39:449–462. https://doi.org/10.1016/S0921-8009(01)00238-5
153. Tilman D, Cassman KG, Matson PA, Naylor R, Polasky S (2002) Agricultural sustainability and intensive production practices. Nature 418(6898):671–677. https://doi.org/10.1038/nature01014
154. Lamichhane JR, Dachbrodt-Saaydeh S, Kudsk P, Messéan A (2016) Toward a reduced reliance on conventional pesticides in European agriculture. Plant Dis 100(1):10–24. https://doi.org/10.1094/PDIS-05-15-0574-FE
155. Reganold JP, Glover JD, Andrews PK, Hinman HR (2001) Sustainability of three apple production systems. Nature 410(6831):926–930. https://doi.org/10.1038/35073574
156. Lechenet M, Bretagnolle V, Bockstaller C, Boissinot F, Petit M-S, Petit S, Munier-Jolain NM (2014) Reconciling pesticide reduction with economic and environmental sustainability in arable farming. PLoS ONE 9(6):e97922. https://doi.org/10.1371/journal.pone.0097922
157. Raliya R, Saharan V, Dimkpa C, Biswas P (2017) Nanofertilizer for precision and sustainable agriculture: current state and future perspectives. J Agric Food Chem 66(26):6487–6503
158. Wang P, Lombi E, Zhao FJ, Kopittke PM (2016) Nanotechnology: a new opportunity in plant sciences. Trends Plant Sci 21(8):699–712
159. Alghuthaymi MA, Kalia A, Bhardwaj K, Bhardwaj P, Abd-Elsalam KA, Valis M, Kuca K (2021) Nanohybrid antifungals for control of plant diseases: current status and future perspectives. J Fungi 7(1):48
160. Kalia A, Bhardwaj K, Bhardwaj P, Abd-Elsalam KA, Valis M, Kuca K (2021) Nanohybrid antifungals for control of plant diseases: current status and future perspectives. J. Fungi 2021(7):48
161. Thakur RK, Shirkot P (2017) Potential of biogold nanoparticles to control plant pathogenic nematodes. J Bioanal Biomed 9(4):220–222. https://doi.org/10.4172/1948-593X.1000182

162. Guzman M, Dile J, Godet S (2009) Synthesis of silver nanoparticles by chemical reduction method and their antibacterial activity. Int J Chem Biol Eng 2(3):104–111
163. Bergeson LL (2010) Nanosilver: US EPA's pesticide office considers how best to proceed. Envir Qual Manage 19(3):79–85
164. Oldenburg SJ (2017) Silver nanoparticles: properties and applications. Merck KGaA, Darmstadt, Germany and/or its affiliates. 14 pp https://www.sigmaaldrich.com/technical-documents/articles/materials-science/nanomaterials/silver-nanoparticles.html
165. El-Shazly MA, Attia YA, Kabil FF, Anis E, Hazman M (2017) Inhibitory effects of salicylic acid and silver nanoparticles on potato virus Y-infected potato plants in Egypt. Middle East J Agric Res 6:835–848
166. Park HJ, Kim SH, Kim HJ, Choi SH (2006) A new composition of nanosized silica-silver for control of various plant diseases. J Plant Pathol 22:295302
167. Jo YK, Kim BH, Jung G (2009) Antifungal activity of silver ions and nanoparticles on phytopathogenic fungi. Plant Dis 93(10):1037–1043
168. Taha RA (2016) Nanotechnology and its application in agriculture. Adv Plants Agricult Res 3(2):15406
169. Chhipa H (2017) Nanofertilizers and nanopesticides for agriculture. Environ Chem Lett 15(1):15–22
170. Kookana RS, Boxall AB, Reeves PT, Ashauer R, Beulke S, Chaudhry Q, Cornelis G, Fernandes TF, Gan J, Kah M, Lynch I (2014) Nanopesticides: guiding principles for regulatory evaluation of environmental risks. J Agric Food Chem 62(19):4227–4240
171. Iavicoli I, Leso V, Beezhold DH, Shvedova AA (2017) Nanotechnology in agriculture: opportunities, toxicological implications, and occupational risks. Toxicol Appl Pharmacol 329:96–111
172. Tong Y, Wu Y, Zhao C, Xu Y, Lu J, Xiang S, Zong F, Wu X (2017) Polymeric nanoparticles as a metolachlor carrier: water-based formulation for hydrophobic pesticides and absorption by plants. J Agric Food Chem 65(34):7371–7378
173. Elmer WH, White JC (2018) The future of nanotechnology in plant pathology. Ann Rev phytopathol 56:6.1–6.23
174. Martins CR, Lopes WA, Andrade JB (2013) Solubility of organic substances. Quim Nova 36:1248–1255
175. Giroto AS, Guimarães GG, Foschini M, Ribeiro C (2017) Role of slow-release nanocomposite fertilizers on nitrogen and phosphate availability in soil. Sci Rep 7:46032
176. Gevao B, Semple KT, Jones KC (2000) Bound pesticide residues in soils: a review. Environ Pollut 108:3–14
177. Hilz E, Veermer AWP (2013) Spray drift review: the extent to which a formulation can contribute to spray drift reduction. Crop Prot 44:75–83
178. Bolognesi C, Carrasquilla G, Volpi S, Solomon KR, Marshall EJ (2009) Biomonitoring of genotoxic risk in agricultural workers from five Colombian regions: association to occupational exposure to glyphosate. J Toxicol Environ Health A 72:986–997
179. Dantas-Torres F, Chomel BB, Otranto D (2012) Ticks and tick-borne diseases: a one health perspective. Trends Parasitol 28(10):437–446
180. Gonfa YH, Tessema FB, Bachheti A, Tadesse MG, Eid EM, Abou Fayssal S, Adelodun B, Choi KS, Širić I, Kumar P, Bachheti RK (2022) Essential oil composition of aerial part of *Pluchea ovalis* (Pers.) DC., silver nanoparticles synthesis, and larvicidal activities against fall armyworm. Sustainability 14(23):15785
181. Green JM, Beestman GB (2007) Recently patented and commercialized formulation and adjuvant technology. Crop Prot 26(3):320–327
182. Hao H, Cheng G, Iqbal Z, Ai X, Hussain HI, Huang L, Dai M, Wang Y, Liu Z, Yuan Z (2014) Benefits and risks of antimicrobial use in food-producing animals. Front Microbiol 5:288
183. Hong P-Y, Al-Jassim N, Ansari MI, Mackie RI (2013) Environmental and public health implications of water reuse: antibiotics, antibiotic resistant bacteria and antibiotic resistance genes. Antibiotics 2:367–399

184. Taghiyari HR, Morrell JJ, Husen A (2022) Emerging nanomaterials (*opportunities and challenges in forestry sectors*). Springer Nature Switzerland AG, Gewerbestrasse 11, 6330 Cham, Switzerland. https://doi.org/10.1007/978-3-031-17378-3
185. Ibrahim SS, Salem NY (2019) Insecticidal efficacy of nano zeolite against *Tribolium confusum* (Col., Tenebrionidae) and *Callosobruchus maculatus* (Col., Bruchidae). Bull Natl Res Centre 43(1):1–8
186. Jin-Chul K, Madhusudhan A, Husen A (2021) Smart nanomaterials in biomedical applications. Springer Nature Switzerland AG, Gewerbestrasse 11, 6330 Cham, Switzerland
187. Kah M, Walch H, Hofmann T (2018) Environmental fate of nanopesticides: durability, sorption and photodegradation of nanoformulated clothianidin. Environ Sci Nano 5(4):882–889. https://doi.org/10.1039/c8en00038g
188. Kashyap PL, Rai P, Sharma S, Chakdar H, Kumar S, Pandiyan K, Srivastava AK (2016) Nanotechnology for the detection and diagnosis of plant pathogens. In: Nanoscience in food and agriculture, vol 2. Springer, Cham, pp 253–276
189. Khodaverdi H, Fowles T, Bick E, Nansen C (2016) Does drought increase the risk of insects developing behavioral resistance to systemic insecticides? J Econ Entomol 109:2027–2031
190. Kim SW, Jung JH, Lamsal K, Kim YS, Min JS, Lee YS (2012) Antifungal effects of silver nanoparticles (AgNPs) against various plant pathogenic fungi. Mycobiology 40:53–58
191. Lechenet M, Dessaint F, Py G, Makowski D, Munier-Jolain N (2017) Reducing pesticide use while preserving crop productivity and profitability on arable farms. Nat Plants 3(3):1–6
192. Lowry GV, Avellan A, Gilbertson LM (2019) Opportunities and challenges for nanotechnology in the agri-tech revolution. Nat Nanotechnol 14(6):517–522. https://doi.org/10.1038/s41565-019-0461-7
193. Murugan K, Dinesh D, Jenil Kumar P, Panneerselvam C, Subramaniam J, Madhiyazhagan P, Suresh U, Nicoletti M, Alarfaj AA, Munusamy MA, Higuchi A, Mehlhorn H, Benelli G (2015) Datura metel-synthesized silver nanoparticles magnify predation of dragonfly nymphs against the malaria vector *Anopheles stephensi*. Parasitol Res 114:4645–4654
194. Nehra M, Dilbaghi N, Hassan AA, Kumar S (2019) Carbon-based nanomaterials for the development of sensitive nanosensor platforms. In: Deep A, Kumar S (eds) Advances in nanosensors for biological and environmental analysis. Elsevier, Amsterdam, pp 1–25
195. Raja MA, Ahmad MA, Daniyal M, Husen A (2023) Management of wastewater and other environmental issues using smart nanomaterials. In: Husen A, Siddiqi KS (Ed) Advances in smart nanomaterials and their applications. Elsevier Inc., 50 Hampshire St., 5th Floor, Cambridge, MA 02139, USA
196. Saiful Islam KBM, Shiraj-Um-Mahmuda S, Hazzaz-Bin-Kabir M (2016) Antibiotic usage patterns in selected broiler farms of Bangladesh and their public health implications. J Public Health Dev Ctries 2:276–284
197. Song B, Gong J, Tang W, Zeng G, Chen M, Xu P, Shen M, Ye S, Feng H, Zhou C, Yang Y (2020) Influence of multi-walled carbon nanotubes on the microbial biomass, enzyme activity, and bacterial community structure in 2, 4-dichlorophenol-contaminated sediment. Sci Total Environ 713:136645
198. Stadler T, López García GP, Gitto JG, Buteler M (2017) Nanostructured alumina: biocidal properties and mechanism of action of a novel insecticide powder
199. Tadros T, Izquierdo P, Esquena J, Solans C (2004) Formation and stability of nano-emulsions. Adv Coll Interface Sci 108:303–318
200. Tian B, Liu J, Liu Y, Wan JB (2022) Integrating diverse plant bioactive ingredients with cyclodextrins to fabricate functional films for food application: a critical review. Crit Rev Food Sci Nutr 1–30
201. You Y, Silbergeld EK (2014) Learning from agriculture: understanding low-dose antimicrobials as drivers of resistome expansion. Front Microbiol 5:284
202. Chaudhary BR, Sharma MD, Shakya SM, Gautam DM (2006) Effect of plant growth regulators on growth, yield and quality of chilli (Capsicum annuum L.) at Rampur, Chitwan. J Inst Agric Anim Sci 27:65–68

Current Applications and Future Perspectives of Nanotechnology for the Preservation and Enhancement of Grain and Seed Traits

Laura Vega-Fernández, Ricardo Quesada-Grosso, María Viñas, Andrea Irías-Mata, Gabriela Montes de Oca-Vásquez, Jose Vega-Baudrit, and Víctor M. Jiménez

Abstract Achieving food security worldwide is a major challenge because nearly 700 million people face hunger and more than 2 billion are affected by mineral and vitamin deficiencies (the so-called "hidden hunger"). Food insecurity can be addressed by implementing sustainable food systems to improve postharvest management of seeds (the propagation structures for the next crop generation) and grains (consumed as food or feed), to which nano-enabled technologies can greatly contribute. This chapter provides an updated description of the use of nanotechnology to improve seed and grain traits. First, the use of nano-priming and nano-coating techniques with different nanomaterials, the mechanisms involved, and their effects on seed germination and seedling growth are described. Furthermore, how to prevent and reduce mycotoxin contamination in grains, where some examples of nanoformulations are addressed (nanoparticles, nanocarriers, nanoliposomes, nanocapsules, nanoemulsions or nanoformulations, and nanoadsorbents) is also described. Under this scope, the challenges and future perspectives of this technology are also described, emphasizing the relevance of involving stakeholders (including consumer perception) and considering human health and environmental impacts.

Keywords Food safety · Food security · Mycotoxin mitigation · Nano-biofortification · Nano-enabled technologies

---

L. Vega-Fernández (✉) · R. Quesada-Grosso · M. Viñas · A. Irías-Mata · V. M. Jiménez
Centro para Investigaciones en Granos y Semillas, Universidad de Costa Rica, Ciudad Universitaria Rodrigo Facio, San Pedro de Montes de Oca, San José, Costa Rica
e-mail: laura.vegafernandez@ucr.ac.cr

G. Montes de Oca-Vásquez · J. Vega-Baudrit
Laboratorio Nacional de Alta Tecnología, Centro Nacional de Alta Tecnología, Pavas, San José, Costa Rica

V. M. Jiménez
Instituto de Investigaciones Agrícolas, Universidad de Costa Rica, Ciudad Universitaria Rodrigo Facio, San Pedro de Montes de Oca, San José, Costa Rica

R. K. Bachheti et al. (eds.), *Nanomaterials for Environmental and Agricultural Sectors*, Smart Nanomaterials Technology,
https://doi.org/10.1007/978-981-99-2874-3_10

1 Introduction

Agriculture is probably the most important source of food and raw materials and one of the essential segments of the world economy [1], with high vulnerability to different hazards, such as droughts, floods, reduction of biodiversity, armed conflicts, and environmental degradation [2, 3]. The challenge of achieving food security worldwide has become a relevant issue, with nearly 700 million people facing hunger and more than 2 billion people worldwide being affected by deficiencies in minerals and vitamins [4]. Therefore, increasing global food production by 50–60% is necessary to supply the demand of a projected global population of 9.8 billion by 2050 [5, 6].

Ensuring an abundant and safe food supply to promote people's health and well-being worldwide requires new social and economic policies and the application of new technologies. The latter will be needed to enable sustainable agricultural practices, achieve higher crop yields, and improve the nutritional value of foods. Moreover, these technologies must be accessible to the population and should have reduced adverse impacts on the environment and health [7].

Nanotechnology has received considerable attention as an option to overcome future challenges in agriculture and the food industry [8]. It has recently emerged as a technology that could transform the agri-food sector to increase the global production and the nutritional value of foods [9, 10]. However, the use of nanotechnology in the agricultural sector is still in the early stages, and the offer for nanoproducts in the market is still low [2, 11]. The diversity of nanomaterials (NMs) used in the agricultural and food industry is quite broad, including NMs based on metals (e.g., Ag, Au, Mg, Zn, ZnO, TiO_2, CuO, and Al_2O_3), polymers [e.g., chitosan (CS), alginate, polylactic acid (PLA), and polyethylene glycol], carbon structures [e.g., carbon dots, graphene, graphene oxides (GOs), single-walled carbon nanotubes, multi-walled carbon nanotubes, and fullerenes], lipids (e.g., nanoliposomes), and minerals (e.g., nano-clays) [1, 11]. These NMs can be applied directly to the soil, by foliar spraying on plants, or in seed treatments [12].

Seeds and grains are essential components of sustainable food systems. While the former constitutes the propagating structures for the next crop generation, grains are mainly used for food or feed. Both can benefit from nanotechnologies by resolving safety issues and enhancing nutritional properties. The use of nanotechnology to improve seed germination and seedling growth, boost grain yield and nutrition [13], and protect them from contaminating agents, such as mycotoxins, have been recently reviewed [14, 15]. However, despite current advances, additional research still needs to be conducted.

The following sections provide an up-to-date description of the use, impact, and applications of nanotechnology in seed and grain production, aiming at describing its use for the preservation and enhancement of seed and grain quality traits, including strategies and alternatives to reduce mycotoxin contamination of the food products (grains).

2 Nanotechnology Strategies for Improving Seed Germination and Grain Quality

Although nanotechnology has several applications in agriculture, its implementation for improving seed germination and grain quality is still under development. While some seeds germinate without delay after imbibition, either directly following harvesting or after a storage period, others enter a latency state that hampers or delays their germination. Thus, physical and chemical treatments must be employed to achieve quick and uniform germination and obtain healthy and vigorous plantlets. On the other side, some grains employed for human or animal nutrition may have suboptimal levels of some minerals or can be prone to accumulate mycotoxins after fungal infection. In both cases, nanotechnological approaches can overcome these disadvantageous situations [16, 17], so they will be described below.

2.1 *Use of Nano-priming and Nano-coating in Seeds and Grains*

The application of seed enhancement technologies (SETs) is a promising area. SETs were initially used in agricultural and horticultural sectors to improve seed handling and germination. However, recent progress has focused on seed-based ecological restoration, long-term conservation (including the conservation of plant genetic resources), quality and yield enhancement or biofortification [18–20]. The most used SETs are priming and coating (Fig. 1). Seed priming involves treating the seeds to regulate water (or solution) uptake before root protrusion occurs. Depending on the strategy, several priming options are available, such as hydro-priming, osmo-priming, hormonal-priming, nutri-priming, bio-priming, among others [16, 18, 21]. When the technique is applied, orthodox primed seeds can then be dried to their regular moisture level for usual handling (planting, packaging, and storage) [16]. Coating, on the other hand, consists of covering the surface of the seeds with specific compounds like micronutrients, soil adjuvants, germination promoters, growth regulators, and symbiotic microorganisms [19, 22].

2.1.1 Metallic Nanoparticles (MNPs) for Seed Nano-priming

MNPs can be categorized as inorganic metal and metal oxide nanoparticles (NPs). Commonly used inorganic metals NPs are aluminum (Al), copper (Cu), silver (Ag), gold (Au), iron (Fe), titanium (Ti), and zinc (Zn). In contrast, metal oxide NPs are the oxidized forms of these metals, such as Al_2O_3, CuO, TiO_2, ZnO, and Fe_3O_4 [23]. Although some of them are essential micronutrients for plant growth (Cu, Fe, and Zn), others, not so frequently used, such as boron (B), manganese (Mn), and molybdenum (Mo), would also need to be considered for adequate germination and

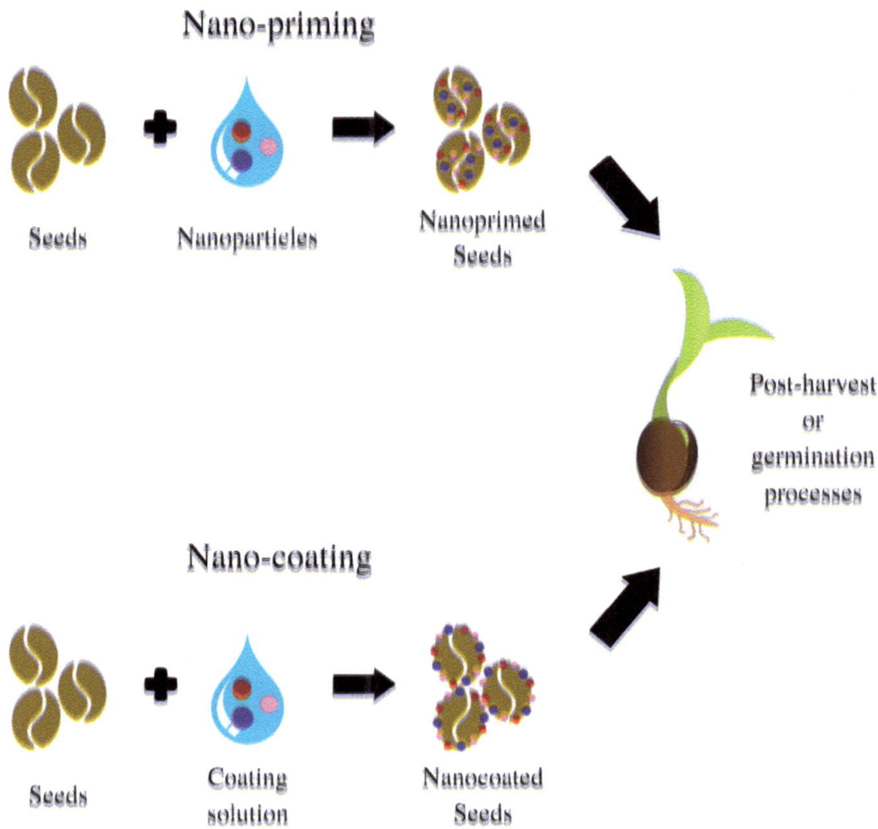

Fig. 1 Comparative representation of nano-priming and nano-coating techniques

plantlet growth. Below, several nano-priming examples with particularly relevant MNPs will illustrate current research tendencies and successful cases.

Itroutwar et al. [24–26] reported the use of Fe_2O_3 and ZnO NPs, synthesized via co-precipitation with the marine macroalgae *Turbinaria ornata*, as priming agents for seed quality improvement in rice (*Oryza sativa*) and maize (*Zea mays*). Priming both seeds with 25 mg/L of the seaweed-based biogenic Fe_2O_3 NPs, and rice seeds with 10 mg/L and maize seeds with 100 mg/L of the equivalent ZnO NPs, significantly enhanced seed germination and seedling parameters (root length, shoot length, seedling vigor, and dry matter production), compared to the hydro-primed control. In another work, ZnO NPs, synthesized by mixing citric acid with Zn acetate, were tested at different concentrations on seeds of rice and pea (*Vigna unguiculata* L.) (100, 500, and 1000 mg/L). While lower concentrations accelerated germination time (two-times faster) and root elongation in both plants, compared to hydro-primed control, higher concentrations showed phytotoxicity symptoms in these crops [27]. Moreover, a lower concentration of ZnO NPs (25 mg/L) used for priming rice seeds increased tolerance to water deficit and improved some plant agronomic traits [28].

Furthermore, TiO_2 NPs were synthesized by mixing titanium isopropoxide with an extract of *Trachyspermum ammi*, a phytochemical-rich herb plant, aiming at enhancing seed germination and seedling growth of mung bean (*Vigna radiata* Wilczek). The antioxidant activity of the NPs was assessed by the 2,2-diphenyl-1-picryl-hydrazyl-hydrate free radical scavenging test, and the results revealed the ability of the TiO_2 NPs to react with and stabilize free radicals, giving them good antioxidant properties. Six concentrations ranging from 25 to 250 mg/L were applied to the seeds, and germination, seedling root, and shoot length were evaluated in vitro (on germination paper) and in vivo (in soil). Results revealed an increase of 80% in the germination percentage of treated seeds compared to control (untreated seeds). Besides, seedling growth increased compared to non-treated control seeds. A dose-dependence was denoted, with better results at 50 mg/L and adverse effects at higher concentrations [29]. Pointing out to a species-specific and not always positive dose–response, Basahi [30] found 50 mg/L TiO_2 NPs to decrease pea (*Pisum sativum* L.) germination rate and mean daily germination, diminish the water absorption, increase the solute leakage, and induce the expression of oxidative stress biomarkers (malondialdehyde and reduced nicotinamide adenine dinucleotide) and antioxidant enzyme activities (guaiacol and catalase). The author hypothesized that TiO_2 NPs treatment triggered reactive oxygen species (ROS) accumulation, disrupting the processes of water and solutes transport and promoting the antioxidant system's activities.

As previously mentioned, besides exerting physiological effects on the seeds that may positively (or negatively) affect germination and plantlet growth, nano-priming, and nano-coating with MNPs can also influence the nutritional properties of treated plant structures (biofortification). Biofortification aims to increase, during plant growth, intrinsic nutrient levels (as opposed to conventional fortification) [16] and nano-biofortification could be a novel approach to enrich crops [17]. Most important biofortified crops include rice, wheat, maize, cassava, sweet potato, and some horticultural crops. Nutrients most relevant in biofortification strategies include B, Cu, Fe, iodine (I), calcium (Ca), selenium (Se), and Zn. Being Zn and Fe crucial nutrients for human health, but with low soil bioavailability because they bind organic and inorganic ligands or could be absorbed or transformed into unavailable forms, thus novel strategies should be used to make them more bioavailable [16, 17, 31]. Phosphorus accumulation in roots, shoots, and, most importantly, grains, was observed in rice plants treated with TiO_2 NPs at a concentration of 750 mg/kg [32].

However, it must also be considered that not all changes in the nutritional status of plants and their consumed parts should be positive after treatment with MNPs. In rice, the use of 500 mg of $n\text{CeO}_2$/kg in the soil reduced the concentration of Fe, S, prolamin, glutelin, lauric and valeric acids, and starch in grains after harvest, while increased the cerium concentration in the grains by 1126% [33], with the negative health implications that this may have [34]. In addition to the recent studies on the use of Fe- and Zn-based MNPs presented above, readers are recommended to look for additional examples, with other metals, in recent review papers (e.g., [23, 31, 35]).

2.1.2 (Bio)polymeric Nanocarriers and Nano-coatings

Active NPs differ from nanocarriers because the first ones cause the biological effects by themselves (such as the MNPs mentioned before). In contrast, nanocarriers are systems where the NPs (active or not) provide a mechanism for the extended release of a loaded active constituent (e.g., nutrients) over time [36]. (Bio)polymeric NPs can be used as delivery systems like nanocarriers or nano-coatings, as described below. According to Dutta et al. [37], choosing a particular nano-enabled carrier to deliver nutrients depends on its ability to hold many different compounds, have a suitable release rate, and minimize nutrient conversion to non-bioavailable forms. Even though there is relevant information on nanocarrier utilization in agriculture (as nano-fertilizers or nano-pesticides), there needs to be more understanding about using the nanocarriers applied directly to seeds.

Merinero et al. [38] reported the use of polyvinyl alcohol (PVA)-tannic acid (TA) NPs containing Fe NPs (PVA-TA-Fe NPs) for biofortification in wheat. The treatment with PVA-TA-Fe NPs showed an increase in the Fe concentration of 35% in the spikes and 215% in the plant compared to the control. Moreover, the Ca concentration increased by 30%. Although the biofortification effect is attributed to the Fe NPs, its association with PVA-TA NPs caused higher stability and lower toxicity. PVA-TA-Fe NPs also promoted the germination rate by 27%.

CS has also been used as a (bio)polymeric NM to enhance other seed traits. CS is produced from chitin, a natural structural compound in crustaceans, and has antimicrobial, antifungal, and antioxidant properties. CS is commonly used because of its high biocompatibility and biodegradability, low immunogenicity, and nontoxic properties [39]. As an example, nano-primed maize seeds with Cu–CS NPs favored germination and seedling growth by inducing α-amylase and protease activities and promoting reserve (starch) mobilization [40]. Moreover, CS-based nanocarriers of gibberellic acid improved initial seedling growth in tomatoes. They caused an increase in fruit yield under field conditions, up to 225.5% for CS/tripolyphosphate and 178.8% for alginate/CS NPs [41]. In addition to its role as a nanocarrier, CS can directly act as NP. Li et al. [42] observed that priming wheat seeds with 5 μg/mL CS NPs promoted seed germination and seedling length, and increased the number of adventitious roots associated with indoleacetic acid (auxin) metabolism. This effect was attributed to the relationship between high surface area and low volume compared to the commercial CS reagent.

Apart from nanocarriers, using (bio)polymers for nano-coating is also known to enhance different seed traits via the controlled and sustained release of compounds of interest [1]. Nano-coating uses nanofibers or nanofibrous mats (from a few nanometers to 500 nm in diameter) synthesized via electrospinning. These nano-coatings are superior in terms of water permeability, gaseous exchange, and lack of residual solvent to other options for seed coating. The seeds can be directly coated with the nanofibers or indirectly coated, wherein the nanofiber mats are prepared and then wrapped around the seeds [22, 43, 44]. This technique is more commonly used for seed preservation and protection [43–47], although in recent years, the use of nano-coating for agronomical enhancement and nutrient delivery has been reported as well [48–54].

2.1.3 Carbon-Based NMs

Like the options described above, carbon-based NMs have been proposed as seed quality enhancers. Nanocarbon can function as a plant growth booster. Safdar et al. [55] described multiple examples of nanocarbon being a coating material for adsorbing nitrogen from ammonia and releasing hydrogen ions, allowing plants to absorb more water and nutrients. Carbon nanotubes (CNTs) have been seen with particular interest because they can be uptaken and transported into the plant tissues, such as observed in tomato seed coats, leading to accelerated seed germination and seedling growth due to improved water uptake. Furthermore, other crops, such as mustard and rice, are being studied in this regard [55, 56].

Sobze et al. [57] summarized information showing that carbon NPs (CNTs and the carbon allotrope fullerene) can improve seed germination of native boreal forest species. Multi-walled CNTs (MWCNTs) can penetrate the seed coat and facilitate water absorption, enhancing seed germination and plant growth. This technology has been proved in the boreal species green alder (*Alnus viridis*), aspen (*Populus tremuloides*), and fireweed (*Chamerion angustifolium*). Aspen and fireweed seeds may lose viability quickly or develop into less vigorous seedlings if seeds are not treated appropriately after collection or during storage, and it was demonstrated that the treatment with CNTs is beneficial to improve germination [57]. In the case of green alder, MWCNTs functionalized with carboxylic acids could improve the germination of dormant bog birch (*Betula pumila* L.) and Labrador tea (*Rododendron groenlandicum* L.) seeds, using stratification as a complement [58].

2.2 Mechanisms Involved in the Nano-treatment of Seeds and Grains

2.2.1 Entry, Translocation, and Interaction of NPs in the Seeds

Nano-priming allows direct interaction of the NPs with the internal seed components (because the NPs enter the tissues). Penetration of NPs into plant structures has been studied in different models, and several mechanisms have been proposed (depending on the composition and size of the NPs, and the plant species and organ or tissue involved [59]). Nonetheless, although NPs of different compositions can be taken into the seeds, the mechanism needs to be better described, especially considering that the seed coat may represent a barrier interacting with the seed's surrounding environment. Evidence shows that rice and maize seeds absorb Zn and

Fe oxides NPs [24–26]. Some MNPs, such as TiO_2, produce holes during seed penetration, thus placing water and nutrients in contact with the embryo, which results in improved germination and seedling growth [29]. Ali et al. [58] observed that MWCNTs act as "nanoneedles" changing the cell membrane phospholipid bilayer's structure and consequently improving germination and vigor by remodeling the membrane lipidome.

Once inside the seed tissues, NPs can exert their effects at different levels (Fig. 2). Evidence shows that one of the mechanisms relies on the modification of gene expression. For example, air-sprayed MWCNTs can induce the expression of aquaporin-coding genes during the imbibition of barley, corn, and soybean seeds, thus increasing water uptake and diffusion of gases, nutrients, and ROS across membranes and then reducing the germination time [60]. Moreover, NPs can also boost the activity of enzymes working on the starch-degradation sites. In this regard, rice nano-primed seeds have increased soluble sugar contents and amylase activity, and modified internal osmotic potential, thus enhancing water uptake [61]. Further information supporting crosstalk between NPs and signaling pathways, nutrient homeostasis, and phytohormones, which could affect plant growth and metabolism associated with environmental stresses (biotic and abiotic), has been recently reviewed [62, 63].

2.2.2 Seed Revitalization and Rejuvenation

During storage, seeds age can deteriorate and progressively lose their vigor; they can also show delayed germination and poor seedling establishment or, in extreme cases, an inability to germinate. The accumulation of ROS and free radicals capable of oxidizing cellular macromolecules in the seed has been proposed as the primary factor negatively affecting seed longevity during storage [64, 65]. However, NPs have been shown to have the potential to be a seed quality enhancer since they can be applied to revitalize and revive aged seeds [23].

ROS production is necessary to trigger germination-related metabolic processes because ROS function as signaling molecules. Oxidative respiration is a metabolic process that up-regulates seed germination and vigor, and ROS are produced during oxidative respiration. Nano-priming influences the ROS homeostasis and antioxidant system in seeds, protecting them from oxidative damage, maintaining seed longevity, and promoting germination. NPs enhance the antioxidant activity of scavenging enzymes, controlling ROS levels in the seed. For instance, its application counteracts seed aging and promotes quality in plant growth, development, and stress resistance. These properties have been observed with MNPs and polymeric NPs; however, it is essential to highlight that the effect depends on the type of NP, crop, and growth environment [23, 66].

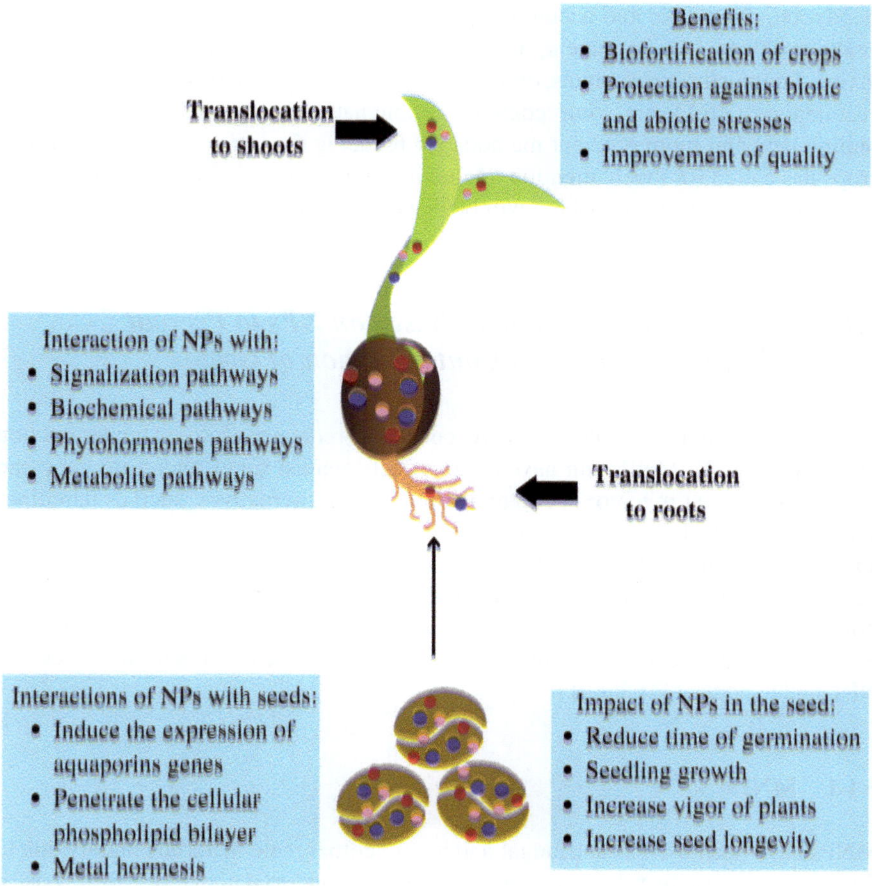

Fig. 2 Interactions of nanoparticles (NPs) with grains, seeds, and plant cycles

3 Use of Nanotechnology to Prevent and Reduce Mycotoxin Contamination in Grains

As mentioned before, NMs can boost the production of food with outstanding quality and high levels of nutrients and help protect plants from pests and diseases. However, to achieve the goal of zero hunger, it is also essential to think about the post-harvest processes, where more than 40% of food is being lost in different ways [10]. One cause of postharvest losses that is not regularly mentioned is mycotoxin contamination. Molds of genera such as *Aspergillus*, *Fusarium,* and *Penicillium* contaminate food products with secondary metabolites they synthesize. These compounds, called mycotoxins, produce global annual losses of nearly USD 930 million with a high impact on food crops globally (with a prevalence of up to 80%) [67]. Maize and wheat grains are two products that are prone to mycotoxin contamination [68].

Mycotoxins intake and accumulation in different human organs lead to serious health problems like immune system affectation, nephrotoxicity, and cancer. Aflatoxins, fumonisins, ochratoxins, deoxynivalenol (DON), trichothecenes, patulin, and zearalenone (ZEA), are some common problematic mycotoxins [67]. Therefore, finding new and more efficient methods for reducing mycotoxins in food and feed crops has called for much attention, including some nanotechnological applications developed in recent years, which will be discussed below.

3.1 Strategies and Alternatives Based on NPs to Prevent and Reduce Mycotoxin Contamination of Grains

There are several nano-strategies to prevent and/or reduce mycotoxin contamination of grains (Fig. 3). MNPs can have direct effects against fungi; therefore, they can be used to prevent mycotoxin contamination. Other alternatives, such as nanoliposomes, nanocapsules, and nanoemulsions, can carry compounds with antifungal and antimycotoxigenic properties, thus when applied to food, they can protect against fungi colonization and growth, reducing the subsequent production and accumulation of mycotoxins. For food already showing mycotoxin contamination, some nano adsorbents could be used to reduce adverse effects (e.g., carbon nanostructures, CS NPs, and nanoclays) [68, 69].

3.1.1 MNPs

MNPs have been used as antifungal and antimycotoxigenic materials mainly due to their capacity to affect the fungal metabolism [70] and to damage or cause lysis of fungal cells [71], either directly because of the metal toxicity or through induction of oxidative stress caused by the generation of ROS in the nanoparticle surface [72]. The most investigated NPs with antifungal effects are Ag and Cu, though other metal-based NPs have also promising antifungal properties, such as Zn, Fe, Se, Ni, and Pb [73–75]. Their effectiveness not only depends on the material (metal and oxidation state) but also on the particle size, with differences also associated with the target microorganism and plant species and structure involved (matrix) [72]. The antifungal effect of Cu and ZnO NPs against mycotoxigenic fungi, such as *Aspergillus flavus*, *Penicillium expansum*, and *Fusarium graminearum*, has been documented [76–79]. Recent research has placed ZnO NPs as one of the most promising antimicrobial NM based on their properties, such as excellent bioactivity, good biocompatibility, easy availability, high efficiency, and cost-effectiveness [80]. The mode of action of these NPs could be related to their photocatalytic properties, which generate oxidative stress leading to fungal cell damage, and in the absence of light, by the direct toxicity of the Zn^{+2} ions, which affect cell metabolism (Table 1), though the exact mechanism remains inconclusive [80].

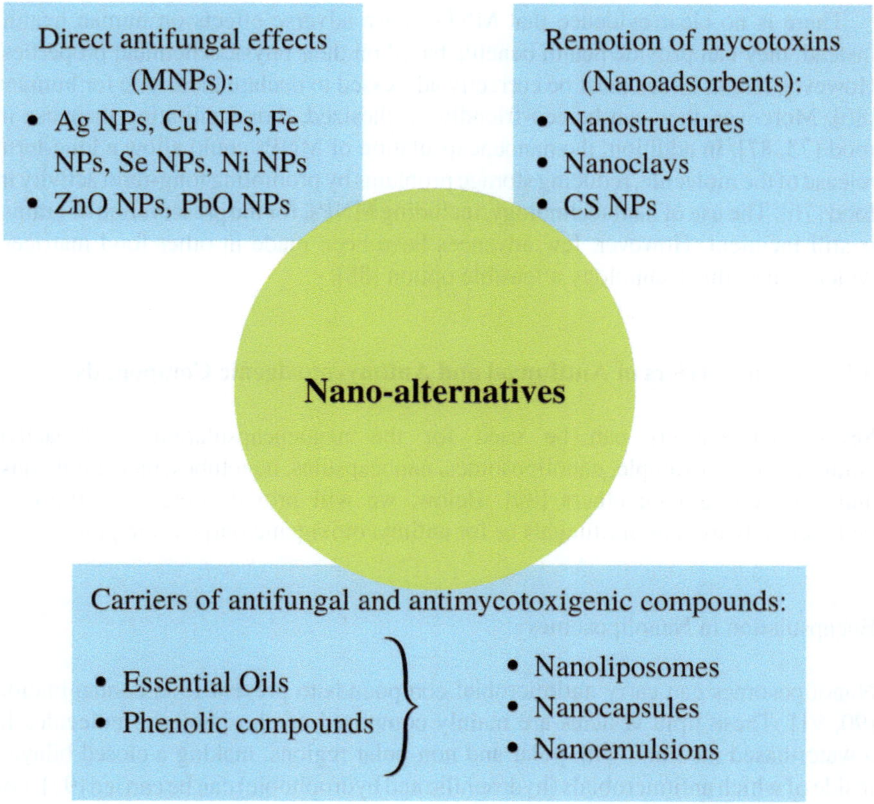

Fig. 3 Nano-alternatives to prevent and/or reduce mycotoxin contamination of grains

Table 1 Mode of action of nanoparticles with antifungal properties

Metal nanoparticle	Mode of action	References
Silver nanoparticles (Ag NPs)	– Alteration of the energy metabolism in the fungi	[81]
	– Disruption of cell walls and membranes in fungi	[82] [83]
	– Changes in gene transcription or enzyme profile in fungi	[70] [84]
Copper nanoparticles (Cu NPs)	– Ergosterol biosynthesis (fungal growth inhibition and changes in colony morphology)	[85]
Zinc oxide nanoparticles (ZnO NPs)	– Photocatalytic properties (ROS in the surface causing oxidative stress leading to fungal cell damage)	[80]
	– Direct toxicity of the Zn^{+2} ions (affect fungal cell metabolism)	

There is no clear evidence that MNPs cause adverse effects on human health; instead, they can provide health benefits based on their physicochemical properties. However, health issues must be correctly addressed to declare them safe for humans [86]. Moreover, they can be eco-friendly synthesized, thus facilitating their use in food [73, 87]. In addition, the nanoencapsulation of MNPs could allow a long-term release of the molecule, reducing storage problems by promoting long-term activity in food [76]. The use of nanotechnology, including MNPs, for the preservation of grains, is still incipient. However, few advances have been made in other food matrices, which makes this technology a feasible option [88].

3.1.2 Nanocarriers of Antifungal and Antimycotoxigenic Compounds

Several nanocarriers can be used for the nanoencapsulation of bioactive compounds, for example, nanoliposomes, nanocapsules, nanotubes, nanoemulsions, and nanogels, among others [89]. Below, we will provide details on those to be potentially used as antifungals or for antimycotoxigenic purposes in grains.

Encapsulation in Nanoliposomes

Nanoliposomes can carry antimicrobial compounds to prevent food contamination [90, 91]. These lipid vesicles are mainly composed of phospholipid molecules in a water-based medium, with polar and non-polar regions, making a closed bilayer, inside of which antimicrobials (hydrophilic and hydrophobic) can be carried [92]. For instance, turmeric extract loaded into nanoliposomes has more significant antimicrobial activity than the free extract [93]. Oregano and clove essential oils (EOs)-loaded nanoliposomes showed mycelial growth inhibition of 82 and 98%, respectively, against the fungus *Trichophyton rubrum* [94, 95]. Atienza et al. [96] demonstrated the antifungal properties of a nano-bio-fungicide composed of plant growth-promoting bacteria loaded into nanoliposomes against *F. oxysporum* f. sp. *cubense* Tropical Race 1.

Despite their application for antifungal purposes is very recent, and that there is no scientific research related to their use for mycotoxin prevention in grains, they are already being used in the food industry with other objectives (to improve texture, taste, and oral sensation of flavors) [97]. Besides, their synthesis using green approaches is making them more attractive for food industry applications. Moreover, they are non-toxic, flexible, biocompatible, biodegradable, and non-immunogenic. However, before they can be routinely used in the grain industry, issues associated with lipid oxidation during storage, which could represent a problem, must be addressed [97].

Encapsulation in Nanocapsules

Nanocapsule is the generic name for a nanoscale capsule containing liquids inside a polymeric membrane. It includes biodegradable and biocompatible polymers such as PLA and poly (DL-lactide-co-glycolide) acid (PLGA) [98]. Not much information about using PLA and PLGA nanocapsules for antimycotoxigenic purposes has been published. Most information relates to their potential as antimicrobial carriers for medical uses, mainly against bacteria. Some of the existing information can set the basis for preventing mycotoxins accumulation in food, including grains. For instance, PLA nanocapsules containing lemongrass EO were applied to apples after harvesting to control phytopathogenic fungi (*Colletotrichum acutatum* and *C. gloeosporioides*). These nanocapsules reduced the development of bitter rot lesions on apples and, more importantly, the effect lasted up to 10 days [99]. Curcumin encapsulated in PLGA NPs with the surface cationic surfactant cetyltrimethylammonium bromide showed fungicidal activity against *Pythium ultimum* var. *ultimum* [100].

Encapsulation in Nanoemulsions

Nanoemulsions are liquid–liquid dispersions that are kinetically stable over an extended time. Typically, they contain oil, water, and an emulsifier (generally a surfactant), and their size ranges from 20 to 500 nm [101]. They can protect some bioactive compounds with antifungal and antimycotoxigenic potential from degradation, thus extending foodstuff shelf life [89, 102]. The mode of action of some bioactive compounds against fungi relies on their adverse effects on enzyme activity, mainly on those related to the biosynthesis or integrity of the cell wall and membranes (including the mitochondrial) [102, 103]. Regarding their anti mycotoxigenic potential, this can be explained by their antioxidant activity, which reduces the fungal oxidative stress responses and mycotoxin production (in fungi, antioxidant molecules are induced together by some gene clusters related to mycotoxin biosynthesis [104]).

Moreover, a direct effect of some plant-derived bioactive compounds on the degradation of mycotoxin molecules has also been proposed [105–107]. Although research about the antifungal and antimycotoxigenic properties of nano encapsulated bioactive compounds is very recent, they have excellent potential in the food industry, while the authorization for their use should not be complex because they are already being used as food additives [108]. As mentioned below, relevant bioactive compounds with antifungal and antimycotoxigenic potential are EOs and phenolic compounds.

Nanoformulations with EOs

EOs are produced by the secondary metabolism of plants and are mainly composed of volatile terpene molecules. Around 3000 EOs have been identified, some with antimicrobial and antioxidant properties [109], making them useful as food additives. They are considered green preservatives in the agricultural and food industries, and many of them have been defined as Generally Recognized As Safe (GRAS) by the United States Food and Drug Administration [110]. Their hydrophobic nature makes them able to interact with fungal membranes, disturbing the structure and leading to cell death [111]. Nanoformulations of EOs can increase their stability to light, heat, pH, moisture, and oxygen, maintaining their antifungal and antimycotoxigenic properties.

Moreover, the negative impact of EOs supplementation on food organoleptic attributes can be countered with their nanoencapsulation [112]. The nanoencapsulation of EOs has been proven to be effective against toxigenic fungi such as *Aspergillus* [113] and *Fusarium* species [114], and more specifically, encapsulation of EOs in nanogels of CS-benzoic, -caffeic, and -cinnamic acids have enhanced antimicrobial activity against the important mycotoxigenic fungus *A. flavus* [115–117]. Regarding a direct effect on the synthesis of mycotoxins, the nanoencapsulation of cinnamaldehyde increases its activity against *A. flavus* and reduces aflatoxin production [118], while a nanoemulsion of EOs from *Origanum majorana* in CS NPs results in a promising antimycotoxigenic agent to prevent aflatoxin B1 (AFB1) contamination of maize samples [111]. Furthermore, nanoencapsulation of EOs and their use as a green preservative for fruits against fungal and mycotoxin deterioration was recently reviewed [119]. Figure 4 presents an overview of research related to EOs nanocarriers, some examples of the main sources of EOs, the mycotoxins and fungi species in which they exerted an in vitro effect, and the main types of nanocarriers and techniques for EO nanoencapsulation [67, 108, 120, 121].

Additional examples showing that nanoformulations increase the positive impact of EOs on reducing fungal infection and mycotoxin accumulation in grains have been recently published. In this regard, EOs nanoemulsions (thyme, lemongrass, cinnamon, peppermint, and clove) enhance specific DON inhibitory activities when applied to rice-based cultures [122]. In this research, the antifungal activity of thyme oil nanoemulsions against *F. graminearum* was particularly effective. Similarly, Wan et al. [123] showed that nanoemulsions significantly increased clove oil's antifungal and antimycotoxigenic potential, and Wan et al. [124] obtained similar results by using food-grade thyme oil nanoemulsions containing natural emulsifiers. Moreover, thymol nanoemulsions sprayed on wheat grains reduced the number of spikelets infected with *Fusarium* head blight [125]. Natural nanoemulsions of *Citrus reticulata* (mandarin) EOs mixed with cinnamon or clove EOs showed antifungal activity against four fungal species including the mycotoxigenic species *Aspergillus niger* [14]. Wan et al. [126] proved that applying clove-oil nanoemulsions during the barley malting process reduces *F. graminearum* growth and the levels of the mycotoxin DON. Moreover, clove oil nanoemulsions can also reduce mycotoxin (fumonisin B1 and B2) contamination during maize storage conditions [15]. Relevant results were published very recently, showing that, by simulating storage conditions, nanoemulsions of *Monarda didyma* and *Neopallasia pectinata* EOs are promising alternatives for protecting peanut grains against AFB1 [127].

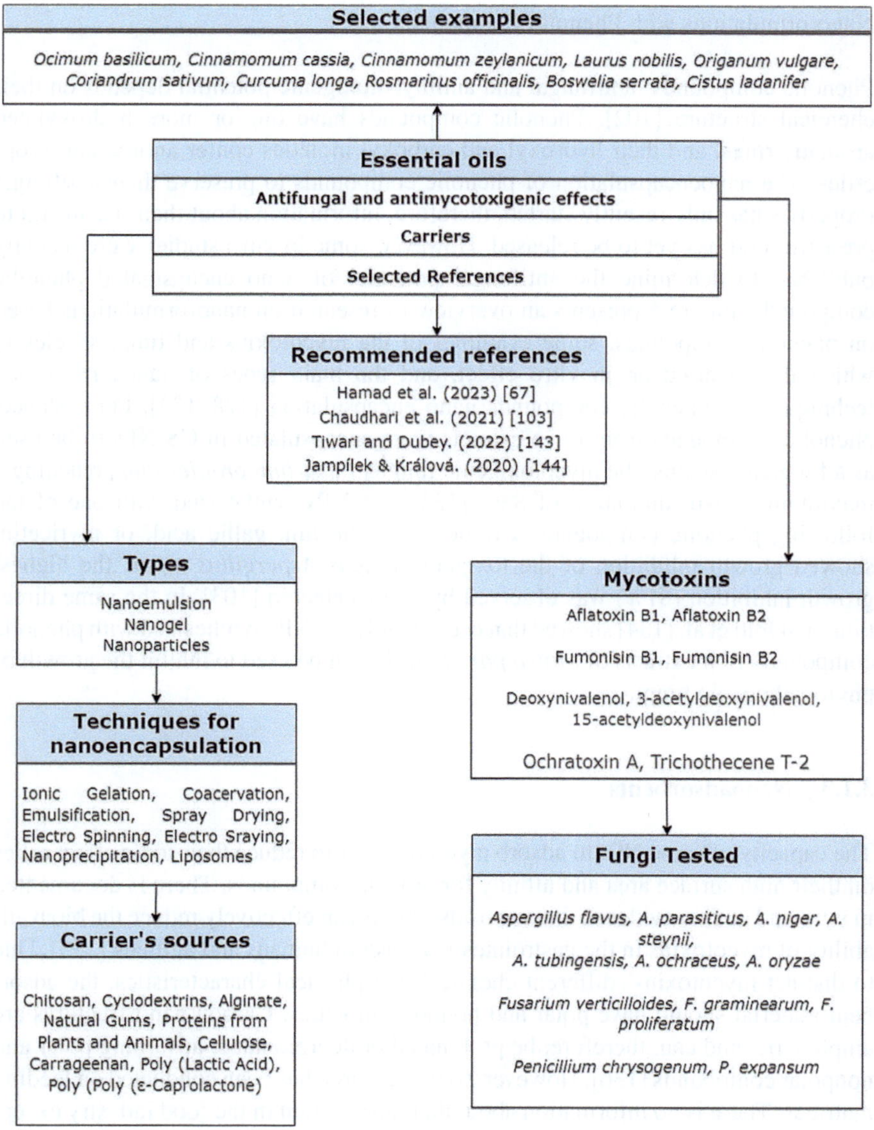

Fig. 4 Overview on research related to essential oil nanocarriers; some examples of the main sources of essential oils, the mycotoxins, and fungi species in which they exerted an in vitro effect and the main types of nanocarriers and techniques for essential oils nanoencapsulation

Nanoformulations with Phenolic Compounds

Phenolic compounds' antifungal and antimycotoxigenic potential depends on their chemical structure [102]. Phenolic compounds have one or more hydroxylated aromatic rings, and their hydroxyl and carboxyl moieties confer antioxidant properties. The nanoencapsulation of phenolic compounds to preserve their antifungal properties has only recently started, therefore, information about their use in situ to preserve food has yet to be released. However, some in vitro studies were recently published to determine the antifungal potential of nano encapsulated phenolic compounds. Figure 5 presents an overview of research on nanoformulations based on phenolic compounds, some examples of the mycotoxins and fungi species in which they exerted an in vitro effect, and the main types of nanocarriers and techniques for phenolic compounds nano encapsulation [128–132]. For instance, phenolic compounds from olive extracts were encapsulated in CS NPs to be used as a fungicide against the mycotoxigenic fungus *Fusarium proliferatum*, reaching a maximum growth inhibition of 88% [133]. Ag NPs synthesized with one of the following phenolic compounds: caffeic acid, catechin, gallic acid, or myricetin, showed growth inhibition of the toxigenic fungus *Aspergillus niger*; the highest growth inhibition (81%) was observed by using catechin [103]. In the same direction, Al-Obiti et al. [134] showed that eco-friendly Ag NPs synthesized with phenolic compounds-rich extracts of *Malva parviflora* L. can be used to inhibit the growth of phytopathogenic fungi.

3.1.3 Nanoadsorbents

The capacity of some NPs to adsorb mycotoxins or to reduce their toxic effect relies on their high surface area and affinity for organic compounds. There is documented in vitro and in vivo evidence that nano adsorbents can effectively reduce the bioavailability of mycotoxins in the gastrointestinal tract in humans and animals [135]. Due to distinct mycotoxins' different chemical and physical characteristics, the adsorbent material should have polar and nonpolar moieties. Carbon nanostructures are amphoteric, and can, therefore, be protonated or deprotonated, adsorbing polar and nonpolar compounds [136]. However, so far, research has been conducted on feeding matrices. There is no information about their application in the food industry except for their indirect application in food packaging [68]. One possible drawback of most nano adsorbents is that, based on their characteristics, they can also bind minerals and vitamins from the food, making them unavailable for human nutrition; thus, these issues will require further research. Below, we will describe some of the most known nano adsorbents as possible options to reduce mycotoxins contamination in grains.

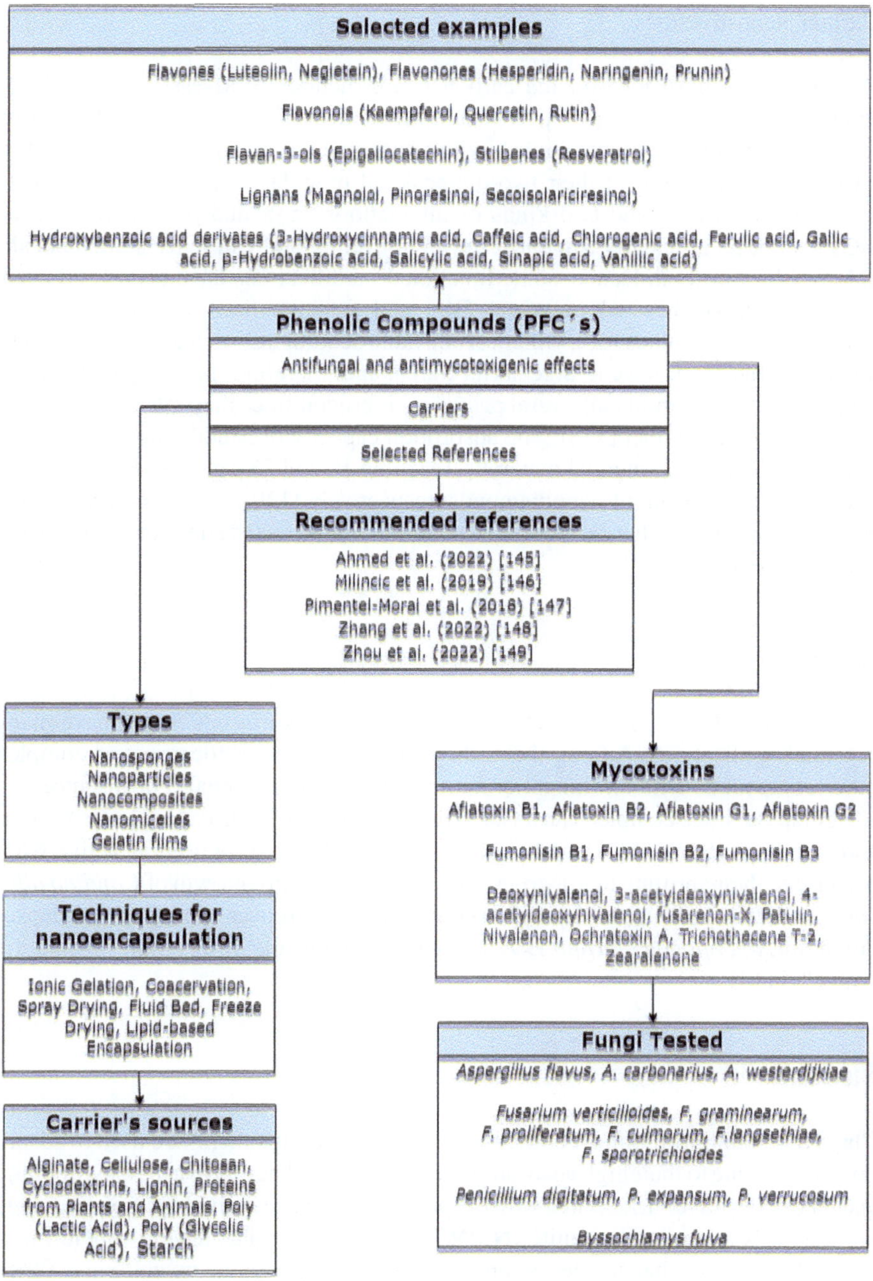

Fig. 5 Overview on research related to nanoformulations based on phenolic compounds, some examples of the mycotoxins and fungi species in which they exerted an effect and the main types of nanocarriers and techniques for phenolic compounds nanoencapsulation

Carbon Nanostructures

Recent studies have investigated carbon nanostructures to assess their feasibility in adsorbing mycotoxins [137]. GOs and reduced GOs (RGOs), characterized by their large surface area and high binding capacity for mycotoxins, are excellent adsorption materials with their two-dimensional monolayer of carbon atoms with an sp^2 hybridization and two kinds of interactions: π–π and p–π. However, the GOs cannot be efficiently separated during recycling and can leave some undesirable residues in food. Hence, some functional modifications have been conducted, for example, synthesized magnetic GOs and functionalized GO systems with amphiphilic compounds [68]. Other carbon nano adsorbents, such as nanodiamonds, are stable and inert, have a large surface area, and, therefore, have high adsorptive properties. Their stability at several pH values is crucial since they must pass through the gastrointestinal tract [138]. The adsorption capacity of nanodiamonds is greater than that of other commercial materials, reaching 10 and 25 μg of AFB1 and ochratoxin A (OTA) per mg of nanodiamonds, respectively [139]. CNTs, fullerenes, and activated carbon are other carbon-based materials that attracted attention as promising antifungal options [137].

CS NPs

As described above, due to functional groups on their surface, CS NPs have many potential applications. Among them, they can be used as nanocarriers to complex other compounds with antifungal or antimycotoxigenic potential. For instance, the CS-glutaraldehyde complex can adsorb some mycotoxins such as AFB1, ZEA, OTA, and fumonisin B1 [135]. In addition, CS NPs can be directly used as active NMs because of their antifungal properties, like inhibiting in vitro growth of *Candida albicans*, *Fusarium solani*, and *Aspergillus niger* [140]. Moreover, the CS nanoencapsulation of *Cymbopogon martinii* EOs stabilizes the molecule by a complex formation with CS [114].

Nanoclays

Clay NPs (montmorillonite, bentonite, and zeolite, among others) are used to adsorb mycotoxins due to their high adsorption capacity [141]. They have a high hydrophilic and electronegative surface area with the properties to be selective in cation exchange processes. Nanomontmorillonite (NMMT), for example, has an ion–dipole interaction between exchangeable cations and carbonyl groups in mycotoxins. When NMMT is used within compatible surfactants, some polar and weak polar mycotoxins have a better chance of being captured [68]. Investigations with NMMT showed a reduction of AFB1, ZEA, and DON by up to 97% [68]. Magnetic nano-zeolite was used in barley as a binder of several mycotoxins reducing >99% of aflatoxins, 50% of OTA, 22% of ZEA, and 1.8% of DON contents [142].

Nanoadsorbents Based on Inorganic Hybrid Composite and Organometallic Frameworks (MFOs)

The inorganic hybrid nano-adsorbents, like Surface Active Maghemite Nanoparticles, Mg–Al/Mg–Fe layered hydroxides, MgO-modified diatomite ceramic membrane, and hydroxyl magnesium silicate, are some examples of materials that were successfully tested to reduce the food contamination with mycotoxins [136, 143]. These are porous materials or coordination polymers, in which metallic ions are bonded to multi-dentate organic ligands to make a multidimensional crystalline network [144]. They have a good adsorption capacity because they have rich binding sites and can form hydrogen bonds with mycotoxins. Therefore, they perform significantly well in properties like accuracy, selectivity, and sensitivity. Particularly promising are the MFOs, because of their positive outcomes in adsorbing AFB1, probably through van der Waals forces, hydrogen bonds, and/or electrostatic interactions [68].

Despite the promising results, there have yet to be nano adsorbents in the market. Most of these materials still need to be tested, and information about the possible risks of their residues in food commodities still needs to be included. The high cost of large-scale application of these materials and the need for further optimization steps are also relevant disadvantages [68].

4 Challenges and Future Perspectives

Projections estimate that food demand will likely increase between 59 and 98% because the world population will reach 9 billion by 2050. Nanotechnology is a promising area to boost safe food production with increased quality and nutrient bioavailability. NMs can enhance food security by promoting better crops from germination to post-harvest processes, in which more than 40% of food losses occur [10]. It is necessary to standardize the design of processes and make them easily up scalable at industrial levels (e.g., type of NPs, dose range, etc.), following the principles of green chemistry and environmental sustainability, and demonstrating fundamental practical advances. Furthermore, it is necessary to achieve risk- and life-cycle assessment studies of NMs, to understand the dynamics of biostimulation, susceptibility, biosafety, and toxicity of NMs in the environment [145–147].

The number of scientific publications and commercial products related to the application of nanotechnology in agriculture and food security is increasing. Most research has been conducted on using NPs as nano-fertilizer and nano-pesticides, particularly with metallic and oxide metallic NPs. Nevertheless, other promising NMs, such as polymeric and carbon-based, have yet to be studied. For these reasons, it is crucial to conduct more studies on the use of NPs on seeds and grains to enhance productivity, food quality, and safety through enhanced germination and vigor biofortification and the prevention and reduction of mycotoxin contamination. This will have to be done without leaving aside potential associated risks, such as environmental contamination due to NPs (through the soil, water, and wind), and their possible adverse effects on human health, which are still poorly studied and less understood.

It is important to keep in mind that, among all nanotechnology applications, food and agriculture are among the most sensitive sectors to public scrutiny. Previous experiences with other emerging agrifood technologies (e.g., the first generation of genetically modified crops) demonstrated the importance of the responsible use of innovation, ensuring transparency and people's confidence. To achieve this, stakeholders must engage without forgetting social and ethical concerns, resulting in wider acceptance and openness of final consumers to NMs [148, 149].

5 Conclusions

Nanotechnology is one of the most promising alternatives to improve agriculture and food industries, in terms of increasing global food production and nutritional value. However, its use for enhancing grain and seed traits is still under development. This technology can improve seed germination and seedling growth and enhance grain yield and nutrition through the direct use of different NPs or nanocarriers to encapsulate biomolecules. Some techniques, such as nano-priming and nanocoating, have enhanced the postharvest quality of seeds. Additionally, research has been conducted on finding more adequate ways to protect seeds and grains from contaminating agents, such as fungi, with less information about their use to prevent and reduce mycotoxins. Metal and metal oxide NPs are the most studied to improve germination, seed quality, and nutritional value. More research needs to be done on using nanocarriers applied directly to seeds. Several nanocarriers can be used for the nanoencapsulation of bioactive compounds, such as EOs and polyphenols, with the potential to be used as antifungals or for antimycotoxigenic purposes. In addition, some NPs (such as carbon nanostructures, clay NPs, and nano adsorbents based on inorganic hybrid composite and organometallic frameworks) can adsorb mycotoxins or reduce their toxic effects due to their high surface area and their affinity for organic compounds. Finally, ensuring the human and environmental safety of the NPs and nanoformulations is of great importance without neglecting the final goal of offering better nutrition and food quality worldwide, considering the importance of engaging all stakeholders and including consumer acceptance of this technology.

Author Contributions **L.V.-F.**: Conceptualization, Methodology, Investigation, Writing—Original Draft, Writing—Review and Editing. **R.Q.-G**: Conceptualization, Methodology, Investigation, Writing—Original Draft, Writing—Review and Editing. **M.V.**: Conceptualization, Methodology, Investigation, Writing—Original Draft, Writing—Review and Editing. **A.I.-M.**: Conceptualization, Methodology, Investigation, Writing—Original Draft, Writing—Review and Editing. **G.MdO.-V.**: Conceptualization, Methodology, Investigation, Writing—Original Draft, Writing—Review and Editing. **J.V.-B.**: Review and Editing. **V.M.J.**: Conceptualization, Investigation, Writing—Review and Editing.

Funding This work was partially funded by the University of Costa Rica under the research project VI-734- C1-453.

Conflict of Interest The authors certify that they have no affiliations with or involvement in any organization or entity with any financial or non-financial interest in the subject matter or materials discussed in this chapter.

References

1. Ioannou A, Gohari G, Papaphilippou P, Panahirad S, Akbari A, Dadpour MR, Krasia-Christoforou T, Fotopoulos V (2020) Advanced nanomaterials in agriculture under a changing climate: the way to the future? Environ Exp Bot 176:104048. https://doi.org/10.1016/j.envexpbot.2020.104048
2. Ali SS, Al-Tohamy R, Koutra E, Moawad MS, Kornaros M, Mustafa AM, Mahmoud YA-G, Badr A, Osman MEH, Elsamahy T, Jiao H, Sun J (2021) Nanobiotechnological advancements in agriculture and food industry: applications, nanotoxicity, and future perspectives. Sci Total Environ 792:148359. https://doi.org/10.1016/j.scitotenv.2021.148359
3. FAO, FIDA, OMS, PMA, UNICEF (2022) Versión resumida de El estado de la seguridad alimentaria y la nutrición en el mundo 2022: Adaptación de las políticas alimentarias y agrícolas para hacer las dietas saludables más asequibles. FAO, Rome, Italy
4. World Health Organization, Food and Agriculture Organization of the United Nations (2018) The nutrition challenge: food system solutions. https://apps.who.int/iris/handle/10665/277440. Accessed 18 Nov 2022
5. Castro-Mayorga JL, Cabrera-Villamizar L, Balcucho-Escalante J, Fabra MJ, López-Rubio A (2020) Applications of nanotechnology in agry-food productions. In: Rajendran S, Mukherjee A, Nguyen TA, Godugu C, Shukla RK (eds) Nanotoxicity. Elsevier, pp 319–340
6. Falcon WP, Naylor RL, Shankar ND (2022) Rethinking global food demand for 2050. Popul Dev Rev 12508. https://doi.org/10.1111/padr.12508
7. Kagan CR (2016) At the nexus of food security and safety: opportunities for nanoscience and nanotechnology. ACS Nano 10:2985–2986. https://doi.org/10.1021/acsnano.6b01483
8. Majumdar S, Keller AA (2021) Omics to address the opportunities and challenges of nanotechnology in agriculture. Crit Rev Environ Sci Technol 51:2595–2636. https://doi.org/10.1080/10643389.2020.1785264
9. Hofmann T, Lowry GV, Ghoshal S, Tufenkji N, Brambilla D, Dutcher JR, Gilbertson LM, Giraldo JP, Kinsella JM, Landry MP, Lovell W, Naccache R, Paret M, Pedersen JA, Unrine JM, White JC, Wilkinson KJ (2020) Technology readiness and overcoming barriers to sustainably implement nanotechnology-enabled plant agriculture. Nat Food 1:416–425. https://doi.org/10.1038/s43016-020-0110-1
10. Neme K, Nafady A, Uddin S, Tola YB (2021) Application of nanotechnology in agriculture, postharvest loss reduction and food processing: food security implication and challenges. Heliyon 7:e08539. https://doi.org/10.1016/j.heliyon.2021.e08539
11. Acharya A, Pal PK (2020) Agriculture nanotechnology: translating research outcome to field applications by influencing environmental sustainability. NanoImpact 19:100232. https://doi.org/10.1016/j.impact.2020.100232
12. Vijayakumar MD, Surendhar GJ, Natrayan L, Patil PP, Ram PMB, Paramasivam P (2022) Evolution and recent scenario of nanotechnology in agriculture and food industries. J Nanomater 2022:1280411. https://doi.org/10.1155/2022/1280411

13. Nile SH, Thiruvengadam M, Wang Y, Samynathan R, Shariati MA, Rebezov M, Nile A, Sun M, Venkidasamy B, Xiao J, Kai G (2022) Nano-priming as emerging seed priming technology for sustainable agriculture—recent developments and future perspectives. J Nanobiotechnol 20:254. https://doi.org/10.1186/s12951-022-01423-8
14. Mahdi AA, Al-Maqtari QA, Mohammed JK, Al-Ansi W, Cui H, Lin L (2021) Enhancement of antioxidant activity, antifungal activity, and oxidation stability of *Citrus reticulata* essential oil nanocapsules by clove and cinnamon essential oils. Food Biosci 43:101226. https://doi.org/10.1016/j.fbio.2021.101226
15. Singh P, Dasgupta N, Singh V, Chandra Mishra N, Singh H, Purohit SD, Srivastava N, Ranjan S, Yadav NP, Mishra BN (2020) Inhibitory effect of clove oil nanoemulsion on fumonisin isolated from maize kernels. LWT Food Sci Technol 134:110237. https://doi.org/10.1016/j.lwt.2020.110237
16. De La Torre-Roche R, Cantu J, Tamez C, Zuverza-Mena N, Hamdi H, Adisa IO, Elmer W, Gardea-Torresdey J, White JC (2020) Seed biofortification by engineered nanomaterials: a pathway to alleviate malnutrition? J Agric Food Chem 68:12189–12202. https://doi.org/10.1021/acs.jafc.0c04881
17. El-Ramady H, Abdalla N, Elbasiouny H, Elbehiry F, Elsakhawy T, Omara AE-D, Amer M, Bayoumi Y, Shalaby TA, Eid Y, Zia-ur-Rehman M (2021) Nano-biofortification of different crops to immune against COVID-19: a review. Ecotoxicol Environ Saf 222:112500. https://doi.org/10.1016/j.ecoenv.2021.112500
18. Brown VS, Erickson TE, Merritt DJ, Madsen MD, Hobbs RJ, Ritchie AL (2021) A global review of seed enhancement technology use to inform improved applications in restoration. Sci Total Environ 798:149096. https://doi.org/10.1016/j.scitotenv.2021.149096
19. Javed T, Afzal I, Shabbir R, Ikram K, Saqlain Zaheer M, Faheem M, Haider Ali H, Iqbal J (2022) Seed coating technology: an innovative and sustainable approach for improving seed quality and crop performance. J Saudi Soc Agric Sci S1658077X22000273. https://doi.org/10.1016/j.jssas.2022.03.003
20. Rao NK, Dulloo ME, Engels JMM (2017) A review of factors that influence the production of quality seed for long-term conservation in genebanks. Genet Resour Crop Evol 64:1061–1074. https://doi.org/10.1007/s10722-016-0425-9
21. Shelar A, Singh AV, Maharjan RS, Laux P, Luch A, Gemmati D, Tisato V, Singh SP, Santilli MF, Shelar A, Chaskar M, Patil R (2021) Sustainable agriculture through multidisciplinary seed nanopriming: prospects of opportunities and challenges. Cells 10:2428. https://doi.org/10.3390/cells10092428
22. Sohail M, Pirzada T, Opperman CH, Khan SA (2022) Recent advances in seed coating technologies: transitioning toward sustainable agriculture. Green Chem 24:6052–6085. https://doi.org/10.1039/D2GC02389J
23. Kaur R, Chandra J, Keshavkant S (2021) Nanotechnology: an efficient approach for rejuvenation of aged seeds. Physiol Mol Biol Plants 27:399–415. https://doi.org/10.1007/s12298-021-00942-2
24. Itroutwar PD, Govindaraju K, Tamilselvan S, Kannan M, Raja K, Subramanian KS (2020) Seaweed-based biogenic ZnO nanoparticles for improving agro-morphological characteristics of rice (*Oryza sativa* L.). J Plant Growth Regul 39:717–728. https://doi.org/10.1007/s00344-019-10012-3
25. Itroutwar PD, Govindaraju K, Tamilselvan S, Kannan M, Vasantharaja R, Chaturvedi S, Shkolnik D (2021) Influence of nanoscale micro-nutrient α-Fe_2O_3 on seed germination, seedling growth, translocation, physiological effects and yield of rice (*Oryza sativa*) and maize (*Zea mays*). Plant Physiol Biochem 162:564–580. https://doi.org/10.1016/j.plaphy.2021.03.023
26. Itroutwar PD, Kasivelu G, Raguraman V, Malaichamy K, Sevathapandian SK (2020) Effects of biogenic zinc oxide nanoparticles on seed germination and seedling vigor of maize (*Zea mays*). Biocatal Agric Biotechnol 29:101778. https://doi.org/10.1016/j.bcab.2020.101778
27. Cyriac J, Melethil K, Thomas B, Sreejit M, Varghese T (2020) Synthesis of biogenic ZnO nanoparticles and its impact on seed germination and root growth of *Oryza sativa* L. and *Vigna unguiculata* L. Mater Today Proc 25:224–229. https://doi.org/10.1016/j.matpr.2020.01.107

28. Mazhar MW, Ishtiaq M, Hussain I, Parveen A, Bhatti K, Azeem M, Thind S, Ajaib M, Maqbool M, Sardar T, Muzammil K, Nasir N (2022) Seed nano-priming with zinc oxide nanoparticles in rice mitigates drought and enhances agronomic profile. PLoS One 17:e0264967. https://doi.org/10.1371/journal.pone.0264967
29. Sunny NE, Mathew SS, Venkat Kumar S, Saravanan P, Rajeshkannan R, Rajasimman M, Vasseghian Y (2022) Effect of green synthesized nano-titanium synthesized from *Trachyspermum ammi* extract on seed germination of *Vigna radiate*. Chemosphere 300:134600. https://doi.org/10.1016/j.chemosphere.2022.134600
30. Basahi M (2021) Seed germination with titanium dioxide nanoparticles enhances water supply, reserve mobilization, oxidative stress and antioxidant enzyme activities in pea. Saudi J Biol Sci 28:6500–6507. https://doi.org/10.1016/j.sjbs.2021.07.023
31. Kapoor P, Dhaka RK, Sihag P, Mehla S, Sagwal V, Singh Y, Langaya S, Balyan P, Singh KP, Xing B, White JC, Dhankher OP, Kumar U (2022) Nanotechnology-enabled biofortification strategies for micronutrients enrichment of food crops: current understanding and future scope. NanoImpact 26:100407. https://doi.org/10.1016/j.impact.2022.100407
32. Zahra Z, Waseem N, Zahra R, Lee H, Badshah MA, Mehmood A, Choi H-K, Arshad M (2017) Growth and metabolic responses of rice (*Oryza sativa* L.) cultivated in phosphorus-deficient soil amended with TiO_2 nanoparticles. J Agric Food Chem 65:5598–5606. https://doi.org/10.1021/acs.jafc.7b01843
33. Rico CM, Morales MI, Barrios AC, McCreary R, Hong J, Lee W-Y, Nunez J, Peralta-Videa JR, Gardea-Torresdey JL (2013) Effect of cerium oxide nanoparticles on the quality of rice (*Oryza sativa* L.) grains. J Agric Food Chem 61:11278–11285. https://doi.org/10.1021/jf404046v
34. Gómez-Aracena J, Riemersma RA, Gutiérrez-Bedmar M, Bode P, Kark JD, Garcia-Rodríguez A, Gorgojo L, van't Veer P, Fernández-Crehuet J, Kok FJ, Martin-Moreno JM (2006) Toenail cerium levels and risk of a first acute myocardial infarction: the EURAMIC and heavy metals study. Chemosphere 64:112–120. https://doi.org/10.1016/j.chemosphere.2005.10.062
35. Khan MK, Pandey A, Hamurcu M, Gezgin S, Athar T, Rajput VD, Gupta OP, Minkina T (2021) Insight into the prospects for nanotechnology in wheat biofortification. Biology 10:1123. https://doi.org/10.3390/biology10111123
36. Pereira A do ES, Oliveira H, Fraceto LF, Santaella C (2021) Nanotechnology potential in seed priming for sustainable agriculture. Nanomaterials 11:267. https://doi.org/10.3390/nano11020267
37. Dutta S, Pal S, Panwar P, Sharma RK, Bhutia PL (2022) Biopolymeric nanocarriers for nutrient delivery and crop biofortification. ACS Omega 7:25909–25920. https://doi.org/10.1021/acsomega.2c02494
38. Merinero M, Alcudia A, Begines B, Martínez G, Martín-Valero MJ, Pérez-Romero JA, Mateos-Naranjo E, Redondo-Gómez S, Navarro-Torre S, Torres Y, Merchán F, Rodríguez-Llorente ID, Pajuelo E (2022) Assessing the biofortification of wheat plants by combining a plant growth-promoting rhizobacterium (PGPR) and polymeric Fe-nanoparticles: allies or enemies? Agronomy 12:228.https://doi.org/10.3390/agronomy12010228
39. Divya K, Jisha MS (2018) Chitosan nanoparticles preparation and applications. Environ Chem Lett 16:101–112. https://doi.org/10.1007/s10311-017-0670-y
40. Saharan V, Kumaraswamy RV, Choudhary RC, Kumari S, Pal A, Raliya R, Biswas P (2016) Cu-chitosan nanoparticle mediated sustainable approach to enhance seedling growth in maize by mobilizing reserved food. J Agric Food Chem 64:6148–6155. https://doi.org/10.1021/acs.jafc.6b02239
41. do Pereira AES, Oliveira HC, Fraceto LF (2019) Polymeric nanoparticles as an alternative for application of gibberellic acid in sustainable agriculture: a field study. Sci Rep 9:7135. https://doi.org/10.1038/s41598-019-43494-y
42. Li R, He J, Xie H, Wang W, Bose SK, Sun Y, Hu J, Yin H (2019) Effects of chitosan nanoparticles on seed germination and seedling growth of wheat (*Triticum aestivum* L.). Int J Biol Macromol 126:91–100. https://doi.org/10.1016/j.ijbiomac.2018.12.118

43. Farias BV, Pirzada T, Mathew R, Sit TL, Opperman C, Khan SA (2019) Electrospun polymer nanofibers as seed coatings for crop protection. ACS Sustain Chem Eng 7:19848–19856. https://doi.org/10.1021/acssuschemeng.9b05200
44. Xu T, Ma C, Aytac Z, Hu X, Ng KW, White JC, Demokritou P (2020) Enhancing agrichemical delivery and seedling development with biodegradable, tunable, biopolymer-based nanofiber seed coatings. ACS Sustain Chem Eng 8:9537–9548. https://doi.org/10.1021/acssuschemeng.0c02696
45. Adak T, Kumar J, Shakil NA, Pandey S (2016) Role of nano-range amphiphilic polymers in seed quality enhancement of soybean and imidacloprid retention capacity on seed coatings. J Sci Food Agric 96:4351–4357. https://doi.org/10.1002/jsfa.7643
46. de Castro e Silva P, Pereira LAS, Lago AMT, Valquíria M, de Rezende ÉM, Carvalho GR, Oliveira JE, Marconcini JM (2019) Physical-mechanical and antifungal properties of pectin nanocomposites/neem oil nanoemulsion for seed coating. Food Biophys 14:456–466. https://doi.org/10.1007/s11483-019-09592-0
47. Tian F, Chen W, Wu C, Kou X, Fan G, Li T, Wu Z (2019) Preservation of *Ginkgo biloba* seeds by coating with chitosan/nano-TiO_2 and chitosan/nano-SiO_2 films. Int J Biol Macromol 126:917–925. https://doi.org/10.1016/j.ijbiomac.2018.12.177
48. Asim M, Ahmad W, Qamar Z, Awais M, Nepal J, Ahmad I (2022) Seed coating with zinc oxide nanofiber (ZnONF) and urea improved zinc uptake; recovery efficiency, growth, and yield of bread wheat (*Triticum aestivum* L.). J Soil Sci Plant Nutr. https://doi.org/10.1007/s42729-022-00978-7
49. Baddar ZE, Unrine JM (2018) Functionalized-ZnO-nanoparticle seed treatments to enhance growth and Zn content of wheat (*Triticum aestivum*) seedlings. J Agric Food Chem 66:12166–12178. https://doi.org/10.1021/acs.jafc.8b03277
50. Krishnamoorthy V, Rajiv S (2018) Tailoring electrospun polymer blend carriers for nutrient delivery in seed coating for sustainable agriculture. J Clean Prod 177:69–78. https://doi.org/10.1016/j.jclepro.2017.12.141
51. Mohanraj J, Subramanian KS, Raja K (2022) Effect of multinutrients loaded electrospun PVA nanofibre on germination and its growth parameters of green gram [*Vigna radiata* (L.) Wilczek]. Legume Res 45:587–593
52. Montanha GS, Rodrigues ES, Marques JPR, de Almeida E, Colzato M, de Carvalho HWP (2020) Zinc nanocoated seeds: an alternative to boost soybean seed germination and seedling development. SN Appl Sci 2:857. https://doi.org/10.1007/s42452-020-2630-6
53. Tondey M, Kalia A, Singh A, Dheri GS, Taggar MS, Nepovimova E, Krejcar O, Kuca K (2021) Seed priming and coating by nano-scale zinc oxide particles improved vegetative growth, yield and quality of fodder maize (*Zea mays*). Agronomy 11:729. https://doi.org/10.3390/agronomy11040729
54. Zhao W, Liu Y, Zhang P, Zhou P, Wu Z, Lou B, Jiang Y, Shakoor N, Li M, Li Y, Lynch I, Rui Y, Tan Z (2022) Engineered Zn-based nano-pesticides as an opportunity for treatment of phytopathogens in agriculture. NanoImpact 28:100420. https://doi.org/10.1016/j.impact.2022.100420
55. Safdar M, Kim W, Park S, Gwon Y, Kim Y-O, Kim J (2022) Engineering plants with carbon nanotubes: a sustainable agriculture approach. J Nanobiotechnol 20:275. https://doi.org/10.1186/s12951-022-01483-w
56. Wang Y, Deng C, Rawat S, Cota-Ruiz K, Medina-Velo I, Gardea-Torresdey JL (2021) Evaluation of the effects of nanomaterials on rice (*Oryza sativa* L.) responses: underlining the benefits of nanotechnology for agricultural applications. ACS Agric Sci Technol 1:44–54. https://doi.org/10.1021/acsagscitech.1c00030
57. Sobze J-M, Galagedara L, Cheema M, Thomas R, Inoue S (2022) The potential of carbon nanoparticles as a stimulant to improve the propagation of native boreal forest species: a mini-review. Front For Glob Change 5:872780. https://doi.org/10.3389/ffgc.2022.872780
58. Ali MdH, Sobze J-M, Pham TH, Nadeem M, Liu C, Galagedara L, Cheema M, Thomas R (2020) Carbon nanotubes improved the germination and vigor of plant species from peatland ecosystem via remodeling the membrane lipidome. Nanomaterials 10:1852. https://doi.org/10.3390/nano10091852

59. Faraz A, Faizan M, Sami F, Siddiqui H, Pichtel J, Hayat S (2019) Nanoparticles: biosynthesis, translocation and role in plant metabolism. IET Nanobiotechnol 13:345–352. https://doi.org/10.1049/iet-nbt.2018.5251
60. Lahiani MH, Dervishi E, Chen J, Nima Z, Gaume A, Biris AS, Khodakovskaya MV (2013) Impact of carbon nanotube exposure to seeds of valuable crops. ACS Appl Mater Interfaces 5:7965–7973. https://doi.org/10.1021/am402052x
61. Mahakham W, Sarmah AK, Maensiri S, Theerakulpisut P (2017) Nanopriming technology for enhancing germination and starch metabolism of aged rice seeds using phytosynthesized silver nanoparticles. Sci Rep 7:8263. https://doi.org/10.1038/s41598-017-08669-5
62. Banerjee A, Roychoudhury A (2022) Explicating the cross-talks between nanoparticles, signaling pathways and nutrient homeostasis during environmental stresses and xenobiotic toxicity for sustainable cultivation of cereals. Chemosphere 286:131827. https://doi.org/10.1016/j.chemosphere.2021.131827
63. Tripathi D, Singh M, Pandey-Rai S (2022) Crosstalk of nanoparticles and phytohormones regulate plant growth and metabolism under abiotic and biotic stress. Plant Stress 6:100107. https://doi.org/10.1016/j.stress.2022.100107
64. Bailly C (2004) Active oxygen species and antioxidants in seed biology. Seed Sci Res 14:93–107. https://doi.org/10.1079/SSR2004159
65. Sano N, Rajjou L, North HM, Debeaujon I, Marion-Poll A, Seo M (2016) Staying alive: molecular aspects of seed longevity. Plant Cell Physiol 57:660–674. https://doi.org/10.1093/pcp/pcv186
66. Kandhol N, Singh VP, Ramawat N, Prasad R, Chauhan DK, Sharma S, Grillo R, Sahi S, Peralta-Videa J, Tripathi DK (2022) Nano-priming: impression on the beginner of plant life. Plant Stress 5:100091. https://doi.org/10.1016/j.stress.2022.100091
67. Hamad GM, Mehany T, Simal-Gandara J, Abou-Alella S, Esua OJ, Abdel-Wahhab MA, Hafez EE (2023) A review of recent innovative strategies for controlling mycotoxins in foods. Food Control 144:109350. https://doi.org/10.1016/j.foodcont.2022.109350
68. Song C, Qin J (2022) High-performance fabricated nano-adsorbents as emerging approach for removal of mycotoxins: a review. Int J Food Sci Technol 57:5781–5789. https://doi.org/10.1111/ijfs.15953
69. Singh BK, Tiwari S, Dubey NK (2021) Essential oils and their nanoformulations as green preservatives to boost food safety against mycotoxin contamination of food commodities: a review. J Sci Food Agric 101:4879–4890. https://doi.org/10.1002/jsfa.11255
70. Pietrzak K, Twarużek M, Czyżowska A, Kosicki R, Gutarowska B (2015) Influence of silver nanoparticles on metabolism and toxicity of moulds. Acta Biochim Pol 62:851–857
71. Kim K-J, Sung WS, Suh BK, Moon S-K, Choi J-S, Kim JG, Lee DG (2009) Antifungal activity and mode of action of silver nanoparticles on *Candida albicans*. Biometals 22:235–242. https://doi.org/10.1007/s10534-008-9159-2
72. Dizaj SM, Lotfipour F, Barzegar-Jalali M, Zarrintan MH, Adibkia K (2014) Antimicrobial activity of the metals and metal oxide nanoparticles. Mater Sci Eng C 44:278–284. https://doi.org/10.1016/j.msec.2014.08.031
73. Asghar MA, Zahir E, Shahid SM, Khan MN, Asghar MA, Iqbal J, Walker G (2018) Iron, copper and silver nanoparticles: green synthesis using green and black tea leaves extracts and evaluation of antibacterial, antifungal and aflatoxin B1 adsorption activity. LWT Food Sci Technol 90:98–107. https://doi.org/10.1016/j.lwt.2017.12.009
74. Cruz-Luna AR, Cruz-Martínez H, Vásquez-López A, Medina DI (2021) Metal nanoparticles as novel antifungal agents for sustainable agriculture: current advances and future directions. J Fungi 7:1033. https://doi.org/10.3390/jof7121033
75. El-Naggar MA, Alrajhi AM, Fouda MM, Abdelkareem EM, Thabit TM, Bouqellah NA (2018) Effect of silver nanoparticles on toxigenic *Fusarium* spp. and deoxynivalenol secretion in some grains. J AOAC Int 101:1534–1541. https://doi.org/10.5740/jaoacint.17-0442
76. Brunel F, El Gueddari NE, Moerschbacher BM (2013) Complexation of copper(II) with chitosan nanogels: toward control of microbial growth. Carbohydr Polym 92:1348–1356. https://doi.org/10.1016/j.carbpol.2012.10.025

77. Ghasemian E, Naghoni A, Tabaraie B, Tabaraie T (2012) In vitro susceptibility of filamentous fungi to copper nanoparticles assessed by rapid XTT colorimetry and agar dilution method. J Mycol Médicale 22:322–328. https://doi.org/10.1016/j.mycmed.2012.09.006
78. He L, Liu Y, Mustapha A, Lin M (2011) Antifungal activity of zinc oxide nanoparticles against *Botrytis cinerea* and *Penicillium expansum*. Microbiol Res 166:207–215. https://doi.org/10.1016/j.micres.2010.03.003
79. Jayaseelan C, Rahuman AA, Kirthi AV, Marimuthu S, Santhoshkumar T, Bagavan A, Gaurav K, Karthik L, Rao KVB (2012) Novel microbial route to synthesize ZnO nanoparticles using *Aeromonas hydrophila* and their activity against pathogenic bacteria and fungi. Spectrochim Acta A Mol Biomol Spectrosc 90:78–84. https://doi.org/10.1016/j.saa.2012.01.006
80. Sun Q, Li J, Le T (2018) Zinc oxide nanoparticle as a novel class of antifungal agents: current advances and future perspectives. J Agric Food Chem 66:11209–11220. https://doi.org/10.1021/acs.jafc.8b03210
81. Shen T, Wang Q, Li C, Zhou B, Li Y, Liu Y (2020) Transcriptome sequencing analysis reveals silver nanoparticles antifungal molecular mechanism of the soil fungi *Fusarium solani* species complex. J Hazard Mater 388:122063. https://doi.org/10.1016/j.jhazmat.2020.122063
82. Lee J, Kim K-J, Sung WS, Kim JG, Lee DG (2010) The silver nanoparticle (nano-Ag): a new model for antifungal agents. In: Pérez-Pozo D (ed) Silver nanoparticles. IntechOpen, pp 295–308
83. Pinto RJB, Almeida A, Fernandes SCM, Freire CSR, Silvestre AJD, Neto CP, Trindade T (2013) Antifungal activity of transparent nanocomposite thin films of pullulan and silver against *Aspergillus niger*. Colloids Surf B Biointerfaces 103:143–148. https://doi.org/10.1016/j.colsurfb.2012.09.045
84. Deabes MM, Khalil WKB, Attallah AG, El-Desouky TA, Naguib KM (2018) Impact of silver nanoparticles on gene expression in *Aspergillus flavus* producer aflatoxin B1. Open Access Maced J Med Sci 6:600–605. https://doi.org/10.3889/oamjms.2018.117
85. Ibarra-Laclette E, Blaz J, Pérez-Torres C-A, Villafán E, Lamelas A, Rosas-Saito G, Ibarra-Juárez LA, de García-Ávila CJ, Martínez-Enriquez AI, Pariona N (2022) Antifungal effect of copper nanoparticles against *Fusarium kuroshium*, an obligate symbiont of *Euwallacea kuroshio* ambrosia beetle. J Fungi 8:347. https://doi.org/10.3390/jof8040347
86. Medici S, Peana M, Pelucelli A, Zoroddu MA (2021) An updated overview on metal nanoparticles toxicity. Semin Cancer Biol 76:17–26. https://doi.org/10.1016/j.semcancer.2021.06.020
87. Roy A, Bulut O, Some S, Kumar Mandal A, Deniz Yilmaz M (2019) Green synthesis of silver nanoparticles: biomolecule-nanoparticle organizations targeting antimicrobial activity. RSC Adv 9:2673–2702. https://doi.org/10.1039/C8RA08982E
88. Silveira MP, de Neves N A, da Silveira JVW, Schmiele M (2019) Nanotechnology applied to cereal grains and cereal-based products and its food safety. In: Molina G, Inamuddin, Pelissari FM, Asiri AM (eds) Food applications of nanotechnology, 1st edn. CRC Press, Boca Raton, Fl, pp 211–223
89. Bazana MT, Codevilla CF, de Menezes CR (2019) Nanoencapsulation of bioactive compounds: challenges and perspectives. Curr Opin Food Sci 26:47–56. https://doi.org/10.1016/j.cofs.2019.03.005
90. Mozafari MR, Johnson C, Hatziantoniou S, Demetzos C (2008) Nanoliposomes and their applications in food nanotechnology. J Liposome Res 18:309–327. https://doi.org/10.1080/08982100802465941
91. Rafiee Z, Barzegar M, Sahari MA, Maherani B (2018) Nanoliposomes containing pistachio green hull's phenolic compounds as natural bio-preservatives for mayonnaise. Eur J Lipid Sci Technol 120:1800086. https://doi.org/10.1002/ejlt.201800086
92. Zarrabi A, Abadi MAA, Khorasani S, Mohammadabadi M-R, Jamshidi A, Torkaman S, Taghavi E, Mozafari MR, Rasti B (2020) Nanoliposomes and tocosomes as multifunctional nanocarriers for the encapsulation of nutraceutical and dietary molecules. Molecules 25:638. https://doi.org/10.3390/molecules25030638

93. Karimi N, Ghanbarzadeh B, Hajibonabi F, Hojabri Z, Ganbarov K, Kafil HS, Hamishehkar H, Yousefi M, Mokarram RR, Kamounah FS, Yousefi B, Moaddab SR (2019) Turmeric extract loaded nanoliposome as a potential antioxidant and antimicrobial nanocarrier for food applications. Food Biosci 29:110–117. https://doi.org/10.1016/j.fbio.2019.04.006
94. Aguilar-Pérez KM, Medina DI, Narayanan J, Parra-Saldívar R, Iqbal HMN (2021) Synthesis and nano-sized characterization of bioactive oregano essential oil molecule-loaded small unilamellar nanoliposomes with antifungal potentialities. Molecules 26:2880. https://doi.org/10.3390/molecules26102880
95. Aguilar-Pérez KM, Medina DI, Parra-Saldívar R, Iqbal HMN (2022) Nano-size characterization and antifungal evaluation of essential oil molecules-loaded nanoliposomes. Molecules 27:5728. https://doi.org/10.3390/molecules27175728
96. Atienza MaTJA, Magpantay MaDA, Santos KLT, Mora NB, Balaraman RP, Gemeinhardt ME, Dela Cueva FM, Paterno ES, Fernando LM, Kohli P (2021) Encapsulation of plant growth-promoting bacterial crude extract in nanoliposome and its antifungal property against *Fusarium oxysporum*. ACS Agric Sci Technol 1:691–701. https://doi.org/10.1021/acsagscitech.1c00188
97. Bondu C, Yen FT (2022) Nanoliposomes, from food industry to nutraceuticals: interests and uses. Innov Food Sci Emerg Technol 81:103140. https://doi.org/10.1016/j.ifset.2022.103140
98. Anderson JM, Shive MS (1997) Biodegradation and biocompatibility of PLA and PLGA microspheres. Adv Drug Deliv Rev 28:5–24. https://doi.org/10.1016/S0169-409X(97)00048-3
99. Antonioli G, Fontanella G, Echeverrigaray S, Longaray Delamare AP, Fernandes Pauletti G, Barcellos T (2020) Poly(lactic acid) nanocapsules containing lemongrass essential oil for postharvest decay control: in vitro and in vivo evaluation against phytopathogenic fungi. Food Chem 326:126997. https://doi.org/10.1016/j.foodchem.2020.126997
100. Khatua A, Prasad A, Priyadarshini E, Virmani I, Ghosh L, Paul B, Meena R, Barabadi H, Patel AK, Saravanan M (2020) CTAB-PLGA curcumin nanoparticles: preparation, biophysical characterization and their enhanced antifungal activity against phytopathogenic fungus *Pythium ultimum*. Chem Sel 5:10574–10580. https://doi.org/10.1002/slct.202002158
101. Gupta A, Eral HB, Hatton TA, Doyle PS (2016) Nanoemulsions: formation, properties and applications. Soft Matter 12:2826–2841. https://doi.org/10.1039/C5SM02958A
102. Loi M, Paciolla C, Logrieco AF, Mulè G (2020) Plant bioactive compounds in pre- and postharvest management for aflatoxins reduction. Front Microbiol 11:243. https://www.frontiersin.org/articles/10.3389/fmicb.2020.00243
103. Scroccarello A, Molina-Hernández B, Della Pelle F, Ciancetta J, Ferraro G, Fratini E, Valbonetti L, Chaves Copez C, Compagnone D (2021) Effect of phenolic compounds-capped AgNPs on growth inhibition of *Aspergillus niger*. Colloids Surf B Biointerfaces 199:111533. https://doi.org/10.1016/j.colsurfb.2020.111533
104. Reverberi M, Ricelli A, Zjalic S, Fabbri AA, Fanelli C (2010) Natural functions of mycotoxins and control of their biosynthesis in fungi. Appl Microbiol Biotechnol 87:899–911. https://doi.org/10.1007/s00253-010-2657-5
105. Iram W, Anjum T, Iqbal M, Ghaffar A, Abbas M (2015) Mass spectrometric identification and toxicity assessment of degraded products of aflatoxin B1 and B2 by *Corymbia citriodora* aqueous extracts. Sci Rep 5:14672. https://doi.org/10.1038/srep14672
106. Ponzilacqua B, Rottinghaus GE, Landers BR, Oliveira CAF (2019) Effects of medicinal herb and Brazilian traditional plant extracts on in vitro mycotoxin decontamination. Food Control 100:24–27. https://doi.org/10.1016/j.foodcont.2019.01.009
107. Velazhahan R, Vijayanandraj S, Vijayasamundeeswari A, Paranidharan V, Samiyappan R, Iwamoto T, Friebe B, Muthukrishnan S (2010) Detoxification of aflatoxins by seed extracts of the medicinal plant, *Trachyspermum ammi* (L.) Sprague ex Turrill—structural analysis and biological toxicity of degradation product of aflatoxin G1. Food Control 21:719–725. https://doi.org/10.1016/j.foodcont.2009.10.014
108. Chaudhari AK, Singh VK, Das S, Dubey NK (2021) Nanoencapsulation of essential oils and their bioactive constituents: a novel strategy to control mycotoxin contamination in food system. Food Chem Toxicol 149:112019. https://doi.org/10.1016/j.fct.2021.112019

109. Ríos J-L (2016) Essential oils: what they are and how the terms are used and defined. In: Preedy VR (ed) Essential oils in food preservation, flavor and safety. Academic Press, San Diego, pp 3–10
110. Falleh H, Ben Jemaa M, Saada M, Ksouri R (2020) Essential oils: a promising eco-friendly food preservative. Food Chem 330:127268. https://doi.org/10.1016/j.foodchem.2020.127268
111. Chaudhari AK, Singh VK, Das S, Deepika, Prasad J, Dwivedy AK, Dubey NK (2020) Improvement of in vitro and in situ antifungal, AFB1 inhibitory and antioxidant activity of *Origanum majorana* L. essential oil through nanoemulsion and recommending as novel food preservative. Food Chem Toxicol 143:111536. https://doi.org/10.1016/j.fct.2020.111536
112. Chaudhari AK, Dwivedy AK, Singh VK, Das S, Singh A, Dubey NK (2019) Essential oils and their bioactive compounds as green preservatives against fungal and mycotoxin contamination of food commodities with special reference to their nanoencapsulation. Environ Sci Pollut Res 26:25414–25431. https://doi.org/10.1007/s11356-019-05932-2
113. Nasseri M, Golmohammadzadeh S, Arouiee H, Jaafari MR, Neamati H (2016) Antifungal activity of *Zataria multiflora* essential oil-loaded solid lipid nanoparticles in-vitro condition. Iran J Basic Med Sci 19:1231–1237
114. Kalagatur NK, Nirmal Ghosh OS, Sundararaj N, Mudili V (2018) Antifungal activity of chitosan nanoparticles encapsulated with *Cymbopogon martinii* essential oil on plant pathogenic fungi *Fusarium graminearum*. Front Pharmacol 9:610. https://doi.org/10.3389/fphar.2018.00610
115. Beyki M, Zhaveh S, Khalili ST, Rahmani-Cherati T, Abollahi A, Bayat M, Tabatabaei M, Mohsenifar A (2014) Encapsulation of *Mentha piperita* essential oils in chitosan–cinnamic acid nanogel with enhanced antimicrobial activity against *Aspergillus flavus*. Ind Crops Prod 54:310–319. https://doi.org/10.1016/j.indcrop.2014.01.033
116. Khalili ST, Mohsenifar A, Beyki M, Zhaveh S, Rahmani-Cherati T, Abdollahi A, Bayat M, Tabatabaei M (2015) Encapsulation of Thyme essential oils in chitosan-benzoic acid nanogel with enhanced antimicrobial activity against *Aspergillus flavus*. LWT Food Sci Technol 60:502–508. https://doi.org/10.1016/j.lwt.2014.07.054
117. Zhaveh S, Mohsenifar A, Beiki M, Khalili ST, Abdollahi A, Rahmani-Cherati T, Tabatabaei M (2015) Encapsulation of *Cuminum cyminum* essential oils in chitosan-caffeic acid nanogel with enhanced antimicrobial activity against *Aspergillus flavus*. Ind Crops Prod 69:251–256. https://doi.org/10.1016/j.indcrop.2015.02.028
118. Li H, Shen Q, Zhou W, Mo H, Pan D, Hu L (2015) Nanocapsular dispersion of cinnamaldehyde for enhanced inhibitory activity against aflatoxin production by *Aspergillus flavus*. Molecules 20:6022–6032. https://doi.org/10.3390/molecules20046022
119. Das S, Ghosh A, Mukherjee A (2021) Nanoencapsulation-based edible coating of essential oils as a novel green strategy against fungal spoilage, mycotoxin contamination, and quality deterioration of stored fruits: an overview. Front Microbiol 12:768414. https://doi.org/10.3389/fmicb.2021.768414
120. Tiwari S, Dubey NK (2022) Nanoencapsulated essential oils as novel green preservatives against fungal and mycotoxin contamination of food commodities. Curr Opin Food Sci 45:100831. https://doi.org/10.1016/j.cofs.2022.100831
121. Jampílek J, Kráľová K (2020) Chapter 14—Impact of nanoparticles on toxigenic fungi. In: Rai M, Abd-Elsalam KA (eds) Nanomycotoxicology. Academic Press, pp 309–348
122. Wan J, Zhong S, Schwarz P, Chen B, Rao J (2019) Physical properties, antifungal and mycotoxin inhibitory activities of five essential oil nanoemulsions: impact of oil compositions and processing parameters. Food Chem 291:199–206. https://doi.org/10.1016/j.foodchem.2019.04.032
123. Wan J, Zhong S, Schwarz P, Chen B, Rao J (2018) Influence of oil phase composition on the antifungal and mycotoxin inhibitory activity of clove oil nanoemulsions. Food Funct 9:2872–2882. https://doi.org/10.1039/C7FO02073B
124. Wan J, Zhong S, Schwarz P, Chen B, Rao J (2019) Enhancement of antifungal and mycotoxin inhibitory activities of food-grade thyme oil nanoemulsions with natural emulsifiers. Food Control 106:106709. https://doi.org/10.1016/j.foodcont.2019.106709

125. Gill TA, Li J, Saenger M, Scofield SR (2016) Thymol-based submicron emulsions exhibit antifungal activity against *Fusarium graminearum* and inhibit *Fusarium* head blight in wheat. J Appl Microbiol 121:1103–1116. https://doi.org/10.1111/jam.13195
126. Wan J, Jin Z, Zhong S, Schwarz P, Chen B, Rao J (2020) Clove oil-in-water nanoemulsion: Mitigates growth of *Fusarium graminearum* and trichothecene mycotoxin production during the malting of *Fusarium* infected barley. Food Chem 312:126120. https://doi.org/10.1016/j.foodchem.2019.126120
127. Song C, Ding G, Dai J, Wang Y, Liu Y, Zhang Y, Zhang Q, Yang J, Qin J (2022) Anti-aflatoxigenic nanoemulsions based on *Monarda didyma* and *Neopallasia pectinata* essential oils as novel green agent for food preservation. Ind Crops Prod 180:114777. https://doi.org/10.1016/j.indcrop.2022.114777
128. Ahmed OS, Tardif C, Rouger C, Atanasova V, Richard-Forget F, Waffo-Téguo P (2022) Naturally occurring phenolic compounds as promising antimycotoxin agents: where are we now? Compr Rev Food Sci Food Saf 21:1161–1197. https://doi.org/10.1111/1541-4337.12891
129. Milinčić DD, Popović DA, Lević SM, Kostić AŽ, Tešić ŽLj, Nedović VA, Pešić MB (2019) Application of polyphenol-loaded nanoparticles in food industry. Nanomaterials 9:1629. https://doi.org/10.3390/nano9111629
130. Pimentel-Moral S, Teixeira MC, Fernandes AR, Arráez-Román D, Martínez-Férez A, Segura-Carretero A, Souto EB (2018) Lipid nanocarriers for the loading of polyphenols—a comprehensive review. Adv Colloid Interface Sci 260:85–94. https://doi.org/10.1016/j.cis.2018.08.007
131. Zhang Z, Li X, Sang S, McClements DJ, Chen L, Long J, Jiao A, Jin Z, Qiu C (2022) Polyphenols as plant-based nutraceuticals: health effects, encapsulation, nano-delivery, and application. Foods 11:2189. https://doi.org/10.3390/foods11152189
132. Zhou H, Sun F, Lin H, Fan Y, Wang C, Yu D, Liu N, Wu A (2022) Food bioactive compounds with prevention functionalities against fungi and mycotoxins: developments and challenges. Curr Opin Food Sci 48:100916. https://doi.org/10.1016/j.cofs.2022.100916
133. Muzzalupo I, Badolati G, Chiappetta A, Picci N, Muzzalupo R (2020) In vitro antifungal activity of olive (*Olea europaea*) leaf extracts loaded in chitosan nanoparticles. Front Bioeng Biotechnol 8:151. https://doi.org/10.3389/fbioe.2020.00151
134. Al-Otibi F, Perveen K, Al-Saif NA, Alharbi RI, Bokhari NA, Albasher G, Al-Otaibi RM, Al-Mosa MA (2021) Biosynthesis of silver nanoparticles using *Malva parviflora* and their antifungal activity. Saudi J Biol Sci 28:2229–2235.https://doi.org/10.1016/j.sjbs.2021.01.012
135. Zhao Z, Liu N, Yang L, Wang J, Song S, Nie D, Yang X, Hou J, Wu A (2015) Cross-linked chitosan polymers as generic adsorbents for simultaneous adsorption of multiple mycotoxins. Food Control 57:362–369. https://doi.org/10.1016/j.foodcont.2015.05.014
136. Horky P, Skalickova S, Baholet D, Skladanka J (2018) Nanoparticles as a solution for eliminating the risk of mycotoxins. Nanomaterials 8:727. https://doi.org/10.3390/nano8090727
137. Abd-Elsalam KA, El-Naggar MA, Ghannouchi A, Bouqellah NA (2020) Nanomaterials and ozonation: safe strategies for mycotoxin management. In: Rai M, Abd-Elsalam KA (eds) Nanomycotoxicology. Academic Press, pp 285–308
138. Gibson N, Shenderova O, Luo TJM, Moseenkov S, Bondar V, Puzyr A, Purtov K, Fitzgerald Z, Brenner DW (2009) Colloidal stability of modified nanodiamond particles. Diam Relat Mater 18:620–626. https://doi.org/10.1016/j.diamond.2008.10.049
139. Gibson NM, Luo TJM, Brenner DW, Shenderova O (2011) Immobilization of mycotoxins on modified nanodiamond substrates. Biointerphases 6:210–217. https://doi.org/10.1116/1.3672489
140. Ing LY, Zin NM, Sarwar A, Katas H (2012) Antifungal activity of chitosan nanoparticles and correlation with their physical properties. Int J Biomater 2012:e632698. https://doi.org/10.1155/2012/632698
141. Zhang N, Han X, Zhao Y, Li Y, Meng J, Zhang H, Liang J (2022) Removal of aflatoxin B1 and zearalenone by clay mineral materials: in the animal industry and environment. Appl Clay Sci 228:106614. https://doi.org/10.1016/j.clay.2022.106614

142. Karami-Osboo R, Maham M, Nasrollahzadeh M (2020) Synthesised magnetic nano-zeolite as a mycotoxins binder to reduce the toxicity of aflatoxins, zearalenone, ochratoxin A, and deoxynivalenol in barley. IET Nanobiotechnol 14:623–627. https://doi.org/10.1049/iet-nbt.2020.0107
143. Lin X, Yu W, Tong X, Li C, Duan N, Wang Z, Wu S (2022) Application of nanomaterials for coping with mycotoxin contamination in food safety: from detection to control. Crit Rev Anal Chem in press:1–34. https://doi.org/10.1080/10408347.2022.2076063
144. Liu C, Wang J, Wan J, Yu C (2021) MOF-on-MOF hybrids: synthesis and applications. Coord Chem Rev 432:213743. https://doi.org/10.1016/j.ccr.2020.213743
145. Fraceto LF, Grillo R, de Medeiros GA, Scognamiglio V, Rea G, Bartolucci C (2016) Nanotechnology in agriculture: Which innovation potential does it have? Front Environ Sci 4:20. https://doi.org/10.3389/fenvs.2016.00020
146. Juárez-Maldonado A, Tortella G, Rubilar O, Fincheira P, Benavides-Mendoza A (2021) Biostimulation and toxicity: the magnitude of the impact of nanomaterials in microorganisms and plants. J Adv Res 31:113–126. https://doi.org/10.1016/j.jare.2020.12.011
147. Singh D, Gurjar BR (2022) Nanotechnology for agricultural applications: facts, issues, knowledge gaps, and challenges in environmental risk assessment. J Environ Manag 322:116033. https://doi.org/10.1016/j.jenvman.2022.116033
148. Grieger K, Merck A, Kuzma J (2022) Formulating best practices for responsible innovation of nano-agrifoods through stakeholder insights and reflection. J Respons Technol 10:100030. https://doi.org/10.1016/j.jrt.2022.100030
149. Pedrini S, Merritt DJ, Stevens J, Dixon K (2017) Seed coating: science or marketing spin? Trends Plant Sci 22:106–116. https://doi.org/10.1016/j.tplants.2016.11.002

Interaction Between Metal Oxide Nanoparticles and PGPR on Plant Growth and Development

Apekshakumari Patel, Nimisha Patel, Ahmad Ali, and Hina Alim

Abstract Plant growth and development depend highly on interaction with the immediate surroundings, mainly the highly complex sphere called the rhizosphere. Rhizosphere is a highly interactive space that comprises plant root's interactions with the diverse soil microorganisms, which is greatly dependent on the physiochemical composition of the soil. The exponential advancement of westernization and overpopulation causes a decrease in fertility soil and limits the land mass for agricultural purposes; this raises concerns when considering the high demand for crop production. Soil microbiota plays a crucial role in maintaining the soil ecosystem and crop productivity. Plant growth-promoting rhizobacteria (PGPR) are free-living microbes essential in sustaining soil fertility and plant development. Metal oxide nanoparticles have increasingly been investigated for their application to increase soil microbiome. Although there is evidence of eco-toxicity of the metal oxide nanoparticles as they have the potential to inhibit the activity of the rhizosphere. This chapter discusses the use of engineered nanoparticles synthesis using aluminum oxide (Al_2O_3), copper oxide (CuO), silicon oxide (SiO_2), iron oxide (Fe_2O_3), titanium dioxide (TiO_2), zinc oxide (ZnO), and other metal oxide nanoparticles to be tested for their eco-friendly effects on PGPR, to facilitate the plant growth and development.

Keywords Metal oxide nanoparticles · PGPR · Plant growth · Plant development · Rhizosphere · Microbiome

A. Patel · A. Ali · H. Alim (✉)
Department of Life Sciences, University of Mumbai, Vidyanagari Campus, Santacruz East, Mumbai 400 098, India
e-mail: hina.alim@lifesciences.mu.ac.in

N. Patel
Department of Life Sciences, J.C. Bose University of Science & Technology, YMCA, Faridabad, Haryana 121 006, India

R. K. Bachheti et al. (eds.), *Nanomaterials for Environmental and Agricultural Sectors*, Smart Nanomaterials Technology,
https://doi.org/10.1007/978-981-99-2874-3_11

1 Introduction

Plants supply food and energy for humans, animals, and poultry and raw materials for the textile, pharmaceutical, timber, and recently established natural biodegradable plastics sectors. However, as the population grows, so does the land accessible for agricultural production. As a result, more natural plantations are destroyed, dependent floras perish, and the most significant catastrophe is biodiversity loss. Various types of agrochemicals and pesticides have exacerbated the situation and resulted in widespread water and soil contamination [41]. According to estimates, global food production will need to expand 70% by 2050, putting additional strain on energy, nutrients, water, land, and labor [78]. Hunger was expected to afflict 702 to 828 million people in 2021. Since the beginning of the COVID-19 pandemic, the number has increased by around 150 million to 103 million more individuals between 2019 and 2020, and 46 million more in 2021, assuming the midpoint of the expected range [20]. At the moment, the production of food is heavily dependent on outside resources like irrigation water and agrichemicals. The natural ecosystem and environment are likely to be further harmed by food production, given the astounding scale and speed of the population shift worldwide. The growth of urbanization, land degradation (such as salinization and erosion), non-food uses of crops and farmland (such as bioenergy, recreational activities, and building of transportation lines), and climate change will all have an impact on how much food can be produced [59]. Abiotic and biotic stressors cause agricultural output losses of up to 30% annually in various parts of the world. These latter ones result from harm done to plants by other living things, including native or cultivated bacteria, fungi, insects, viruses, plants, weeds, and parasites [77]. Prior research has mostly emphasized increasing the food supply while preserving the environment. Interdisciplinary research should boost production while practicing environmental conservation in relation to food security and the environment, however, on its own, these recommendations are unlikely to keep the ecosystem healthy, given the ongoing growth in the world's population [52].

Soil is the most common home for practically all microorganisms, and it is where microbes interact with their biotic components, such as the rhizosphere, and with one another. The microbial world of soil is the Earth's biggest untapped reserve of microorganisms. It has become a significant frontier in biology in recent decades, as it performs several roles for the biosphere, including nutrient cycling and plant growth stimulation [74]. Furthermore, microbial populations in rhizospheric soil are physiologically more active than in non-rhizospheric soil because plant roots influence soil-borne microbial communities through various mechanisms, including organic compound excretion, nutrient competition, and providing a solid surface for attachment [12]. According to research, the physicochemical composition of rhizospheric soil influences bacterial population densities in the rhizosphere. Bacteria such as those from the genera *Azotobacter, Azospirillum, Bacillus, Burkholderia, Enterobacter,* and *Pseudomonas* can colonize plant rhizosphere, root, and shoot interior or the surface and can aid in the growth of plants. These microorganisms are known as plant growth-promoting rhizobacteria (PGPR) [29, 49]. PGPR is a group of beneficial bacteria that are found in the rhizosphere of soil and are known to have a range of positive effects on plant growth and development. PGPR can improve plant growth and development by increasing root growth, improving nutrient uptake, and increasing resistance to stress [5, 58]. PGPR can also produce various compounds beneficial to plants, including hormones, enzymes, and antibiotics [24, 27]. PGPR can be applied to plants as a coating or included in the soil and has been shown to have a variety of positive effects on plant growth and development, including increased yield, improved root growth, and increased resistance to stress [1].

The environmental matrix contains the most naturally occurring nanoparticles as primary particles and as aggregates or agglomerates in the soil [51]. This is caused by ongoing physical or chemical weathering, reorganization of its geogenic components, and intense biological activity that changes dead organic matter and minerals. Humic compounds, nano-, and micron-scale particles give soils (and sediments) a high porosity and extremely high specific surface areas (tens to hundreds of square meters per gram) [50].

In agriculture, metal oxide nanoparticles have been studied for their potential to enhance plant growth and development. These nanoparticles can be applied to plants as a coating or included in the soil and have been shown to have a variety of effects on plant growth and development, including increased photosynthetic efficiency, increased growth rate and biomass production, and improved stress tolerance [28, 79, 80], PGPR can also interact with metal oxide nanoparticles in various ways, including by mitigating the adverse effects of metal oxide nanoparticles on plant growth and development or by enhancing the positive effects of these nanoparticles [53]. However, metal oxide nanoparticles can also have negative effects on plants, including reduced growth and biomass production, impaired root development, and decreased nutrient uptake [67, 68]. These negative effects are due to the toxicity of metal oxide nanoparticles to plants and their ability to interfere with the uptake of nutrients and other essential elements. This chapter highlights the cumulative effect of the interaction between the metal oxide nanoparticles and PGPR on plant growth and development.

2 Metal Oxide Nanoparticles

Nanotechnology has made numerous applications possible, and it has been seen that adding nanoparticles, such as metallic nanoparticles, to the biological sciences produces the intended results. Additionally, these particles have been used in soil transportation and environmental cleanup systems [56]. Nanoparticles are well known for their size, shape, inertness, and surface coating materials. They are stable, less dangerous, and resistant to surface oxidation under certain circumstances. They are candidates of significant interest in a range of disciplines due to their unusual characteristics, including their surface energy, increased surface area to volume ratio, chemical reactivity, and various other related qualities [7, 8, 31, 55]. They have a

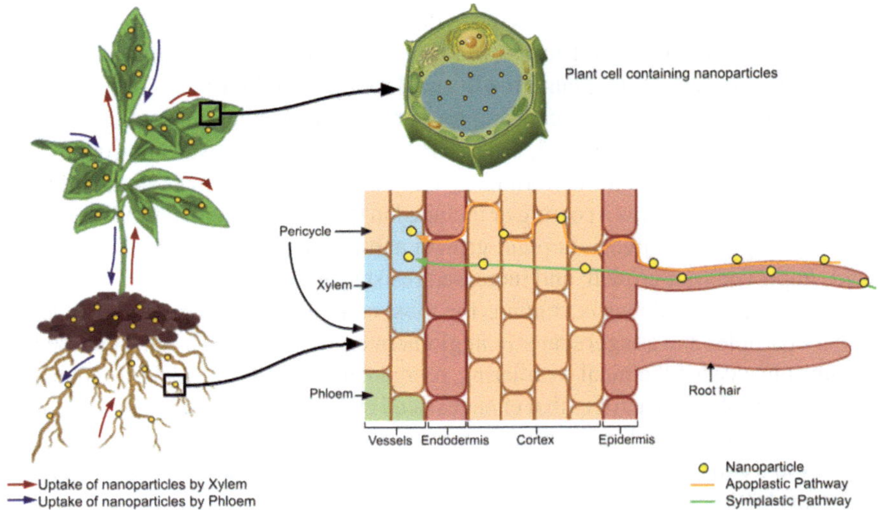

Fig. 1 Translocation of the nanoparticles through various cellular pathways

detrimental effect on plant morphology, physiology, proteome, metabolome, and genetics, endangering the sustainability of ecosystems [32–37, 82]. The soil pH, humic acid, and organic natural matter are only a few factors that affect nanoparticles' toxicity and fate [81]. Additionally, depending on their size, nanoparticles are introduced into the cell via several cellular pathways, such as the apoplastic pathway and symplastic pathway, as shown in Fig. 1 [64, 71].

Metal oxide nanoparticles are extremely small particles (typically less than 100 nm) composed of metal ions and oxygen atoms [23]. Metal oxide nanoparticles are used widely in various fields due to their properties, which are shown in Fig. 2. Aluminum oxide (Al_2O_3), copper oxide (CuO), silicon oxide (SiO_2), iron oxide (Fe_2O_3), calcium oxide (CaO), magnesium oxide (MgO), titanium dioxide (TiO_2), and zinc oxide (ZnO) are among the nano metal-oxides that are advantageous for several applications [25]. The mechanisms by which metal oxide nanoparticles improve plant growth and development are not fully understood but are thought to involve a variety of mechanisms, including:

Photosynthesis is the process by which plants, algae, and some bacteria convert light energy into chemical energy in glucose and other sugars.

(1) Photosynthesis enhancement: Metal oxide nanoparticles, such as titanium dioxide and zinc oxide, can absorb light and reflect it onto the plant's leaves, which can improve the efficiency of photosynthesis. This can lead to an increased growth rate and biomass production in plants. Metal oxide nanoparticles, such as TiO_2 and ZnO, have been shown to have the potential to enhance photosynthesis in plants by absorbing light and transferring the absorbed energy to photosynthetic pigments, such as chlorophyll [60]. There are several ways in which metal oxide nanoparticles can be used to enhance photosynthesis. One

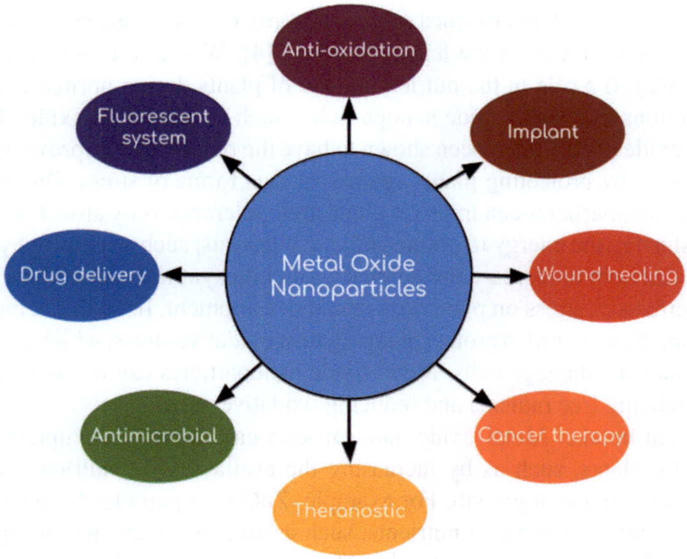

Fig. 2 Metal oxide nanoparticles application in various health sectors

way is by incorporating them into the plant's leaves or adding them to the soil around it. Metal oxide nanoparticles can also be added to the water plants grown in or applied as a coating to the surface of the leaves. There is some evidence to suggest that metal oxide nanoparticles can increase the efficiency of photosynthesis in plants, leading to enhanced growth and productivity. With all dependent physiological markers, the TiO_2 nanoparticles enabled unexpectedly early plant maturation. The most unexpected thing about these discoveries is that, although being more photo-stable than other plant enhancers, TiO_2 nanoparticles are possibly hazardous and offer less nutrition. In contrast, the variety treated with photo-corrosive and less resistant ZnO nanoparticles predicted and better reflected the physiological characteristics of the sunflower with rather good quantitative and nutritional impacts. Its improved zinc bioavailability due to ZnO nanoparticle transformation, which supports the sunflower's physiological and metabolic pathways, is most likely the cause of its superiority [44].

Plant stress refers to the negative effects of environmental conditions, such as extreme temperatures, drought, and soil conditions, on plant growth and development.

(2) Plant stress tolerance: Metal oxide nanoparticles can improve the stress tolerance of plants, such as by protecting them from oxidative stress or by increasing the efficiency of photosynthesis [66]. For example, TiO_2 nanoparticles have been shown to improve the stress tolerance of plants under high light intensity and high temperatures and decrease oxidative damage and lipid peroxidation during drought. ZnO nanoparticles have also been shown to improve the stress tolerance of plants under drought and salt stress, as they decrease oxidative stress and

increase antioxidant enzymes such as superoxide dismutase (SOD) and peroxidase (POD) levels in the leaves of plants [4]. Whereas, Fe_2O_3 nanoparticles have played a role in the nutrient uptake of plants during normal and drought conditions [3]. Metal oxide nanoparticles, such as titanium dioxide (TiO2) and zinc oxide (ZnO), have been shown to have the potential to improve plant stress tolerance by protecting plants against various forms of stress. One way metal oxide nanoparticles can improve plant stress tolerance is by absorbing light and transferring the energy to photosynthetic pigments, such as chlorophyll [4, 61]. This can help to increase the efficiency of photosynthesis and reduce the negative effects of stress on plant growth and development. In addition, metal oxide nanoparticles can also protect plants against oxidative stress, which occurs when free radicals damage cells. Metal oxide nanoparticles can act as antioxidants, neutralizing free radicals and reducing oxidative stress.

(3) Nutrient uptake: Metal oxide nanoparticles can improve the uptake of nutrients by plants, such as by increasing the availability of nutrients in the soil or stimulating root growth. For example, ZnO nanoparticles have been shown to improve the uptake of nutrients, such as zinc, nitrogen, and phosphorus, in plants [14, 62]. There are a number of ways in which metal oxide nanoparticles could improve nutrient uptake in plants. For example, metal oxide nanoparticles have been shown to have the ability to absorb nutrients onto their surface, which could make these nutrients more available to plants. In addition, metal oxide nanoparticles have been shown to have the ability to alter the soil microbiome, which could affect the availability of nutrients to plants [17, 47].

(4) Antimicrobial activity: Metal oxide nanoparticles can inhibit the growth of harmful bacteria and fungi in the soil, improving plant growth and development by reducing the risk of diseases. Several different metal oxide nanoparticle types have antimicrobial activity against certain soil microorganisms, such as Fe_2O_3, ZnO, and CuO [16, 72].

It is important to note that the antimicrobial activity of metal oxide nanoparticles can vary depending on the specific type of metal oxide and the conditions under which they are used. In addition, much is still not known about the potential impacts of metal oxide nanoparticles on soil microorganisms and the ecosystem as a whole, and more research is needed to fully understand the risks and benefits of using these particles in agriculture.

3 Plant Growth and Development

Plant growth and development refers to how plants grow and mature over time. Plant growth refers to a plant's size and mass over time, driven by cell division, cell expansion, and cell differentiation. These processes are complex and involve various factors, including genetics, the environment, and the interactions between plants and other organisms. These processes are regulated by a variety of hormones and signaling pathways, which are influenced by both internal and external factors.

Plant development refers to the progression of a plant from a seedling to a mature plant and involves a series of stages, including germination, vegetative growth, flowering, and senescence. Each stage is characterized by specific developmental changes, such as the emergence of roots, leaves, and flowers, and is influenced by a range of environmental factors, including temperature, light, and nutrient availability. Plant growth and development are important for plants' survival and reproductive success and are also important for various human activities, including agriculture, forestry, and horticulture. The study of plant growth and development is a multidisciplinary field involving various disciplines, including biology, genetics, ecology, and agriculture.

4 Plant Growth-Promoting Rhizobacteria

Plant growth-promoting bacteria (PGPR) are rhizosphere soil bacteria that promote plant growth and development by releasing different regulatory chemicals. They can be located near the roots (rhizosphere), on the leaves (phyllosphere), or within the plant itself (endosphere). Endophytes (PGPE) are often the most effective in promoting development because they can connect with the host plant and exert their positive impact considerably more efficiently. Furthermore, because they are shielded from the outside environment, the PGPE are significantly less susceptible to the soil's regular chemical-physical biotic and abiotic fluctuations. Endophytic bacteria originate in the rhizosphere habitat surrounding the roots and enter plant tissues mostly through natural cracks formed in the roots during growth. Plant exudates and radical metabolites are important for choosing and recruiting beneficial microorganisms. The beneficial function of bacteria associated with plants mostly manifests by directly enhancing nutrient absorption by modifying plant hormone levels. PGPRs are a helpful class of rhizosphere microorganisms that can boost plant growth through a number of processes, such as nutrient uptake, siderophore synthesis, phytohormone synthesis, nitrogen fixation by living organisms, solubilization of insoluble phosphorus, the introduction of systemic tolerance genes, synthesis of ACC deaminase, and the production of various volatile organic compounds. The most significant and well-studied direct processes include nitrogen fixation, the solubilization of inorganic phosphate, auxins (especially 3-indole acetic acid, IAA), cytokinins, and gibberellins; the enzyme 1-aminocyclopropane-1-carboxylate (ACC) deaminase; and the synthesis of siderophore molecules [13, 75, 76]. In contrast, indirect mechanisms are described as the inhibitory actions of the PGPR that prevent pathogenic organisms from causing harm. The most prevalent indirect processes include the formation of hydrogen cyanide, antibiotics, and enzymes capable of attacking and destroying pathogen cell walls [54].

Using PGPR can potentially increase sustainable farming because of its environmentally benign and helpful nature in substituting the increased use of synthetic fertilizers and pesticides. A wide variety of compounds produced by rhizobacteria either directly or indirectly support plant growth. The importance of PGPR research is increasing due to the rapid expansion of commercial biofertilizers, including the best PGPR strains [10]. PGPR are categorized as phytostimulators, biopesticides, and biofertilizers, as shown in Table 1.

The mechanisms by which plant growth-promoting rhizobacteria (PGPR) improve plant growth and development are not fully understood but are thought to involve a variety of mechanisms, including PGPR can be applied to plants as a coating or included in the soil and has been shown to have a variety of positive effects on plant growth and development, including increased yield, improved root growth, and increased resistance to stress. They can also interact with metal oxide nanoparticles in various ways, including by mitigating the negative effects of metal oxide nanoparticles on plant growth and development or by enhancing the positive effects of these nanoparticles. They are a promising tool for improving plant growth and development and have potential applications in agriculture and horticulture. However, further research is needed to fully understand the mechanisms underlying the effects of PGPR on plants and to optimize their use in different plant species and environments.

(1) Nutrient acquisition: PGPR can improve plant growth and development by increasing the availability of nutrients to plants, such as nitrogen, phosphorus, and potassium [19]. PGPR can do this through a variety of mechanisms, including the production of enzymes that help plants to access nutrients in the soil, the nitrogen-fixing ability of certain strains, and the production of hormones that stimulate root growth and improve nutrient uptake [9, 24, 38].
(2) Stress tolerance: PGPR can help plants tolerate stress, such as drought, salinity, and high temperatures, by producing a variety of compounds that help plants to cope with these stresses. For example, PGPR can produce enzymes that help plants to detoxify harmful substances, hormones that stimulate root growth and improve water uptake, and antioxidants that protect plants from oxidative stress [18, 43].

Table 1 Various mechanism performed by PGPR for plant growth and development

Types of PGPR	Function	Mechanism	References
Phytostimulator	These microbes can produce or induce the plant growth regulators	Synthesis of indole acetic acid (IAA), gibberellins, cytokinins and ethylene	[21, 22, 40, 46]
Biopesticides	These microbes release phytopathogenic agents to prevent the growth of plant pathogens	Release of hydrolytic enzymes, synthesis of antibiotics and HCN	[6, 54, 57]
Biofertilizers	These microbes colonize the root of the plant in a rhizosphere to enhance the availability of plant nutrients	Nitrogen fixation, solubilization of insoluble phosphate and synthesis of siderophore	[39, 45, 73]

(3) Antagonism: PGPR can inhibit the growth of harmful bacteria and fungi in the soil, improving plant growth and development by reducing the risk of disease [24]. PGPR can do this through various mechanisms, including the production of antibiotics and other antimicrobial compounds, the competition for nutrients, space, and the induction of plant defense responses [15, 30, 69].
(4) Induced systemic resistance: PGPR can induce systemic resistance in plants, which is the plant's ability to mount an immune response against a variety of pathogens. Specific PGPR strains cause induce systemic resistance against several diseases attacking the same crop. In addition to disease control, using PGPR decreases insect and worm damage. The broad spectrum of control provided by PGPR strains can provide an efficient, cheap, and practical method of plant protection. Because of their capacity to colonize and remain in the intercellular space of epidermal cells, certain PGPR are appropriate for use in vegetatively propagated crops, decreasing the need for subsequent treatment if the same vegetative portions are employed as propagative material. Furthermore, many PGPR strain mixes show synergistic activity in growth promotion and plant protection, showing that various mechanisms are engaged in disease management. This can help plants better cope with stress and disease and improve plant growth and development [63, 65].

5 Interaction Between Metal Oxide Nanoparticles and PGPR

Plant growth-promoting rhizobacteria (PGPR) and metal oxide nanoparticles are two substances studied for their potential to enhance plant growth and development. Both PGPR and metal oxide nanoparticles can be applied to plants as a coating or included in the soil and have been shown to have various effects on plant growth and development. Haris and Ahmad [26] examined the PGPR strains *Pseudomonas aeruginosa, Pseudomonas fluorescens,* and *Bacillus amyloliquefaciens* with the nanoparticles ZnO and TiO_2. They elucidated that the nanoparticles had a dose-dependent impact on PGPR strains, indicating the toxicity of these nanoparticles in the environment. ZnO nanoparticles were discovered to be more hazardous than TiO_2 nanoparticles. ZnO nanoparticles cause more cell wall damage in bacteria than TiO_2-NPs, showing their relative toxicity.

While in another study conducted by Shafiq et al. [71] the synergistic effect of *Bacillus pumilus* with ZnO and TiO_2 nanoparticles on decreasing the Cd stress in maize. The study showed that such synergistic effects enhanced the plant biomass, proline concentration, antioxidants, phytohormones and nutrition content, increasing defense mechanism against Cd stress.

The research on ZnO, SiO_2, CeO_2, and TiO_2 effects on *Azotobacter*, P-solubilizing and K-solubilizing bacteria by Chai et al. [11] showed different results. ZnO, CeO_2, and TiO_2 nanoparticles considerably reduced the number of *Azotobacter*, P- and K-solubilizing bacteria, however SiO_2 nanoparticles had no effect. Enzymatic activity reductions were more substantial in reaction to ZnO, CeO_2, and TiO_2 nanoparticle treatments than in response to SiO_2 nanoparticle treatments. This was most likely due to microbial cell absorption or ingestion of free metals liberated in soil by nanoparticles. In the $CHCl_3$-labile metal and H_2O extractions, they found Zn, Ce, and Ti, but not Si. A considerable portion of the nanoparticles may be transported or absorbed by microbial cells, releasing ions into the soil. In organic farm soil polluted with ZnO and CeO_2 nanoparticles, nitrogen fixation of soybean nodules was inhibited. ZnO and TiO_2 nanoparticles reduced the number of taxa involved with the nitrogen cycle, according to the DNA fingerprinting analysis (*Rhizobiales, Bradyrhizobiaceae, Bradyrhizobium*). ZnO and TiO_2 nanoparticles substantially reduced the activities of protease, catalase, and peroxidase in wheat soil, which might be attributed to dissolved ions in the soil.

In a review by Shang et al. [72], TiO_2 and SiO_2 have been demonstrated to be the potential for directly inhibiting crop diseases via antibacterial action. Recently ZnO nanoparticles have been demonstrated to inhibit the growth of bacteria and fungus such as, *Aspergillus flavus, Alternaria alternate, Fusarium graminearum, Fusarium oxysporum, Penicillium expansum, Rhizopus stolonifer* and *Mucor plumbeus*, including the pathogenic bacteria, *Pseudomonas aeruginosa*. Metal oxide nanoparticles limit the formation of fungal conidia and conidiophores, resulting in the death of fungal hyphae. A study found that germination, root-shoot lengths, fresh-dry weights, and photosynthetic pigments with total protein were all reduced in weeds treated to SiO_2 nanoparticles. Commercial herbicides often suppress or destroy the above-ground sections of weeds without damaging the below-ground parts, such as rhizomes or tubers. As a result, weeds sprout,nevertheless, nano herbicides inhibit weed regeneration. Thus, nanoparticles in insecticides, fungicides, and herbicides have enormous potential for sustainable agricultural growth.

Under stress conditions, it has been previously discussed that metal oxide nanoparticles and PGPR act on the plant mechanism to assist plant survival and enhance plant growth. One such study on Zn-Fe oxide and *Azotobacter* was conducted by Seyed Sharifi et al. [70], demonstrating their synergistic effect on grain crops under water stress conditions. Their research showed that water constraint reduced grain yield, photosynthetic pigments, photosystem II efficiency, and relative water content while increasing proline, soluble sugars, and catalase, peroxidase, and polyphenol oxidase activities. Biofertilizers and nanooxides were used to boost proline, soluble sugars, and enzyme activity. Under severe water scarcity, plants treated with Zn-Fe oxide nanoparticles showed 17% lower electrical conductivity, 27% greater peroxidase activity, and 61% higher polyphenol oxidase activity than control plants. Under severe water scarcity, the combination of *Azotobacter* and nano Zn-Fe oxide increased grain production by 88% when compared to the control. These findings suggest combining biofertilizer and nano Zn-Fe oxide boosted wheat production under drought conditions.

Karunakaran et al. [42] studied the PGPR strains *Bacillus megaterium, Azotobacter vinelandii, Pseudomonas fluorescens* and *Bacillus brevis* with TiO_2 and ZrO_2 (Zirconium) nanoparticles. TiO_2 nanoparticles are extremely toxic to the PGPR strains than their bulk equivalents. They depicted that the size of the particles, hydrophobic potential, and zeta potential are the key determinants of bacterial toxicity. ZrO_2 is discovered to be non-toxic in both its bulk and nano forms. The PGPR, the soil, and its nutrients have been proven harmful to TiO_2 nanoparticles. Thus, it is clear that antimicrobial metal oxides will have a toxic effect on soil and are a significant problem that impacts the soil's health.

Another study involved the green synthesized ZnO nanoparticles from *Eucalyptus globulus* leaves. Results showed that at 36 h, ZnO nanoparticles and PGPR (*Pseudomonas*) strains had greater antibacterial activity on *Escherichia coli* strain than eucalyptus leaf extract. With increased particle dose and treatment period, ZnO nanoparticles became more effective. *Pseudomonas* showed a high antagonistic response to the *Escherichia coli* strain during testing. However, the antagonistic effects are amplified when *Pseudomonas* and ZnO nanoparticles are applied together. The study uses in-silico techniques to determine the binding affinity of ZnO nanoparticles with the appropriate bacterial pathogen receptor. The penicillin-binding protein 6 (PBP 6) receptor's conserved binding site is where the ligand ZnO binds, according to the binding site conformations. ZnO nanoparticles can significantly contribute to the inhibition of PBP 6 since they exhibit a similar pattern of interactions as other reported ligands. This study also discovered that combining ZnO nanoparticles and *Pseudomonas* was an innovative and successful method for treating pathogenic bacteria, including multidrug-resistant bacteria, but not advantageous for the development of plants [48].

The study conducted by Mushtaq et al. [53] was to determine whether iron oxide nanoparticles and *Bacillus subtilis* could reduce arsenic (As) stress in *Cucurbita moschata*. In seedlings exposed to As stress, lower growth factors, such as photosynthetic pigments, photosynthesis rates, and gas exchange characteristics, were seen. However, in supplemented *Cucurbita moschata* seedlings, iron oxide nanoparticles, and *Bacillus subtilis* enhanced growth characteristics and proline levels. Seedlings infected with *Bacillus subtilis* demonstrated increased POD and SOD activity under the toxicity of As. The simultaneous administration of iron oxide nanoparticles and *Bacillus subtilis also enhanced the activity of these antioxidative enzymes*. Reduced levels of hydrogen peroxide, malondialdehyde, and electrolyte leakage in iron oxide nanoparticles and *Bacillus subtilis* treated plants reduce the As stress in inoculated *Cucurbita moschata* seedlings. Additionally, in As-stressed seedlings, a synergistic interaction between plant growth-promoting bacteria (PGPB) and iron oxide nanoparticles improved the production of stress-relieving polyamines such as spermidine and putrescine. According to recent studies, using iron oxide nanoparticles and *Bacillus* subtilis in combination is an efficient, environmentally friendly method for reducing As stress in *Cucurbita moschata* seedlings.

The current study of Ahmed et al. [2] demonstrated that a free-living, N_2-fixing, and nanoparticle-tolerant *Azotobacter salinestris* strain recovered from metal-polluted soil interferes with tomato plant-metal oxide nanoparticles interactions (ZnO, CuO, Al_2O_3, and TiO_2) in a sandy clay loam soil system with bulk materials as control. With and without seed biopriming and *Azotobacter salinestris* root inoculation, tomato plants were cultivated to full maturity in soils supplemented with metal oxide nanoparticles. Under the influence of nanoparticles, *Azotobacter salinestris* was discovered to be metabolically active, producing noticeably high levels of bioactive indole-3-acetic acid, morphologically unchanged, and with minute change in cell membrane permeability. ZnO nanoparticles modestly change the permeability of bacterial membranes. Additionally, even when exposed to nanoparticles, *Azotobacter salinestris* released noticeably more extracellular polymeric substance (EPS), which could trap the nanoparticles and form metal-EPS complexes. Adsorbed nanoparticles were also discovered on bacterial biomass. The nanoparticles were stabilized by EPS, which also gave them a negative zeta potential. Following soil treatment, *A. salinestris* increased tomato fruit output, lycopene concentration, and plant performance even in nanoparticle-stressed soils. It is interesting to note that *Azotobacter salinestris* inoculation increased the production of photosynthetic pigment, flower characteristics, plant and fruit biomass, and decreased proline level. Additionally, bacterial inoculation drastically decreased nanoparticles uptake and accumulation in vegetative organs and fruits. In conclusion, *Azotobacter salinestris* inoculation might be a different option to boost tomato output in metal-oxide nanoparticles.

In order to prevent the toxicity of metal oxide nanoparticles from causing the loss of both structure and functions of agronomically significant microorganisms, the release of nanoparticles into the environment needs to be closely regulated. The toxicity of nanoparticles is impacted by the reaction's medium and procedure, including the size of the particles, hydrophobic potential, and zeta potential. However, both PGPR and metal oxide nanoparticles can also have negative effects on plant growth and development, depending on the specific strains, the plant species, and the concentration and duration of exposure. For example, high concentrations of metal oxide nanoparticles can be toxic to plants, while some PGPR strains may negatively affect plant growth and development under certain conditions. Therefore, optimizing the concentration of metal oxide nanoparticles and PGPR strains should be scrutinized for future aspects.

6 Conclusion

Metal oxide nanoparticles, such as titanium dioxide and zinc oxide, have been shown to have various effects on plant growth and development when applied as a coating or included in the soil. These effects can be positive and negative, depending on the specific metal oxide, the plant species concentration, and the duration of exposure. Positive effects of metal oxide nanoparticles on plant growth and development have been observed in some studies, including increased photosynthetic efficiency, increased growth rate, biomass production, and improved stress tolerance. For example, titanium dioxide nanoparticles have been shown to increase the chlorophyll content and photosynthetic efficiency of lettuce and tomato plants, while zinc oxide nanoparticles have been shown to improve the growth and antioxidant activity of plants but at higher concentrations showed toxicity. Metal oxide nanoparticles can also have negative effects on plants, including reduced growth and biomass production, impaired root development, and decreased nutrient uptake. For example, high concentrations of titanium dioxide nanoparticles have been shown to inhibit the growth of lettuce and tomato plants, while zinc oxide nanoparticles have been shown to reduce the growth and biomass of rice plants.

Plant growth-promoting rhizobacteria (PGPR) are a group of beneficial bacteria that are found in the soil and are known to have a range of positive effects on plant growth and development, including increased root growth, improved nutrient uptake, and increased resistance to stress. When applied to plants, PGPR can improve plant growth, development and may also interact with metal oxide nanoparticles in various ways. For example, PGPR has been shown to mitigate the negative effects of metal oxide nanoparticles on plant growth and development, such as by reducing the toxicity of metal oxide nanoparticles or increasing the uptake of beneficial nutrients. In addition, PGPR can also enhance the positive effects of metal oxide nanoparticles on plants, such as by increasing the efficiency of photosynthesis or by improving stress tolerance.

Overall, the interaction between metal oxide nanoparticles and PGPR on plant growth and development is complex and can be affected by various factors, including the specific metal oxide and PGPR strains, the plant species, concentration, and duration of exposure. Further research is needed to fully understand the mechanisms underlying these interactions and to optimize the use of metal oxide nanoparticles and PGPR in agriculture and horticulture.

References

1. Adesemoye AO, Egamberdieva D (2013) Beneficial effects of plant growth-promoting rhizobacteria on improved crop production: prospects for developing economies. In: Bacteria in agrobiology: crop productivity. Springer, Berlin, Heidelberg, pp 45–63. https://doi.org/10.1007/978-3-642-37241-4_2
2. Ahmed B, Syed A, Rizvi A, Shahid M, Bahkali AH, Khan MS, Musarrat J (2021a) Impact of metal-oxide nanoparticles on growth, physiology and yield of tomato (*Solanum lycopersicum* L.) modulated by *Azotobacter salinestris* strain ASM. Environ Pollut 269:116218 https://doi.org/10.1016/j.envpol.2020.116218
3. Ahmed T, Noman M, Manzoor N, Shahid M, Abdullah M, Ali L, Wang G, Hashem A, Al-Arjani AB, Alqarawi AA, Abd-Allah EF (2021b) Nanoparticle-based amelioration of drought stress and cadmium toxicity in rice via triggering the stress responsive genetic mechanisms and nutrient acquisition. Ecotoxicol Environ Safety 209:111829. https://doi.org/10.1016/j.ecoenv.2020.111829
4. Alabdallah NM, Hasan MM, Hammami I, Alghamdi AI, Alshehri D, Alatawi HA (2021) Green synthesized metal oxide nanoparticles mediate growth regulation and physiology of crop plants under drought stress. Plants 10:1730. https://doi.org/10.3390/plants10081730
5. Asif M, Pervez A, Ahmad R (2019) Role of melatonin and plant-growth-promoting rhizobacteria in the growth and development of plants. CLEAN–Soil, Air, Water 47:1800459. https://doi.org/10.1002/clen.201800459
6. Bajracharya AM (2019) Plant growth promoting rhizobacteria (PGPR): biofertiliser and biocontrol agent-review article. J Balkumari Coll 8:42–45. https://doi.org/10.3126/jbkc.v8i0.29304

7. Bachheti RK, Konwarh R, Gupta V, Husen A, Joshi A (2019) Green synthesis of iron oxide nanoparticles: cutting edge technology and multifaceted applications. In: Nanomaterials and plant potential. Springer, Cham, pp 239–259. https://doi.org/10.1007/978-3-030-05569-1_9
8. Bachheti RK, Godebo Y, Bachheti A, Yassin MO, Azamal Husen (2020) Root-based fabrication of metal/metal-oxide nanomaterials and their various applications. In: Husen A, Jawaid M (eds) Nanomaterials for agriculture and forestry applications. Elsevier, pp 135–166. https://doi.org/10.1016/B978-0-12-817852-2.00006-8
9. Bechtaoui N, Raklami A, Benidire L, Tahiri AI, Göttfert M, Oufdou K (2020) Effects of PGPR co-inoculation on growth, phosphorus nutrition and phosphatase/phytase activities of faba bean under different phosphorus availability conditions. Pol J Environ Stud 29:1557–1565. https://doi.org/10.15244/pjoes/110345
10. Bhattacharyya PN, Jha DK (2012) Plant growth-promoting rhizobacteria (PGPR): emergence in agriculture. World J Microbiol Biotechnol 28:1327–1350. https://doi.org/10.1007/s11274-011-0979-9
11. Chai H, Yao J, Sun J, Zhang C, Liu W, Zhu M, Ceccanti B (2015) The effect of metal oxide nanoparticles on functional bacteria and metabolic profiles in agricultural soil. Bull Environ Contam Toxicol 94:490–495. https://doi.org/10.1007/s00128-015-1485-9
12. Choudhary KK, Pandey D, Agrawal SB (2013) Deterioration of rhizospheric soil health due to elevated ultraviolet-B. Arch Agron Soil Sci 59:1419–1437. https://doi.org/10.1080/03650340.2012.713473
13. Choudhary M, Chandra P, Arora S (2019) Soil-plant-microbe interactions in salt-affected soils. In: Research developments in saline agriculture. Springer, Singapore, pp 203–235. https://doi.org/10.1007/978-981-13-5832-6_6
14. Dimkpa CO, Singh U, Bindraban PS, Elmer WH, Gardea-Torresdey JL, White JC (2019) Zinc oxide nanoparticles alleviate drought-induced alterations in sorghum performance, nutrient acquisition, and grain fortification. Sci Total Environ 688:926–934. https://doi.org/10.1016/j.scitotenv.2019.06.392
15. Dorjey S, Dolkar D, Sharma R (2017) Plant growth promoting rhizobacteria Pseudomonas: a review. Int J Curr Microbiol App Sci 6:1335–1344. https://doi.org/10.20546/ijcmas.2017.602.160
16. Duhan JS, Kumar R, Kumar N, Kaur P, Nehra K, Duhan S (2017) Nanotechnology: the new perspective in precision agriculture. Biotechnol Rep 15:11–23. https://doi.org/10.1016/j.btre.2017.03.002
17. Duncan E, Owens G (2019) Metal oxide nanomaterials used to remediate heavy metal contaminated soils have strong effects on nutrient and trace element phytoavailability. Sci Total Environ 678:430–437. https://doi.org/10.1016/j.scitotenv.2019.04.442
18. Enebe MC, Babalola OO (2018) The influence of plant growth-promoting rhizobacteria in plant tolerance to abiotic stress: a survival strategy. Appl Microbiol Biotechnol 102:7821–7835. https://doi.org/10.1007/s00253-018-9214-z
19. Etesami H, Adl SM (2020) Plant growth-promoting rhizobacteria (PGPR) and their action mechanisms in availability of nutrients to plants. Phyto-Microbiome in stress regulation. Springer, Singapore, pp 147–203 https://doi.org/10.1007/978-981-15-2576-6_9
20. FAO, IFAD, UNICEF, WFP, WHO (2022) The state of food security and nutrition in the world 2022. Repurposing food and agricultural policies to make healthy diets more affordable. FAO, Rome. https://doi.org/10.4060/cc0639en. Accessed 5 Dec 2022
21. Goswami D, Pithwa S, Dhandhukia P, Thakker JN (2014) Delineating Kocuria turfanensis 2M4 as a credible PGPR: a novel IAA-producing bacteria isolated from saline desert. J Plant Interact 9:566–576. https://doi.org/10.1080/17429145.2013.871650
22. Govindasamy V, Senthilkumar M, Mageshwaran V, Annapurna K (2009) Detection and characterization of ACC deaminase in plant growth promoting rhizobacteria. J Plant Biochem Biotechnol 18:71–76. https://doi.org/10.1007/BF03263298
23. Gupta AK, Gupta M (2005) Synthesis and surface engineering of iron oxide nanoparticles for biomedical applications. Biomaterials 26:3995–4021. https://doi.org/10.1016/j.biomaterials.2004.10.012

24. Gupta G, Parihar SS, Ahirwar NK, Snehi SK, Singh V (2015) Plant growth promoting rhizobacteria (PGPR): current and future prospects for development of sustainable agriculture. J Microb Biochem Technol 7:96–102. https://doi.org/10.4172/1948-5948.1000188
25. Hamuda HE (2015) Influence of engineered metal oxide nanoparticles on seed germination, seedling development and chlorophyll content. Obuda Univ e-Bull 5:79
26. Haris Z, Ahmad I (2017) Impact of metal oxide nanoparticles on beneficial soil microorganisms and their secondary metabolites. Int J Life Sci Sci Res 3:1020–1030. https://doi.org/10.21276/ijlssr.2017.3.3.10
27. Hashem A, Tabassum B, Abd_Allah EF (2019) Bacillus subtilis: a plant-growth promoting rhizobacterium that also impacts biotic stress. Saudi J Biol Sci 26:1291–1297. https://doi.org/10.1016/j.sjbs.2019.05.004
28. Hassanisaadi M, Barani M, Rahdar A, Heidary M, Thysiadou A, Kyzas GZ (2022) Role of agrochemical-based nanomaterials in plants: Biotic and abiotic stress with germination improvement of seeds. Plant Growth Regul 4:1–44. https://doi.org/10.1007/s10725-021-00782-w
29. Hayat R, Ali S, Amara U, Khalid R, Ahmed I (2010) Soil beneficial bacteria and their role in plant growth promotion: a review. Ann Microbiol 60:579–598. https://doi.org/10.1007/s13213-010-0117-1
30. Hol WG, Bezemer TM, Biere A (2013) Getting the ecology into interactions between plants and the plant growth-promoting bacterium Pseudomonas fluorescens. Front Plant Sci 4:81. https://doi.org/10.3389/fpls.2013.00081
31. Husen A, Rahman QI, Iqbal M, Yassin MO, Bachheti RK (2019) Plant-mediated fabrication of gold nanoparticles and their applications. In: Nanomaterials and plant potential. Springer, Cham, pp 71–110. https://doi.org/10.1007/978-3-030-05569-1_3
32. Husen A, Siddiqi KS (2014) Carbon and fullerene nanomaterials in plant system. J Nanobiotechnol 12(1):1–16
33. Husen A (2020a) Interactions of metal and metal-oxide nanomaterials with agricultural crops: an overview. In: Husen A, Jawaid M (eds) Nanomaterials for agriculture and forestry applications. Elsevier Inc. 50 Hampshire St., 5th Floor, Cambridge, MA 02139, USA, pp 167–197. https://doi.org/10.1016/B978-0-12-817852-2.00007-X
34. Husen, A (2020b) Carbon-based nanomaterials and their interactions with agricultural crops. In: Husen A, Jawaid M (eds). Elsevier Inc. 50 Hampshire St., 5th Floor, Cambridge, MA 02139, USA, pp 199–218. https://doi.org/10.1016/B978-0-12-817852-2.00008-1
35. Husen A, Siddiqi KS (2023) Advances in smart nanomaterials and their applications. Elsevier Inc., 50 Hampshire St., 5th Floor, Cambridge, MA 02139, USA
36. Husen, A (2022) Engineered nanomaterials for sustainable agricultural production, soil improvement and stress management. Elsevier Inc., 50 Hampshire St., 5th Floor, Cambridge, MA 02139, USA
37. Husen A, Iqbal M (2019) Nanomaterials and plant potential. Springer International Publishing AG, Gewerbestrasse 11, 6330 Cham, Switzerland. https://doi.org/10.1007/978-3-030-05569-1
38. Islam F, Yasmeen T, Ali Q, Ali S, Arif MS, Hussain S, Rizvi H (2014) Influence of Pseudomonas aeruginosa as PGPR on oxidative stress tolerance in wheat under Zn stress. Ecotoxicol Environ Saf 104:285–293. https://doi.org/10.1016/j.ecoenv.2014.03.008
39. Jha CK, Saraf M (2015) Plant growth promoting rhizobacteria (PGPR). J Agric Res Dev 5:108–119. https://doi.org/10.13140/RG.2.1.5171.2164
40. Kang SM, Khan AL, Hamayun M, Hussain J, Joo GJ, You YH, Kim JG, Lee IJ (2012) Gibberellin-producing Promicromonospora sp. SE188 improves Solanum lycopersicum plant growth and influences endogenous plant hormones. J Microbiol 50:902–909. https://doi.org/10.1007/s12275-012-2273-4
41. Karimi H, Mahdavi S, Asgari Lajayer B, Moghiseh E, Rajput VD, Minkina T, Astatkie T (2022) Insights on the bioremediation technologies for pesticide-contaminated soils. Environ Geochem Health 44:1329–1354. https://doi.org/10.1007/s10653-021-01081-z
42. Karunakaran G, Suriyaprabha R, Manivasakan P, Yuvakkumar R, Rajendran V, Kannan N (2013) Impact of nano and bulk ZrO_2, TiO_2 particles on soil nutrient contents and PGPR. J Nanosci Nanotechnol 13:678–685. https://doi.org/10.1166/jnn.2013.6880

43. Khoshru B, Mitra D, Khoshmanzar E, Myo EM, Uniyal N, Mahakur B, Mohapatra PK, Panneerselvam P, Boutaj H, Alizadeh M, Cely MV (2020) Current scenario and future prospects of plant growth-promoting rhizobacteria: an economic valuable resource for the agriculture revival under stressful conditions. J Plant Nutr 43:3062–3092. https://doi.org/10.1080/01904167.2020.1799004
44. Kolenčík M, Ernst D, Urík M, Ďurišová Ľ, Bujdoš M, Šebesta M, Dobročka E, Kšiňan S, Illa R, Qian Y, Feng H (2020) Foliar application of low concentrations of titanium dioxide and zinc oxide nanoparticles to the common sunflower under field conditions. Nanomaterials 10:1619. https://doi.org/10.3390/nano10081619
45. Kuan KB, Othman R, Abdul Rahim K, Shamsuddin ZH (2016) Plant growth-promoting rhizobacteria inoculation to enhance vegetative growth, nitrogen fixation and nitrogen remobilisation of maize under greenhouse conditions. PLoS ONE 11:e0152478. https://doi.org/10.1371/journal.pone.0152478
46. Liu F, Xing S, Ma H, Du Z, Ma B (2013) Cytokinin-producing, plant growth-promoting rhizobacteria that confer resistance to drought stress in Platycladus orientalis container seedlings. Appl Microbiol Biotechnol 97:9155–9164. https://doi.org/10.1007/s00253-013-5193-2
47. Lowry GV, Avellan A, Gilbertson LM (2019) Opportunities and challenges for nanotechnology in the agri-tech revolution. Nat Nanotechnol 14:517–522. https://doi.org/10.1038/s41565-019-0461-7
48. Masood K, Yasmin H, Batool S, Ilyas N, Nosheen A, Naz R, Khan N, Hassan MN, Aldhahrani A, Althobaiti F (2021) A strategy for mitigating avian colibacillosis disease using plant growth promoting rhizobacteria and green synthesized zinc oxide nanoparticles. Saudi J Biol Sci 28:4957–4968. https://doi.org/10.1016/j.sjbs.2021.06.100
49. Mekonnen H, Kibret M (2021) The roles of plant growth promoting rhizobacteria in sustainable vegetable production in Ethiopia. Chem Biol Technol Agric 8:1–11. https://doi.org/10.1186/s40538-021-00213-y
50. Mishra VK, Kumar A (2009) Impact of metal nanoparticles on the plant growth promoting rhizobacteria. Dig J Nanomater Biostruct 4:587–592
51. Montano MD, Lowry GV, von der Kammer F, Blue J, Ranville JF (2014) Current status and future direction for examining engineered nanoparticles in natural systems. Environ Chem 11:351–366. https://doi.org/10.1071/EN14037
52. Munirah A, Norfarizan-Hanoon NA (2022) Interrelated of food safety, food security and sustainable food production. Food Res 6:304–310. https://doi.org/10.26656/fr.2017.6(1).696
53. Mushtaq T, Shah AA, Akram W, Yasin NA (2020) Synergistic ameliorative effect of iron oxide nanoparticles and Bacillus subtilis S4 against arsenic toxicity in Cucurbita moschata: polyamines, antioxidants, and physiochemical studies. Int J Phytorem 22:1408–1419. https://doi.org/10.1080/15226514.2020.1781052
54. Odoh CK (2017) Plant growth promoting rhizobacteria (PGPR): a bioprotectant bioinoculant for sustainable agrobiology. a review. Int J Adv Res Biol Sci 4:123–142. https://doi.org/10.22192/ijarbs.2017.04.05.014
55. Painuli S, Semwal P, Bacheti A, Bachheti RK, Husen A (2020) Nanomaterials from nonwood forest products and their applications. In: Husen A, Jawaid M (eds) Nanomaterials for agriculture and forestry applications. Elsevier, pp 15–40. https://doi.org/10.1016/B978-0-12-817852-2.00002-0
56. Pan B, Xing B (2012) Applications and implications of manufactured nanoparticles in soils: a review. Eur J Soil Sci 63:437–456. https://doi.org/10.1111/j.1365-2389.2012.01475.x
57. Panicker S, Sayyed RZ (2022) Hydrolytic enzymes from PGPR against plant fungal pathogens. Inantifungal metabolites of Rhizobacteria for sustainable agriculture. Fungal biology. Springer, Cham, pp 211–238. https://doi.org/10.1007/978-3-031-04805-0_10
58. Paul D, Lade H (2014) Plant-growth-promoting rhizobacteria to improve crop growth in saline soils: a review. Agron Sustain Dev 34:737–752. https://doi.org/10.1007/s13593-014-0233-6
59. Pindral S, Kot R, Hulisz P (2022) The influence of city development on urban pedodiversity. Sci Rep 12:1–13. https://doi.org/10.1038/s41598-022-09903-5

60. Poddar K, Sarkar D, Sarkar A (2020) Nanoparticles on photosynthesis of plants: effects and role. In: Green nanoparticles. Nanotechnology in the life sciences. Springer, Cham. pp 273–287 https://doi.org/10.1007/978-3-030-39246-8_13
61. Rai-Kalal P, Jajoo A (2021) Priming with zinc oxide nanoparticles improve germination and photosynthetic performance in wheat. Plant Physiol Biochem 160:341–351. https://doi.org/10.1016/j.plaphy.2021.01.032
62. Raliya R, Tarafdar JC, Biswas P (2016) Enhancing the mobilization of native phosphorus in the mung bean rhizosphere using ZnO nanoparticles synthesized by soil fungi. J Agric Food Chem 64:3111–3118. https://doi.org/10.1021/acs.jafc.5b05224
63. Ramamoorthy V, Viswanathan R, Raguchander T, Prakasam V, Samiyappan R (2001) Induction of systemic resistance by plant growth promoting rhizobacteria in crop plants against pests and diseases. Crop Prot 20:1–11. https://doi.org/10.1016/S0261-2194(00)00056-9
64. Rauch J, Kolch W, Laurent S, Mahmoudi M (2013) Big signals from small particles: regulation of cell signaling pathways by nanoparticles. Chem Rev 113:3391–3406. https://doi.org/10.1021/cr3002627
65. Raupach GS, Liu L, Murphy JF, Tuzun S, Kloepper JW (1996) Induced systemic resistance in cucumber and tomato against cucumber mosaic cucumovirus using plant growth-promoting rhizobacteria (PGPR). Plant Dis 80:891–894. https://doi.org/10.1094/PD-80-0891
66. Rizwan M, Ali S, Ali B, Adrees M, Arshad M, Hussain A, ur Rehman MZ, Waris AA (2019) Zinc and iron oxide nanoparticles improved the plant growth and reduced the oxidative stress and cadmium concentration in wheat. Chemosphere 214:269–277https://doi.org/10.1016/j.chemosphere.2018.09.120
67. Rizwan M, Ali S, Qayyum MF, Ok YS, Adrees M, Ibrahim M, Zia-ur-Rehman M, Farid M, Abbas F (2017) Effect of metal and metal oxide nanoparticles on growth and physiology of globally important food crops: a critical review. J Hazard Mater 322:2–16. https://doi.org/10.1016/j.jhazmat.2016.05.061
68. Rui M, Ma C, White JC, Hao Y, Wang Y, Tang X, Yang J, Jiang F, Ali A, Rui Y, Cao W (2018) Metal oxide nanoparticles alter peanut (Arachis hypogaea L.) physiological response and reduce nutritional quality: a life cycle study. Environ Sci: Nano 5:2088–2102. https://doi.org/10.1039/C8EN00436F
69. Santoyo G, Urtis-Flores CA, Loeza-Lara PD, Orozco-Mosqueda MD, Glick BR (2021) Rhizosphere colonization determinants by plant growth-promoting rhizobacteria (PGPR). Biology 10:475. https://doi.org/10.3390/biology10060475
70. Seyed Sharifi R, Khalilzadeh R, Pirzad A, Anwar S (2020) Effects of biofertilizers and nano zinc-iron oxide on yield and physicochemical properties of wheat under water deficit conditions. Commun Soil Sci Plant Anal 51:2511–2524. https://doi.org/10.1080/00103624.2020.1845350
71. Shafiq T, Yasmin H, Shah ZA, Nosheen A, Ahmad P, Kaushik P, Ahmad A (2022) Titanium oxide and zinc oxide nanoparticles in combination with cadmium tolerant bacillus pumilus ameliorates the cadmium toxicity in maize. Antioxidants 11:2156. https://doi.org/10.3390/antiox11112156
72. Shang Y, Hasan MK, Ahammed GJ, Li M, Yin H, Zhou J (2019) Applications of nanotechnology in plant growth and crop protection: a review. Molecules 24:2558. https://doi.org/10.3390/molecules24142558
73. Singh R, Pandey DK, Kumar A, Singh M (2017) PGPR isolates from the rhizosphere of vegetable crop Momordica charantia: characterization and application as biofertilizer. Int J Curr Microbiol App Sci. 6:1789–1802. https://doi.org/10.20546/ijcmas.2017.603.205
74. Suyal DC, Joshi D, Debbarma P, Soni R, Das B, Goel R (2019). Soil metagenomics: unculturable microbial diversity and its function. In: Varma A, Choudhary D (eds) Mycorrhizosphere and pedogenesis. Springer, Singapore, pp 355–362 https://doi.org/10.1007/978-981-13-6480-8_20
75. Velivelli SL, Sessitsch A, Prestwich BD (2014) The role of microbial inoculants in integrated crop management systems. Potato Res 57:291–309. https://doi.org/10.1007/s11540-014-9278-9
76. Vocciante M, Grifoni M, Fusini D, Petruzzelli G, Franchi E (2022) The role of plant growth-promoting rhizobacteria (PGPR) in mitigating plant's environmental stresses. Appl Sci 12:1231. https://doi.org/10.3390/app12031231

77. Walli MH, Al-Jubouri Z, Madumarov MM, Margaryta M, Aldibe AA (2022) Genetic and environment diversity to improve wheat (Triticum spp.) productivity: a review. Res Crops 23:295–306. https://doi.org/10.31830/2348-7542.2022.041
78. Wang X (2022) Managing land carrying capacity: key to achieving sustainable production systems for food security. Land 11:484. https://doi.org/10.3390/land11040484
79. Wu H, Li Z (2022) Recent advances in nano-enabled agriculture for improving plant performance. Crop J 10:1–12. https://doi.org/10.1016/j.cj.2021.06.002
80. Wu H, Shabala L, Shabala S, Giraldo JP (2018) Hydroxyl radical scavenging by cerium oxide nanoparticles improves Arabidopsis salinity tolerance by enhancing leaf mesophyll potassium retention. Environ Sci Nano 5:1567–1583. https://doi.org/10.1039/C8EN00323H
81. Zhai Y, Wang Z, Wang G, Peijnenburg WJ, Vijver MG (2020) The fate and toxicity of Pb-based perovskite nanoparticles on soil bacterial community: impacts of pH, humic acid, and divalent cations. Chemosphere 249:126564. https://doi.org/10.1016/j.chemosphere.2020.126564
82. Zhang LJ, Qian L, Ding LY, Wang L, Wong MH, Tao HC (2021) Ecological and toxicological assessments of anthropogenic contaminants based on environmental metabolomics. Environ Sci Ecotechnol 5:100081. https://doi.org/10.1016/j.ese.2021.100081

Recent Application and Future Prospects of Nanoparticles-Based Colorimetric Sensors for Residual Pesticides Detection

Selvaraj Mohana Roopan, Murugesan Shobika, and Gunabalan Madhumitha

Abstract All over the world, there is consistent and regular use of pesticides, resulting in their accumulation in food, water, and soil. Numerous significant problems have been connected to pesticide exposure, including cancer, pregnancy, various other adverse health effects, and birth defects. As a result, there is a squeezing requirement for quick and easy methods to find pesticides in foodstuffs to protect human health. Nanoparticle-based colorimetric diagnostics have received a lot of interest for rapid and efficient pesticide residue detection. Pesticides can be detected with the naked eye due to a chemical interaction between the analytes and the surface of nanoparticles, which causes periodic optical alterations. When metal nanoparticles cluster show various linear and non-linear optical characteristics in the visible spectral range, Localized Surface Plasmon Resonance, or LSPR, is the term used to describe this phenomenon. These methodologies are used for on-site detection and are both precise and cost-effective. Furthermore, the modification and functionalization of nanoparticles have significantly increased the selectivity and sensitivity of such assays. We summarised and presented the most current uses of SMNPs for the identification of leftover pesticides in this chapter.

Keywords Colorimetric sensors · Metal nanoparticles · Residual pesticides · Surface-modified nanoparticles

Abbreviations

DLS analysis	Dynamic light scattering
DM	Dimethoate
IR spectroscopy	Infrared spectroscopy
LSPR	Localized surface plasmon resonance

S. M. Roopan (✉) · M. Shobika · G. Madhumitha
Chemistry of Heterocycles & Natural Product Research Laboratory, Department of Chemistry, School of Advanced Sciences, Vellore Institute of Technology, Vellore, Tamil Nadu 632014, India
e-mail: mohanaroopan.s@vit.ac.in

R. K. Bachheti et al. (eds.), *Nanomaterials for Environmental and Agricultural Sectors*, Smart Nanomaterials Technology,
https://doi.org/10.1007/978-981-99-2874-3_12

NWs	Nanowires
OP	Organophosphorus
SEM	Scanning electron microscopy
SMNPs	Surface-modified nanoparticles
TCh	Thiocholine
TEM	Transmission electron microscopy
TMB	3,3′,5,5′-Tetramethylbenzidine
UV–vis spectroscopy	Ultraviolet-visible spectroscopy

1 Introduction

Hazardous levels of chemical compounds and poisons in foodstuffs have been a major source of concern in food safety and public health communities [21, 63]. Every year, roughly 600 million people are affected, resulting in 4,20,000 deaths [9, 31]. This type of catastrophe greatly impacts the nation's health and prosperity and creates barriers to worldwide trade [2]. Pesticides are a major source of food contamination and safety. Modern agriculture is causing food poisoning now. If not utilized carefully and under regulated conditions, their hazardous qualities constitute a significant danger to humans. Since the mid-twentieth century, there has occurred too much unjustifiable need to use pesticides, resulting in substantial environmental degradation. Its plays a major role not only in agriculture but also in other sources of agriculture. In addition to being employed in the agricultural industry, they are also utilized in non-agricultural areas like disease vectors in both industrial and residential settings and pest management. They undergo metabolic alterations. It can be found in all ecosystem elements, including food, river, groundwater, and air.

We now needed an accurate and straightforward approach to checking pesticide residue in food products. Various traditional and modern techniques, including IR spectroscopy [13], spectrophotometry [5, 10], chromatography [45, 49], polarography [1], and GC–MS [55], HPLC [56, 62], have been used for this purpose over the years. These approaches offer strong sensitivity, stability, and reproducibility; however, they come with several drawbacks, such as difficult sample preparation processes, expensive and insufficient real-time analysis, and long analysis times. As a result, these limitations underscore the importance of developing facile and quick strategies for tracking and detecting leftover pesticides in food and environmental items that do not require expensive equipment [52].

Nanoparticles have been extensively employed in designing inexpensive and fast sensors that use colour since the dawn of nanotechnology [51, 58]. The nanoparticles-based sensor has garnered a lot of interest for detecting residual pesticides owing to its easy operation, minimal cost, and great sensitivity [22]. Because of their tiny size, high extinction coefficients, and vast surface area, nanoparticles are ideal for detecting leftover pesticides [6, 41, 47]. Furthermore, SMNPs, including an additional ligand, improve binding potential to desired compounds [33, 43, 48]. There have been several published publications describing the relevance of nanoparticle-based colorimetric approaches for detecting proteins[64], heavy metal ions [7, 46], biological agents [57] and organic compounds [11], and melamine [17].

Colorimetric sensing methods represent one of the most efficient and precise methods for identifying a variety of Pesticides, which including Fungicides, insecticides etc. [18]. The major focus of this investigation was the use of nanoparticle-based SMNPs to find the presence of leftover pesticides using colorimetry tactics in foodstuff and ecology. This chapter summarises the tactics for rapid identification of leftover pesticides in foodstuff or items and the environment using nanoparticle-based colorimetry sensors.

2 Nanoparticle-Based Colorimetric Approaches for the Mechanism of Pesticide Detection

The colour change is used to quantify nanoparticle-based colorimetric experiments. The essential premise for colorimetric tests is interactions between the chemical and molecules between the surrounding nanoparticles and target analytes. Donor–acceptor reactions based on coordination chemistry, including electrostatic, Van der Waal, and hydrogen interactions, have undoubtedly been utilized to selectively and sensitively find various analytes [34]. When the desired sample components come into touch with NPs with essential unique features, they produce agglomeration and a colour change, confirming the presence of analytes [33, 50]. comparable methods have also been used to assess pesticide residues.

Surface-modified nanoparticles have also been utilized in a comparable manner to evaluate pesticide residues in food products and the environment. In ideal circumstances, pesticides recovered from food samples are frequently examined by affixing them to a colloidal solution with minute concentrations of nanoparticles. Surface-modified nanoparticles interact with pesticides to aggregate, which alters the nanoparticles' colour. This interaction is clearly visible to the unaided eye and is independently confirmed by TEM, DLS analysis, and UV–vis spectroscopy.

3 Current Use of Nanoparticles in Assessing Leftover Pesticides in Surrounding and the Foodstuff

For the low-cost monitoring for organophosphorus (OP) pesticides in foodstuffs, Wei et al. [54] synthesised a novel composite of 3DRGO-$NiFe_2O_4$/NiO. It was regarded as one of the easiest and quickest ways of detection. The high-performance mimetic peroxidase activity in these nanoparticles also allows for the colorimetric detection of OPs. The reference chemical for evaluating detection capacity was dichlorvos. The linear detection range for dichlorvos is 50 mg/mL to 2.5. The detection threshold is set at 10 mg/mL at 104 mg/mL^{-1}.

Dichlorvos may be converted into digital signals, which a smartphone can use to recognise quantitative data and the paper sensor's image data. Notably, this method may be used to locate dichlorvos in actual environments. Additionally, a test paper was created using 3DRGO-$NiFe_2O_4$/NiO NPs.

Ag_2O in cube form has an oxidase-like activity that can be used a unique colorimetric strategy for the recognition of dimethoate (DM) was established by [60]. The experiment demonstrated that dimethoate might facilitate the chromogenic substrate oxidation under catalytic conditions by accelerating the discharge of oxygen radicals and $^1O_2$ and promoting electron transfer from dissolved oxygen to Ag_2O. Individual calculations yielded the colorimetry's LOD and LOQ to be 14 gL^{-1} and 46 gL^{-1}, respectively, which are significantly less than the upper limit residues of DM allowed in various plants. With a minimum detection threshold of 14 gL^{-1}, the colorimetric approach demonstrated excellent sensitivity. Further evidence that this technology has a bright future for studying pesticide residues in cuisine comes from its high selectivity compared to rival pesticides and robust recovery in several common vegetables.

Liu et al. created boron carbon oxynitride quantum dots in a one-step hydrothermal procedure by combining boric acid and ethanolamine [23]. Thiocholine (TCh) may reductively break down MnO_2 nanosheets to form Mn^{2+}, restoring BCNO QDs' fluorescent properties produced by the breakdown of acetylcholine with the assistance of the catalyse Acetylcholinesterase. Organophosphorus insecticides have inhibited AChE enzyme activity, TCh synthesis and MnO_2 nanosheet disintegration causing the "switch-off" of fluorescence. Consequently, by observing the fluctuation of fluorescence potency of ACHE-ATCH-MnO_2-BCNO-QDs system, the OP concentration may be estimated. Under optimum experimental circumstances, Paraoxon's dynamic detection range was 0.1–250 ng mL^{-1}, whereas the 0.03 ng mL^{-1} limit of detection was used. Then they identified concentration-dependent visible colour changes in the reaction system. Additionally, they have been developed a BCNO QD-enabled test paper that is disposable. Finally, they performed digital picture chromaticity analysis utilizing a smartphone with RGB values of the test paper and reaction solution, lowered the identification cost and time and offered a practical way to quickly identify OPs on the spot.

The non-contaminating detection of OPs remains a complicated topic. As a result, Huang et al. [12] devised a steady and sensitive approach. They developed a simple colour paper sensor utilizing 3,3′,5,5′-tetramethylbenzidine (TMB) and γ-MnOOH nanowires (NWs) and for quickly and sensitively screened OPs and acetylcholine function. Utilizing variations in absorbance at 652 nm or the blue colour of oxTMB preparations, the dose of OPs was calculated to act as an AChE inhibitor. As a result of this detection platform, considerable LODs for the activity of AChE was established at 0.007 mU mL^{-1}, 0.14 ng mL^{-1}, and 0.35 ng mL^{-1} for dichlorvos and omethoate, respectively. After the portable construction of this type of AChE-nanozyme tandem response was attained on the test paper, LODs for AChE, omethoate, and dichlorvos were 0.1, 10, and 3 ng mL^{-1}, respectively. Finally, they concluded that this sensing system has excellent selectivity and interference resistance in both solid and solution phases and stated that they performed admirably in vegetable samples and actual serum.

Shefali Singh and his colleagues [44] have published a straightforward, accurate, and focused label-free colorimetric test for malathion detection employing a palladium-gold nanorod as a nanozyme. The study examined the nanozyme's capacity for peroxidase activity on colorimetric stimuli and explored the impact of several OPs on this activity. The Pd-Au nanozyme exhibited outstanding peroxidase-mimicking activity with O-phenylenediamine in the presence of H_2O_2. It is a superior enzyme because it has better kinetic characteristics than horseradish peroxidase, namely K_m and K_{cat}. Nanozyme has strong peroxidase activity between 2 and 6 pH and is enduring throughout temperatures vary from 4 to 70 °Celsius. The fundamental idea behind the assay was the selective quenching nanozyme peroxidase activity with enhancing malathion concentration. The test has a detection limit of 60 ng/ml and did not exhibit cross-reaction with similar metal salts or organophosphates. Validation on tap water samples enhanced with various levels of malathion showed excellent restoration in the 80–106% range. The intra- and inter-assay precision of a test was good and ranged from 2.7 to 6.1% and 3.2 to 5.9%, respectively. This work showed the palladium-gold nanorods' catalytic capability, which may be used as a nanozyme to provide extremely sensitive detection techniques.

Tracer antibodies have been used to create the spatially-resolved multianalyte immunoassay (ICA) for the first time. g-C_3N_4/$BiFeO_3$ NCs were used as a chemiluminescent/colorimetric double ICA for the measurement of leftover pesticides using carbaryl and chlorpyrifos are the model analytes by Hui Ouyang [32]. The tracer antibodies were created by attaching g-C_3N_4/$BiFeO_3$ NCs to the carbaryl and chlorpyrifos antibodies. The generation of G-C_3N_4/$BiFeO_3$ NCs-induced CL signal was also captured as a sensitive, quantitative signal after starting the luminol-H_2O_2 reaction in the test lines. Both carbaryl and chlorpyrifos had LODs of 0.033 ng/mL^{-1} under ideal circumstances. Injections of carbaryl and chlorpyrifos were made into ambient specimens and water were effectively detected using the dual-readout ICA with 80–119% and 90–118% are reasonable recovery levels. The innovative ICA demonstrated tremendous potential in various fields, including medication safety, environmental control, and medical medicine, due to several benefits, including cheap cost, speed of delivery, fast response, and superior mobility.

3.1 Cu Nanoparticle-Based Colorimetric Pesticide Sensing

Cu is a vital component of the bioactive components crucial to biological systems [3]. Along with Ag and Au NPs, copper nanoparticles are receiving great interest as colorimetric assays for the pesticide detection process [3, 14, 29, 36]. Cu NPs are easier and less costly to synthesize than Ag and Au NPs. Cu nanostructure materials have various drawbacks, such as easy agglomeration and low stability, that restrict their wider use when detecting trace levels of analytes using colorimetry. This is true even if the manufacture of Cu NPs is less expensive than that of Au and Ag NPs. Only a few papers have recently discussed the functionalization of the Cu NPs allows for colorimetric pesticide tracking.

Identifying chlorpyrifos calorimetrically by using a Cu-MOF sensor. The Fe_3O_4-aptamer surfaces of the Cu-MOF were changed with c-DNA before being constructed, creating a magnetically controllable optical detector for chlorpyrifos testing [25]. It is obvious that the TMB/H_2O_2 colour shifts previous and after the Cu MOF-Fe_3O_4-aptasensor interaction with chlorpyrifos served as the basis for the investigation of chlorpyrifos. By using UV–visible spectrometry to monitor the colour of solutions at 650 nm, the colorimetric test for chlorpyrifos was carried out. The results showed dynamic linear range with a 4.4 ng/mL^{-1} LOD from 0 to 1250 ng/mL^{-1}.Notably, even in the presence of other pesticides, the new analytical technique demonstrated strong selectivity for chlorpyrifos. The method's potential was evaluated for measuring the presence of chlorpyrifos in food items like vegetables and fruits. The results showed high recoveries, indicating that the approach might be employed as a promising alternative to GC–MS.

In juice and ambient samples, CTAB-Cu nanoparticles were effectively used as a Nanosensor for colorimetry for sensing insecticides maneb ziram, and, zineb, that include Dithiocarbamates [8]. By injecting pesticides (maneb, ziram, and zineb), the red tint of these NPs transformed to a yellow tint, enabling the development of straightforward colorimetric techniques for each pesticide's detection. With greater dynamic values of maneb ziram and, zineb, the prepared NPs served as a colour-based sensor device for pesticide detection. Three types of pesticides were identified using the analytical application of the technique in the water, mango juice, and tomato extract, with excellent rebounds between 95.8 and 108.5%. These findings show that incorporating functionalized Cu nanoparticles and a smartphone, UV–Visible spectrophotometry for visual readout on-site detection of trace amounts of pesticides in water and food items has been effective.

The creation of a sensor made of paper for specific phenthoate detection in food and beverage also a ligand called citrate was utilized to synthesize core–shell of Cu@Ag NPs [42]. It is possible to construct a colorimetric sensor which made of paper for specific phenthoate sensing as a consequence of the way that phenthoate had contact with the outer layers of core–shell of Cu@Ag nanostructure. This interaction resulting in a shift in colour because of agglomeration. The newly created sensor had an outstanding LOD of 15 mg L^{-1}, a larger linear range of 50 to 1500 mg L^{-1}, and ranging from 92.6 to 97.4% strong recovery. Phenthoate concentrations in edible samples and water were measured using a colorimetric instrument with a paper-based medium linked to a smartphone.

In order to find dimethoate, a portable nano colorimetric equipment that could be linked to a smartphone was created. Cu NPs were added to a stimuli-responsive hydrogel to selectively and sensitively find dimethoate [16]. This approach demonstrated a potential accessible analysis approach for dimethoate on-site screening in edible items and has a 1.0 mg L^{-1} detection limit.

In order to detect acetamiprid, Cu NP surfaces were combined with a biopolymer called DNA with an AT-rich double strand (ds) [4]. The development of a compound combining acetamiprid and aptamer was the foundation for the detection theory. Acetamiprid may be detected by the designed probe as low as 2.37 nm. The Cu NPs with biopolymer caps worked well as a probe to find acetamiprid in fruits like apples.

3.2 Colorimetric Pesticide Sensing Using Au–Ag Nanoparticle

AChE-based colorimetric assays utilising NPs have been employed in several investigations to detect AChE inhibitors quickly and directly in aqueous solutions. Due to their unique colorimetric traits, AgNPs and AuNPs were used in biological and ecological systems for OP pesticide detection [30].

Considering their robust plasmon resonance and high extinction coefficients, gold and silver have received much interest for detecting residues of pesticides in the surroundings and groceries matrices [59, 61]. Despite significant advances in metal nanoparticle-based colorimetric techniques for the identification of different sample matrix such as pathogens chemical toxins, and heavy metals, only just a few studies that describe SMNPs being used as colorimetric indicators for the sensing of the leftover pesticides in packaged foods [35, 40].

A quick and easy Au NPs-based colorimetric sensor was created by Sun and colleagues to find OP pesticides. The tracking approach is founded on AchE catalytic reaction in the ubiety of OP pesticides that led to the aggregation and colour transformation of lipoic acid-capped Au-NPs to steely blue from red. The suggested sensor system is adequate for identifying OP chemicals at lower concentrations in contrast to the presence of other insecticides, according to the results of the TEM and UV–vis investigation. Consequently, the created Colorimetric test might develop into an effective on-site. Because this sensor technology has great potential, it can detect organophosphate chemicals to deliver quick and precise results without utilising any specialised instrumentation.

A colorimetric sensor for tricyclazole fungicide detection using SAADTC-Ag NPs was developed by Rohit and Kailasa [38]. The detecting mechanism is based on SAADTC-Ag NPs tricyclazole-induced aggregation via donor–acceptor of electrons interaction, which results in a color change of SAADTC-Ag NPs. The proposed NPs-based sensor was incredibly biased for TCZ fungicide in contrast to the additional presence of pesticides that potentially interfere with a LOD of 1.8 10^{-7} M for tricyclazole fungicide. Additionally, the analysis of actual rice samples is successfully utilised for this NPs-based sensor.

Employing melamine-modified Au NPs, Kang et al. [15] suggested a trustworthy and extremely confidential approach to identifying pymetrozine. The effective integration of melamine onto the AuNPs surface was confirmed by zeta potential, X-ray microscopy, and EDX methods after AuNPs were created utilizing a chemical synthesis process. A new SPR peak was discovered at 680 nm when pymetrozine pesticides were administered to M-Au NPs, which is owing to the h—bonding between M-AuNPs and pymetrozine, that produced accumulation and resulting color shifts to dark-blue from red. The sensing system is very sensitive, with a LOD of 10 nM for UV–vis spectroscopy and 80 nM for the unaided eye. Pymetrozine pesticides were effectively detected using the sensor technology in green tea and apple juice samples. The outcomes revealed good concordance with those found by HPLC.

Dipstick immunoassay was used by Malarkodi et al. [27] to provide a straightforward, quick technique for the identification of KOP in vegetables and fruits. The bioconjugate Au- NPs were added to an IgG antibody, which was then employed as a kitazine detection colorimetric probe. Using DIA, various levels of organophosphates in the food samples were found by developing a purple colour on the membrane strips. Other interferant compounds were used to test the method's sensitivity and selectivity, and the findings revealed that kitazine and other pesticides had a strong affinity for the DIA. As a result, our strategy offers a unique, economical, and environmentally benign way for kitazine's quick detection in usual food specimens.

Menon and colleagues describe employing pSC4R-Ag NPs to detect dimethoate pesticides simply and extremely sensitively in industrial wastewater [28]. Alteration of Silver NPs surfaces by pSC4R was identified using a slight peak shift obtained from UV–Vis spectroscopy. The pSC4R specifically binds via electrostatic interactions and π–π interactions to dimethoate, causing those NPs to aggregate in the availability of dimethoate and change color to red from yellow. UV–vis spectroscopy was employed to measure dimethoate's concentration and quantify the colour shift visible to the naked eye. Additionally, pSC4-Ag NPs were used to analyze several pesticides, including imidacloprid, 2,4-D, monocrotophos, parathion, dimethoate, dichlorvos, and hexaconazole, to evaluate the uniqueness of the suggested sensors. The observations from UV–vis spectroscopy show that the dimethoate-containing solution changed colour from yellow to red, whereas no other pesticides demonstrated any discernible colour change. The new assay has consequently shown a lot of promise for analysing dimethoate in actual analytes.

A lauryl sulphate-capped gold nanoparticle-based assay method was created by Lakade et al. [19] for the quantitative determination of Mangoes that have been artificially ripened contain calcium carbide. Utilizing lauryl sulphate for the synthesis, UV–vis spectroscopy, DLS, and SEM methods were used to characterize the Au NPs. Gold nanoparticles with lauryl sulphate are red in colour and has exhibit maximum absorption at 520 nm. After adding arsenic standard solution, the absorption of lauryl sulphate-capped Au NPs decreased, and a new absorption peak was discovered at 620 nm. This resulted from the accumulation of Au NPs with lauryl sulphate when arsenic is present and altered the red to blue colour. Calcium carbamide may be found using arsenic in both natural conditions and unnaturally torn mangoes. The detection concept is based on the evidence that when Au NPs interact with arsenic that has lauryl sulphate, the arsenic causes them to aggregate. The method may be used to find carbide in fruits and vegetables at the point of care.

One of the deadliest pyrethroid pesticides is deltamethrin. It is frequently used to eradicate insects and lice from fruits and vegetables. Wang and coworkers reported employing MNBT-Au NPs as a clearer vision to detect deltamethrin [53]. The fast adhesion of deltamethrin to the top of altered Au NPs and changes color became visible after inducing aggregation that was further assessed by UV–vis spectrophotometry, forming the basis of the detection mechanism. Nano-based sensors have been developed to detect deltamethrin in tomato and cherry samples. The claimed method's LOD was 0.25 M by bare eye & 0.05 M by UV–vis spectrophotometer. These findings demonstrated that nano-based sensor are suitable for a quick and precise examination.

CS-Au NPs derived colorimetric technique for finding glyphosate in ecological and aquatic specimens was created by Zheng et al. [65]. Based on the electrostatic attraction that causes these NPs to aggregate in the vicinity of glyphosate, the detection approach. After 1.5 ml of 20 mM Acetic acid- sodium acetate buffer with pH = 4.0 and 1.2 mL solution of the CS-Au NPs were mixed for detection, the amount of glyphosate was measured by UV–vis spectrophotometry. This caused the CS-Au NPs to shift color, which is visible to the unaided eye. Other pesticides did not affect these NPs; the only aggregate found was glyphosate. With a LOD of 5.88 108 M, this technique performed analytically quite well. This method demonstrates excellent sensitivity for detecting glyphosate in actual water samples.

Besides that, for the colorimetric assay utilized in mancozeb measurement, Rohit and his coworkers employed DDTC-Ag NPs [39]. They employed mancozeb, which possesses a secondary amine group that facilitates the enamine reactions and Michael additional reaction to connect the phenolic groups of DDTC to it easily. According to the researchers, only these NPs aggregated in the presence of Mancozeb, and no other pesticides were found to be absorbed. Mancozeb was quickly detected using this NPs-based colorimetric sensor down to concertation of 21.1×10^{-6} M. The results indicated a fair degree of agreement with those obtained by gas chromatography. As a result, it shows that the created colorimetry probe is extremely sensitive and selective for mancozeb.

A very sensitive sensor based on Ag NPs was created by Ma et al. [26] in the case of colorimetric identification of triazophos pesticides in specimens of water and food. Utilizing tri-sodium citrate and sodium borohydride, silver nanoparticles have been created. By adding triazophos insecticide, the absorption peak of the as-prepared nanoparticles of silver, which was used to be analysed by UV–vis spectroscopy, shifted to red at 525 nm. The detection phenomena are confirmed utilizing dynamic light scattering methods and TEM. The yellow colour changes to red because of π–π interactions between triazophos and citrate and H-bonding in the detecting mechanism. Using a 5 nm detection limit, the suggested sensor system demonstrated exceptional sensitivity. Additionally, water and fruit samples were examined using the suggested technique to judge the constructed sensor's usability and reliability. Moreover, the outcome demonstrates that the sensing system is reliable, easy to use, and capable of quickly detecting triazophos in a genuine food sample.

A team of researchers has developed a sensor that can detect the pesticide quinalphos even if other pesticides are present. Additionally, p-NADTC-Au NPs were employed by Rohit and his team to perform colorimetry identification of quinalphos in food and water specimens [37]. It was discovered at 522 nm, as compared to 520 nm, for the surface plasmon resonance of bare Au NPs. The gold nanoparticles' colour altered to blue from dark red because of electrostatic interactions between quinalphos and the modified NPs, which were the basis for the detecting technique. By evaluating how the technique responded to various pesticides such 2,4-D, chlorpyrifos, hexaconazole, monocrotophos, atrazine, indoxacarb, and acetamiprid, its selectivity was determined. This method's LOD was 0.35 M, which is less than that of previous approaches that have been published. It was also effectively detected in real samples of tomatoes, rice, wheat, and water.

The most common application of atrazine is to control grassy weeds that grow in various species. Using gold nanoparticles treated with melamine, Liu and colleagues [24] established a sensitive colorimetric detection technique. Melamine has been shown to alter the Au NPs surface, changing its colour from red to blue. When an atrazine sample is exposed to M-Au NPs, the most absorption occurs at redshifts and at 640 nm, a new absorption peak is observed. With a LOD of 0.0165 M, the established strategy has been effectively employed to evaluate atrazine in aquatic specimens. The assay's selectivity was also evaluated by including a few achievable pesticide interventions. According to the results, atrazine was the assay system's most sensitive test material out of all the pesticides.

Malathion was recently detected using a nanoparticle-based colorimetric approach based on Au NPs' anti-aggregation properties [20]. Trisodium citrate was used as a reducing and moderating agent to create the wine-red spherical Au NPs. Typically, citrate-stabilized gold nanoparticles caused aggregation at alkaline circumstances (pH > 9) because of electrostatic interactions, which changed the colour from red to grey. Because malathion is easily hydrolysed in an alkaline environment, the detection concept of this approach was based on the observation that when aggregated nanoparticles were treated with hydrolysed malathion, their grey colour changed to red. According to the data, this proposed method's analytical performance under ideal circumstances was 11.8 nm. The detection of malathion in samples of water, fruit, and vegetables using an anti-aggregation sensor has demonstrated this detector's usefulness for studying genuine specimens.

4 Future Outcomes and Conclusions

A technology that has promise for the advancement of colorimetric techniques for identifying leftover pesticides in specimens of ecological and foodstuffs is the surface modification of nanoparticles. It is now necessary to constantly test the environment and food supply for the presence of residual pesticides. In terms of ease of use, accuracy, speed, and lack of specialised equipment, colorimetric techniques based on nanotechnology are superior to more traditional ones for detecting pesticides. This kind of test might eventually take the place of certain current techniques for detecting pesticides. The molecular and chemical interactions between desired pesticides and SMNPs are the basis for nanoparticle-based colorimetric detectors, implying a unique and delicate procedure. Therefore, surface-modified nanoparticles are a practical tool for quickly and economically identifying leftover pesticides in various environmental and food matrices without the need for expensive and specialized apparatus due to their multifunctional properties.

Despite their great promise, the commercialization of nanotechnology-based colorimetric approaches presents many difficulties for the scientific community to overcome.

1. Although many researchers have simplified sample extraction for the colorimetric approach, most of them have used water as a sample. Creating an extraction technique for complicated food samples continues to be a significant problem.
2. To investigate surface modifiers that can tolerate the potential interference from other pollutants, additional research is needed to improve the specificity and repeatability of the colorimetric approach.
3. The point-of-care use of these procedures can be enhanced by creating strip-based techniques employing colloidal nanoparticles with modified surfaces.

The authors write in their conclusion that the new study will inspire more research on this subject. The authors are optimistic that the difficulties in the colorimetric identification of leftover pesticides in challenging environmental and food samples will be addressed by the efforts of researchers working together.

References

1. Affam AC, Chaudhuri M (2013) Degradation of pesticides chlorpyrifos, cypermethrin and chlorothalonil in aqueous solution by TiO_2 photocatalysis. J Environ Manage 130:160–165. https://doi.org/10.1016/j.jenvman.2013.08.058
2. Bülbül G, Hayat A, Andreescu S (2015) Portable nanoparticle-based sensors for food safety assessment. Sensors 15:30736–30758. https://doi.org/10.3390/s151229826
3. Dhavle V, Kateshiya MR, Park T-J, Kailasa SK (2021) Functionalization of silver nanoparticles with carbohydrate derivative for colorimetric assay of thiram. J Electron Mater 50:3676–3685. https://doi.org/10.1007/s11664-021-08875-y
4. Fan K, Kang W, Qu S, Li L, Qu B, Lu L (2019) A label-free and enzyme-free fluorescent aptasensor for sensitive detection of acetamiprid based on AT-rich dsDNA-templated copper nanoparticles. Talanta 197:645–652. https://doi.org/10.1016/j.talanta.2019.01.069
5. Farajzadeh MA, Sattari Dabbagh M, Yadeghari A (2017) Deep eutectic solvent based gas-assisted dispersive liquid-phase microextraction combined with gas chromatography and flame ionization detection for the determination of some pesticide residues in fruit and vegetable samples. J Sep Sci 40:2253–2260. https://doi.org/10.1002/jssc.201700052
6. Fu C, Zhao G, Zhan H et al (2013) Evaluation and characterization of reduced graphene oxide nanosheets as anode materials for lithium-ion batteries. 6269–6280. Int J Electrochem Sci 8:6269–6280
7. Gao Y, Li X, Li Y, Li T, Zhao Y, Wu A (2014) A simple visual and highly selective colorimetric detection of Hg^{2+} based on gold nanoparticles modified by 8-hydroxyquinolines and oxalates. Chem Commun 50:6447. https://doi.org/10.1039/c4cc00069b
8. Ghoto SA, Khuhawar MY, Jahangir TM, Mangi JUD (2019) Applications of copper nanoparticles for colorimetric detection of dithiocarbamate pesticides. J Nanostruct Chem 9:77–93. https://doi.org/10.1007/s40097-019-0299-4
9. Godfray HCJ, Beddington JR, Crute IR, Haddad L, Lawrence D, Muir JF, Pretty J, Robinson S, Thomas SM, Toulmin C (2010) Food security: the challenge of feeding 9 billion people. Science 327(80–):812–818. https://doi.org/10.1126/science.1185383

10. Grahovac ZM, Mitić SAS, Pecev ET, Pavlović AN (2010) Development of new kinetic-spectrophotometric method for determination insecticide dimethoate in milk and water. J Chinese Chem Soc 57:1027–1034. https://doi.org/10.1002/jccs.201000143
11. Hu J, Ni P, Dai H, Sun Y, Wang Y, Jiang S, Li Z (2015) A facile label-free colorimetric aptasensor for ricin based on the peroxidase-like activity of gold nanoparticles. RSC Adv 5:16036–16041. https://doi.org/10.1039/C4RA17327A
12. Huang L, Sun D-W, Pu H, Wei Q, Luo L, Wang J (2019) A colorimetric paper sensor based on the domino reaction of acetylcholinesterase and degradable γ-MnOOH nanozyme for sensitive detection of organophosphorus pesticides. Sens Actuators B Chem 290:573–580. https://doi.org/10.1016/j.snb.2019.04.020
13. Jia D, Gao J, Wang L, Gao Y, Ye B (2015) Electrochemical behavior of the insecticide pymetrozine at an electrochemically pretreated glassy carbon electrode and its analytical application. Anal Methods 7:9100–9107. https://doi.org/10.1039/C5AY01987G
14. Kailasa SK, Koduru JR, Desai ML, Park TJ, Singhal RK, Basu H (2018) Recent progress on surface chemistry of plasmonic metal nanoparticles for colorimetric assay of drugs in pharmaceutical and biological samples. TrAC Trends Anal Chem 105:106–120. https://doi.org/10.1016/j.trac.2018.05.004
15. Kang J, Zhang Y, Li X, Dong C, Liu H, Miao L, Low PJ, Gao Z, Hosmane NS, Wu A (2018) Rapid and sensitive colorimetric sensing of the insecticide pymetrozine using melamine-modified gold nanoparticles. Anal Methods 10:417–421. https://doi.org/10.1039/C7AY02658G
16. Kong D, Jin R, Wang T, Li H, Yan X, Su D, Wang C, Liu F, Sun P, Liu X, Gao Y, Ma J, Liang X, Lu G (2019) Fluorescent hydrogel test kit coordination with smartphone: robust performance for on-site dimethoate analysis. Biosens Bioelectron 145:111706. https://doi.org/10.1016/j.bios.2019.111706
17. Kumar N, Seth R, Kumar H (2014) Colorimetric detection of melamine in milk by citrate-stabilized gold nanoparticles. Anal Biochem 456:43–49. https://doi.org/10.1016/j.ab.2014.04.002
18. Kushwaha A, Singh G, Sharma M (2020) Colorimetric sensing of chlorpyrifos through negative feedback inhibition of the catalytic activity of silver phosphate oxygenase nanozymes. RSC Adv 10:13050–13065. https://doi.org/10.1039/C9RA10719C
19. Lakade AJ, Sundar K, Shetty PH (2018) Gold nanoparticle-based method for detection of calcium carbide in artificially ripened mangoes (*Magnifera indica*). Food Addit Contam Part A 35:1078–1084. https://doi.org/10.1080/19440049.2018.1449969
20. Li D, Wang S, Wang L, Zhang H, Hu J (2019) A simple colorimetric probe based on anti-aggregation of AuNPs for rapid and sensitive detection of malathion in environmental samples. Anal Bioanal Chem 411:2645–2652. https://doi.org/10.1007/s00216-019-01703-7
21. Li F, Feng Y, Zhao C, Li P, Tang B (2012) A sensitive graphene oxide–DNA based sensing platform for fluorescence "turn-on" detection of bleomycin. Chem Commun 48:127–129. https://doi.org/10.1039/C1CC15694B
22. Li Z, Wang Y, Ni Y, Kokot S (2014) Unmodified silver nanoparticles for rapid analysis of the organophosphorus pesticide, dipterex, often found in different waters. Sens Actuators B Chem 193:205–211. https://doi.org/10.1016/j.snb.2013.11.096
23. Liu F, Lei T, Zhang Y, Wang Y, He Y (2021) A BCNO QDs-MnO_2 nanosheets based fluorescence "off-on-off" and colorimetric sensor with smartphone detector for the detection of organophosphorus pesticides. Anal Chim Acta 1184:339026. https://doi.org/10.1016/j.aca.2021.339026
24. Liu G, Yang X, Li T, Yu H, Du X, She Y, Wang J, Wang S, Jin F, Jin M, Shao H, Zheng L, Zhang Y, Zhou P (2015) Spectrophotometric and visual detection of the herbicide atrazine by exploiting hydrogen bond-induced aggregation of melamine-modified gold nanoparticles. Microchim Acta 182:1983–1989. https://doi.org/10.1007/s00604-015-1531-7
25. Liu Q, He Z, Wang H, Feng X, Han P (2020) Magnetically controlled colorimetric aptasensor for chlorpyrifos based on copper-based metal-organic framework nanoparticles with peroxidase mimetic property. Microchim Acta 187:524. https://doi.org/10.1007/s00604-020-04499-x

26. Ma S, He J, Guo M, Sun X, Zheng M, Wang Y (2018) Ultrasensitive colorimetric detection of triazophos based on the aggregation of silver nanoparticles. Colloids Surf A Physicochem Eng Asp 538:343–349. https://doi.org/10.1016/j.colsurfa.2017.11.030
27. Malarkodi C, Rajeshkumar S, Annadurai G (2017) Detection of environmentally hazardous pesticide in fruit and vegetable samples using gold nanoparticles. Food Control 80:11–18. https://doi.org/10.1016/j.foodcont.2017.04.023
28. Menon SK, Modi NR, Pandya A, Lodha A (2013) Ultrasensitive and specific detection of dimethoate using a p-sulphonato-calix[4]resorcinarene functionalized silver nanoprobe in aqueous solution. RSC Adv 3:10623. https://doi.org/10.1039/c3ra40762d
29. Mu J, Peng Y, Shi Z, Zhang D, Jia Q (2021) Copper nanocluster composites for analytical (bio)-sensing and imaging: a review. Microchim Acta 188:384. https://doi.org/10.1007/s00604-021-05011-9
30. Nanda Kumar D, Alex SA, Chandrasekaran N, Mukherjee A (2016) Acetylcholinesterase (AChE)-mediated immobilization of silver nanoparticles for the detection of organophosphorus pesticides. RSC Adv 6:64769–64777. https://doi.org/10.1039/C6RA13185A
31. Otte J, Roland-Holst DPD et al (2007) People food safety
32. Ouyang H, Tu X, Fu Z, Wang W, Fu S, Zhu C, Du D, Lin Y (2018) Colorimetric and chemiluminescent dual-readout immunochromatographic assay for detection of pesticide residues utilizing g-C_3N_4/$BiFeO_3$ nanocomposites. Biosens Bioelectron 106:43–49. https://doi.org/10.1016/j.bios.2018.01.033
33. Polavarapu L, Pérez-Juste J, Xu Q-H, Liz-Marzán LM (2014) Optical sensing of biological, chemical and ionic species through aggregation of plasmonic nanoparticles. J Mater Chem C 2:7460. https://doi.org/10.1039/C4TC01142B
34. Priyadarshini E, Pradhan N (2017) Gold nanoparticles as efficient sensors in colorimetric detection of toxic metal ions: a review. Sens Actuators B Chem 238:888–902. https://doi.org/10.1016/j.snb.2016.06.081
35. Qian S, Lin H (2015) Colorimetric sensor array for detection and identification of organophosphorus and carbamate pesticides. Anal Chem 87:5395–5400. https://doi.org/10.1021/acs.analchem.5b00738
36. Qing Z, Bai A, Xing S, Zou Z, He X, Wang K, Yang R (2019) Progress in biosensor based on DNA-templated copper nanoparticles. Biosens Bioelectron 137:96–109. https://doi.org/10.1016/j.bios.2019.05.014
37. Rohit JV, Basu H, Singhal RK, Kailasa SK (2016) Development of p-nitroaniline dithiocarbamate capped gold nanoparticles-based microvolume UV–vis spectrometric method for facile and selective detection of quinalphos insecticide in environmental samples. Sens Actuators B Chem 237:826–835. https://doi.org/10.1016/j.snb.2016.07.019
38. Rohit JV, Kailasa SK (2014) 5-Sulfo anthranilic acid dithiocarbamate functionalized silver nanoparticles as a colorimetric probe for the simple and selective detection of tricyclazole fungicide in rice samples. Anal Methods 6:5934–5941. https://doi.org/10.1039/C3AY42092B
39. Rohit JV, Solanki JN, Kailasa SK (2014) Surface modification of silver nanoparticles with dopamine dithiocarbamate for selective colorimetric sensing of mancozeb in environmental samples. Sens Actuators B Chem 200:219–226. https://doi.org/10.1016/j.snb.2014.04.043
40. Saien J, Khezrianjoo S (2008) Degradation of the fungicide carbendazim in aqueous solutions with UV/TiO_2 process: optimization, kinetics and toxicity studies. J Hazard Mater 157:269–276. https://doi.org/10.1016/j.jhazmat.2007.12.094
41. Saini A, Kumar P, Ravelo B, Lallechere S, Thakur A, Thakur P (2016) Magneto-dielectric properties of doped ferrite based nanosized ceramics over very high frequency range. Eng Sci Technol Int J 19:911–916. https://doi.org/10.1016/j.jestch.2015.12.008
42. Shrivas K, Monisha, Patel S, Thakur SS, Shankar R (2020) Food safety monitoring of the pesticide phenthoate using a smartphone-assisted paper-based sensor with bimetallic Cu@Ag core–shell nanoparticles. Lab Chip 20:3996–4006. https://doi.org/10.1039/D0LC00515K
43. Singh P (2016) SPR biosensors: historical perspectives and current challenges. Sens Actuators B Chem 229:110–130. https://doi.org/10.1016/j.snb.2016.01.118

44. Singh S, Tripathi P, Kumar N, Nara S (2017) Colorimetric sensing of malathion using palladium-gold bimetallic nanozyme. Biosens Bioelectron 92:280–286. https://doi.org/10.1016/j.bios.2016.11.011
45. Souza Tette PA, Rocha Guidi L, de Abreu Glória MB, Fernandes C (2016) Pesticides in honey: a review on chromatographic analytical methods. Talanta 149:124–141. https://doi.org/10.1016/j.talanta.2015.11.045
46. Taghdisi SM, Danesh NM, Lavaee P, Emrani AS, Ramezani M, Abnous K (2015) A novel colorimetric triple-helix molecular switch aptasensor based on peroxidase-like activity of gold nanoparticles for ultrasensitive detection of lead (<scp>ii</scp>). RSC Adv 5:43508–43514. https://doi.org/10.1039/C5RA06326D
47. Thakur P, Sharma P, Mattei J-L, Queffelec P, Trukhanov AV, Trukhanov SV, Panina LV, Thakur A (2018) Influence of cobalt substitution on structural, optical, electrical and magnetic properties of nanosized lithium ferrite. J Mater Sci Mater Electron 29:16507–16515. https://doi.org/10.1007/s10854-018-9744-2
48. Unser S, Bruzas I, He J, Sagle L (2015) Localized surface plasmon resonance biosensing: current challenges and approaches. Sensors 15:15684–15716. https://doi.org/10.3390/s150715684
49. Vickers NJ (2017) Animal communication: When I'm Calling You, Will You Answer Too? Curr Biol 27:R713–R715. https://doi.org/10.1016/j.cub.2017.05.064
50. Vilela D, González MC, Escarpa A (2012) Sensing colorimetric approaches based on gold and silver nanoparticles aggregation: chemical creativity behind the assay. A review. Anal Chim Acta 751:24–43. https://doi.org/10.1016/j.aca.2012.08.043
51. Walekar L, Dutta T, Kumar P, Ok YS, Pawar S, Deep A, Kim K-H (2017) Functionalized fluorescent nanomaterials for sensing pollutants in the environment: a critical review. TrAC Trends Anal Chem 97:458–467. https://doi.org/10.1016/j.trac.2017.10.012
52. Wang G, Bai X, Chen X, Ren Y, Pang X, Han J (2022) Detection of adulteration and pesticide residues in Chinese patent medicine Qipi pill using KASP technology and GC-MS/MS. Front Nutr. https://doi.org/10.3389/fnut.2022.837268
53. Wang Z, Huang Y, Wang D, Sun L, Dong C, Fang L, Zhang Y, Wu A (2018) A rapid colorimetric method for the detection of deltamethrin based on gold nanoparticles modified with 2-mercapto-6-nitrobenzothiazole. Anal Methods 10:1774–1780. https://doi.org/10.1039/C8AY00137E
54. Wei Z, Li H, Wu J, Dong Y, Zhang H, Chen H, Ren C (2020) 3DRGO-$NiFe_2O_4$/NiO nanoparticles for fast and simple detection of organophosphorus pesticides. Chin Chem Lett 31:177–180. https://doi.org/10.1016/j.cclet.2019.05.031
55. Wierucka M, Biziuk M (2014) Application of magnetic nanoparticles for magnetic solid-phase extraction in preparing biological, environmental and food samples. TrAC Trends Anal Chem 59:50–58. https://doi.org/10.1016/j.trac.2014.04.007
56. Wong JW, Hennessy MK, Hayward DG, Krynitsky AJ, Cassias I, Schenck FJ (2007) Analysis of organophosphorus pesticides in dried ground ginseng root by capillary gas chromatography−mass spectrometry and−flame photometric detection. J Agric Food Chem 55:1117–1128. https://doi.org/10.1021/jf062774q
57. Yan J, Huang Y, Zhang C, Fang Z, Bai W, Yan M, Zhu C, Chen A (2017) Aptamer based photometric assay for the antibiotic sulfadimethoxine based on the inhibition and reactivation of the peroxidase-like activity of gold nanoparticles. Microchim Acta 184:59–63. https://doi.org/10.1007/s00604-016-1994-1
58. Yue G, Su S, Li N, Shuai M, Lai X, Astruc D, Zhao P (2016) Gold nanoparticles as sensors in the colorimetric and fluorescence detection of chemical warfare agents. Coord Chem Rev 311:75–84. https://doi.org/10.1016/j.ccr.2015.11.009
59. Zeng S, Baillargeat D, Ho H-P, Yong K-T (2014) Nanomaterials enhanced surface plasmon resonance for biological and chemical sensing applications. Chem Soc Rev 43:3426. https://doi.org/10.1039/c3cs60479a
60. Zhan X, Tang Y, Liu Y, Tao H, Wu Y (2022) A novel colorimetric strategy for rapid detection of dimethoate residue in vegetables based on enhancing oxidase-mimicking catalytic activity of cube-shape Ag_2O particles. Sens Actuators B Chem 361:131720. https://doi.org/10.1016/j.snb.2022.131720

61. Zhang D, Liu Q (2016) Biosensors and bioelectronics on smartphone for portable biochemical detection. Biosens Bioelectron 75:273–284. https://doi.org/10.1016/j.bios.2015.08.037
62. Zhang K, Wong JW, Yang P, Tech K, DiBenedetto AL, Lee NS, Hayward DG, Makovi CM, Krynitsky AJ, Banerjee K, Jao L, Dasgupta S, Smoker MS, Simonds R, Schreiber A (2011) Multiresidue pesticide analysis of agricultural commodities using acetonitrile salt-out extraction, dispersive solid-phase sample clean-up, and high-performance liquid chromatography-tandem mass spectrometry. J Agric Food Chem 59:7636–7646. https://doi.org/10.1021/jf2010723
63. Zhang X, Zhao H, Xue Y, Wu Z, Zhang Y, He Y, Li X, Yuan Z (2012) Colorimetric sensing of clenbuterol using gold nanoparticles in the presence of melamine. Biosens Bioelectron 34:112–117. https://doi.org/10.1016/j.bios.2012.01.026
64. Zhang Y, Leng Y, Miao L, Xin J, Wu A (2013) The colorimetric detection of Pb^{2+} by using sodium thiosulfate and hexadecyl trimethyl ammonium bromide modified gold nanoparticles. Dalt Trans 42:5485. https://doi.org/10.1039/c3dt32532f
65. Zheng J, Zhang H, Qu J, Zhu Q, Chen X (2013) Visual detection of glyphosate in environmental water samples using cysteamine-stabilized gold nanoparticles as colorimetric probe. Anal Methods 5:917–924. https://doi.org/10.1039/C2AY26391B

Applications, Opportunities and Challenges of Nanotechnology in the Food Industry

Anteneh Kindu Mersha, Bilisuma Fekadu Finina, and Gebrehiwot Gebreslassie

Abstract Nanotechnology becomes the frontier technological advancement in food industries, especially to overcome the current food security concern. It has diverse applications from crop production to table ranging from nanoscale delivery system of agrochemicals, improved crop production, nanotechnology-based quality control; reduced environmental waste, improved consumer health, and enhanced safety of consumable products. The progress in nanotechnology is mainly driven by the continuous innovation and advancement of new nanomaterials. Nanomaterials are used as food additives, safe delivery of micronutrients, antimicrobials, fillers for structural modification of foods, and as sensors to provide fast and reliable food information. Moreover, nanotechnology has wide application in developing active food packagings, nanoscale enzymatic reactor, nanofltration systems, heat and mass transfer nanofabrication and nanocapsules. On the other side, utilization of nanomaterials in the food sector faces consumer concerns like health effects and environmental damage, demanding responsible manufacture and governing regulations.

Keywords Nanotechnology · Nanomaterials · Nanofoods · Food processing · Antimicrobial packaging · Nutrient delivery

A. K. Mersha (✉) · B. F. Finina · G. Gebreslassie
Department of Industrial Chemistry, Addis Ababa Science and Technology University, Addis Ababa, Ethiopia
e-mail: anteneh.kindu@aastu.edu.et

A. K. Mersha · G. Gebreslassie
Nanotechnology Center of Excellence, Addis Ababa Science and Technology University, Addis Ababa, Ethiopia

B. F. Finina
Department of Chemistry, Kotebe University of Education, Addis Ababa, Ethiopia

R. K. Bachheti et al. (eds.), *Nanomaterials for Environmental and Agricultural Sectors*, Smart Nanomaterials Technology,
https://doi.org/10.1007/978-981-99-2874-3_13

1 Introduction

Food security has become an increasing concern due to alarming global population growth. Since the end of the Second World War, the growth in agricultural productivity has outpaced population growth. However, several challenges still exist. First, the food system is already exceeding global boundaries from a resource perspective and is producing significant waste. Second, there is an increasing food safety concern associated with premature mortality and the growing chronic and infectious diseases globally, including in developing nations. Third, food insecurity continues to be linked with vast inequalities, including political instabilities and the struggle for decent life [60].

The existing interrelationships among population growth, food security, health and sustainability demand more than a mere sufficiency of food supply [60]. It is, thus, imperative to innovative technological advances in deploying transformative solutions to effectively address the challenges of sustainable agriculture and food systems [56]. Nowadays, utilization of nanotechnology in food industries is thought to be the prominent scientific way to overcome the aforementioned problems.

Nanotechnology is multidisciplinary science, which deals with development and use of materials whose at least one of its dimensions is in nanoscale ranging from 1 to 100 nm. While nanomaterial (NM) has one or more of its dimensions in the nanoscale, nanoparticle (NP) has all three dimensions in the nanoscale range [23]. NMs and NPs can be manufactured in either bottom-up or top-down approaches [5], and encompass a variety of nanoforms, including nanotubes, nanofibres, nanorods, nanofilms, nanolayers, nanocoatings and nanosheets [22, 42, 43]. These materials are also characterized by high reactivity, high aspect ratio, tunable pore size and particle morphology. Due to such interesting properties, they are widely applicable in many manufacturing sectors including but not limited to food industries [6, 8, 9, 29, 30, 32].

NMs are reported to enhance and maintain food safety by improving the efficacy of food packaging, food shelf-life, delivery system, flavoring, pollution reduction, providing reliable quality control system and enhancing nutritional value—without changing the taste and physical characteristics of food products [11, 32]. In addition, nanotechnology has revealed many promising applications in the food and agriculture sectors. Nanotechnology-enabled delivery system of agrochemicals, improved crop production, environmental management, consumer health, and safety of consumable products are main contributions of nanotechnology [10, 29, 32, 52]. In particular, nanotechnology is desirable to optimize food production systems—to produce more food, incorporate functional materials in food, and increase bioavailability of material that have beneficial nutritional effects, improve target delivery, reduce food waste, and safely and equitably distribute food without damaging the environment for future generations [41]. Ideally, nanomaterials used in food industry should be food-grade [21]. They shall have or enhance specific characteristics, such as biocompatibility, biodegradability and bioavailability, adequate shelf life, water/lipid solubility, dependent to pH and temperature changes [54]. In addition, food-packaging materials need to enhance properties like oxygen barrier, resistance to moisture, CO_2 and UV [54, 64].

Thus, although nanofoods are reported to have health and environmental risk and the properties of nanomaterials which are being used in the food industry are not fully understood [21, 41], the use of nanotechnology in various food and agricultural processing systems and in end-products is believed to contribute positively to the food ecosystem.

2 Applications of Nanotechnology in the Food Industry

Nanotechnology is applicable in all areas of food industry, including development and processing of new functional foods, food packaging, food preservation and safety monitoring, nutrient delivery, as well as water purification as summarized in Fig. 1. Recent examples of the key nanostructured materials applicable in the food sector are summarized in Table 2 along with the possible designs, synthesis methods and specific applications.

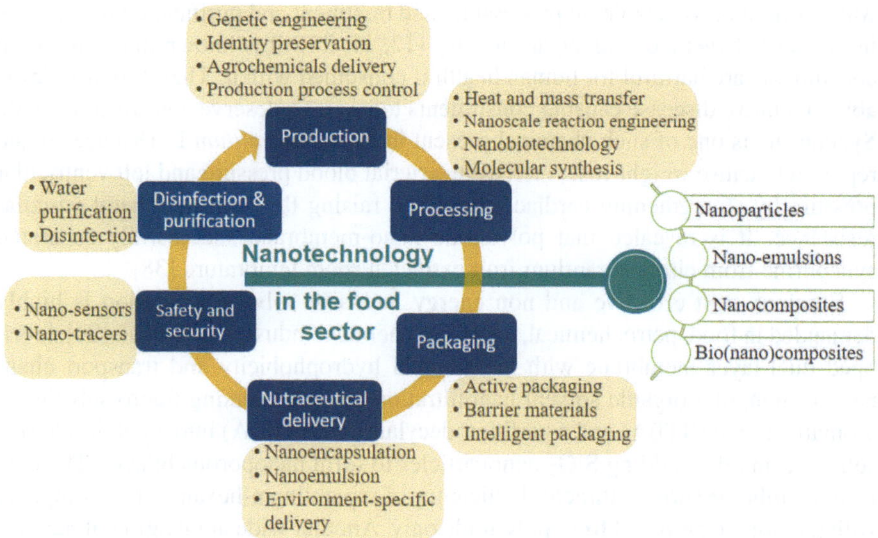

Fig. 1 Applications of nanotechnology in the food industry

2.1 Food Processing

Owing to their promising size related properties, NMs are widely applicable in food processing industries to manufacture or provide food products of improved quality, which are convenient to consumers. Well-controlled manufacturing process could result fresh and flavored food products, authentic, convenient food with prolonged shelf life. In food processing, these NMs can be used as food additives, carriers for smart delivery of nutrients, anticaking agents, antimicrobial agents, fillers and as sensors to provide fast and reliable information (like detection and quantification of contaminants or various target molecules) related to quality of the product [16, 28]. Nanotechnology is also fundamentally applicable in food processing in facilitating nanoscale enzymatic reactor [34], nanofltration [15], heat and mass transfer nanofabrication [27], and nanocapsulation [53].

Nanofiltration in food processing

Nanofiltration is a pressure driven separation technology which works on mild condition of low pressure and temprature. It is evolving technology having multiple advantage and improved feature compared to conventional separation techniques in food processing industries including but not limited to wastewater treatment, fractionation, water softening, vegetable oil processing, and treatment and purification of products in the dairy, beverage, and sugar industry [12, 15, 33]. It is known that some food constituents are harmful for human health if consumed with it. Therefore it is desirable to remove diseease causing constituents to improve/deserve consumer's health. Synephrine is one of such chemical present in *Citrus aurantium* L. (Rutaceae) and reported to cause weight-lose, increasing arterial blood pressure and left ventricular pressure by strengthening cardiac output and raising the total peripheral vascular resistance. It is revealed that polyamide nano-membrane successfully separated synephrine from citrus aurantium fruit extract at room temprature [38].

Efficient, cost effective and non energy intensive solvent separation is highly demanded in food, petrochemical, and other chemical industries. Fu et al. [24] developed dual-layer membrane with hierarchical hydrophobicity and transport channels for nonpolar organic solvent nanofiltration by incorporating fluoro substituted aromatic amine (BTBA) and a perfluorodecylamine (HTFDA) into a polyimide (PI) substrate and then adding SiO_2 nanoparticles to form nanoporous bilayer. The dual layer membrane showed threefold efficiency of separating n-hexane when compared with membrane prepared from polyimide only. Another wide application of nanofiltration is water treatment [25]. It is frequently reported that nanofiltration, one of the membrane filtration methods, is widely employed in water treatment (including food processing wastewater) owing to its advantages of low energy consumption, low environmental footprint, high efficiency and system confinment. Experimental comparision of nanofiltration based on cellulose nanofibers on porous filter paper consisting of ultralong hydroxyapatite nanowires and cellulose fibers (CF-CNF/HAPNW) revealed that the membrane ensures water transport nanochannels for rapid separation, and relatively high rejection rates of impurities which makes it ideal for water filtration and remediation [65]. Another report also indicated polymer of intrinsic microporosity (PIM-1) modified by metal-organic frameworks (MOFs) for the removal of heavy metals and dyes from wastwater. The composite showed an increased water permiability, increased metal ion rejection coefficients and and increased rejection of food dyes than PIM-1 alone [36].

Nanosensors

Foodborne diseases are the immense threat to public health. Generally, these diseases are caused from incomplete food quality which may be due to bacterial toxins in food and environment [3, 11]. For decades, spectroscopic and chromatographic techniques of high cost, tremendous analysis time and huge apparatus were used to get food information to judge the quality of foods. Nowadays, technological advancement employed nanotechnology to fabricate tiny sensing devices that can give information about food composition, its quality, freshness and food safety quickly and cheaply without requiring specialized operator skill. Inded, nanosensors can be incorporated into packaging systems or integrated to smart phones so it helps consumers to determine the quality attributes of foods rapidly [41].

One of the interesting example of food quality assurance is the detection of trace Botulinum toxin [18]. Botulinum neurotoxin (BoNT) is the bacterial toxin produced by Clostridium botulinum bacteria. It is the most prevalent cause of foodborne diseases over the globe with the lethal dose of approximetly 1 ng/kg in unvaccinated persons. Therefore appropriate tool to detect trace Botulinum toxin at lower concentration in food like nano-immunosensors are desirable. Recently, Cheng HP and Chuang HS presented a bead-based diffusometric nanosensor for simple and rapid, yet quantitative detection of biological botulinum toxins [18]. Their result indicates nano-immunosensors achieved lower limit of detection of less than 1 ng/mL when compared to microplate colorimetric Enzyme-Linked ImmunoSorbent Assays (ELISA) which only resolved toxin concentration of over 6.9 ng/mL. ELISA assays with multiwell microplates are the FDA recommended techniques and are routinely applied in the identification and detection of toxins or proteins.

Food freshness is another fastly growing consumer concern. Food contamination and microbial growth are the main cause of food spoilage that the later is proved to produce volatile amine during food processing/transportation and/or storage [2]. For example, Wang et al. developed superhydrophobic pH-Sensing Colorimetric Coating by loading anthocyanin-rich extract on SNPs@PDMS to monitor food freshness. The result indicated that this pH-sensor coating can be reused in fresh food storage containing high moisture content to monitor food freshness without damage [62].

2.2 *Food Packaging*

Convenient food packaging is strongly important to reduce or prevent food waste and/or contamination [17]. For instance, loss of water could cause an undesirable drying of the food while moisture absorption from surroundings may lead to microbial propagation, food rancidity, and reduced shelf-life. Suitable packaging is essential in order to alleviate these problems. In contrast to low flexibility and high density of metals and non-renewability of plastic packaging materials, ideal packaging materials are required to be flexible, tough, resistant to chemical, low density and formable [46]. Nanotechnology is playing an emerging role in demonstrating the ideal engineering of the packaging materials.

Active packaging materials, which have antimicrobial and antioxidant activities, are among the promising applications of nanotechnology in the area [4]. An overview of the various systems employed in active food packaging are presented in Table 1. The corresponding active agents that can be incorporated in the packaging systems are shown in Fig. 2. Furthermore, the characteristic features of packaging materials, possible nanostructure design and enhancement options are discussed in this section.

Poly(lactic acid) (PLA) is one of the widely used biodegradable and environmentally safe polymers in food packaging applications. However, it has limitation in preventing oxidation and UV light, which may cause food degradation. Lignin nanoparticle grafted PLA is reported to have good antioxidant and UV barrier properties. Comparison of different properties of PLA, lignin blended PLA and lignin nanoparticle grafted PLA revealed that lignin nanoparticles grafted PLA showed significantly reduced oxygen permeability, water vapour permeability and UV light transmittance. In addition, incorporation of lignin nanoparticles into PLA showed high radical scavenging activity; thus improved antioxidant activity than pure PLA and lignin blended PLA [13].

Furthermore, Munteanu et al. developed multifunctional PLA nanofibers containing Ag-NP and Vitamin E having enhanced antimicrobial and antioxidant activity. The nanocomposite is reported to inhibit growth of Escherichia coli, Listeria monocytogenes, and Salmonella typhymurium up to 100% whereas the antioxidant activity of PLA/AgNP/Vitamin E nanofibers was determined as 94% efficient [44].

Polypropylene (PP) is another widely applicable material in food packaging. It is inexpensive and has high clarity, chemically inert, and excellent water barrier. However, PP has inherent limitation of poor oxygen and carbon dioxide barrier properties [17, 64]. But nanotechnology assisted modification is proved to overcome this drawback. Cellulose nanocrystal and cellulose nanocrystal-citric acid coated PP showed enhanced O_2 and CO_2 barrier performance by 92%, while maintaining the moisture barrier property. The improved performance is probably resulted from the occupied existing free volume of the substrate system. Additionally, the coated PP is observed to be highly cross-linked that may lead to lower humidity sensitivity of the modified PP packaging material [64].

Polyhydroxyalkanoates, particularly poly(3-hydroxybutyrate) (PHB), being fully degradable and renewable, is another widely used food packaging material. However, the use of PHB has been limited due to several drawbacks such as low thermal degradation temperature, high melting temperature, brittleness and relatively high water vapour permeability. Nanotechnological advancement enabled incorporation of functionalized cellulose nanocrystals methyl ester (CNC-me) and 3-hydroxyvalerate into PHB matrix. This nanocomposite showed enhanced thermal stability, mechanical

Table 1 Potential active packaging applications in the food industry [63]

Type of active packaging	Type of food	Expectation
Active releasing systems (emitter)		
Antimicrobial packaging systems	Fresh and processed meat, fresh and smoked fish, fresh seafood, dairy products, fresh and processed fruits and vegetables, grain, cereals and bakery products, ready to-eat meals	Inhibition or retardation of bacterial growth, extension of the shelf-life
Antioxidant releaser	Fresh fatty fish and meat; fat-containing instant powders; seeds, nuts and oils; fried products	Improvement of oxidative stability
CO_2 emitter	Fresh fish and meat	Extension of microbiological shelf life, reduction in head space volume of modified atmosphere packaging
Active scavenging systems (absorber)		
Oxygen scavenger	Oked meat products	Prevention of discoloration
	Grated cheese, bakery products	Prevention of mold growth
	Fruit and vegetable juices	Retention of vitamin C content, prevention of browning
	Seeds, nuts and oils; fat-containing instant powders, fried snacks; dried meat products	Prevention of rancidity
Moisture scavenger	Mushrooms, tomatoes, strawberries, maize, grains, seeds, fresh fish and meat	Extension of shelf life through maintaining moisture content, decrease in moisture condensation in the packaging, positive impact on the appearance, reduction in browning or discoloratiion
Ethylene absorber	Climacteric fruits and vegetables	Reduction in ripening and senescence, theraby enhancing quality and prolonging shelf-life

strength, interesting barrier and migration properties due to good dispersion of CNC-me in PHB, as well as improved interfacial adhesion and increased crystallinity. These are attributes of good food packaging materials and green plastics which could replace traditional petrochemical materials [64].

Furthermore, Yaoqi Tian fabricated superhydrophobic coating by combining starch nanoparticles (SNPs) with biodegradable, nontoxic, and tasteless poly(dimethylsiloxane) (PDMS) which can be applied to the inner surfaces of food packaging to reduce waste [62]. The developed starch based nanomaterial demonstrated high water repellency level (high hydrophobicity), good stability

Table 2 Recent examples of the key nanostructured designs applicable in the food sector along with their synthesis methods

Class of functional material	Nanostructure design	Synthesis method	Specific application	References
Dual layer nanoporous membrane	BTBA/HTFDA/SiO_2/polyimide	Co-casting method	Separation of n-hexane	[24]
Nanofiltration filter paper	CF-CNF/HAPNW	Papermaking process (filtration, hot pressing, and drying)	Removal of dyes and Na_2SO_4	[65]
Nanofiltration membrane	PIM-1/MOFs	Solution casting method	Removal of heavy metals and food dyes from wastwater	[36]
Nanosensor	Anthocyanin/Chitosan/PVA Film	Solution casting method	Detection of biological botulinum toxins	[18]
pH-sensor coating; Food packaging	Anthocyanin/SNPs@PDMS	Spray coating	Food freshness monitoring; waste reduction	[62]
Active packaging material	PLA/Ag-NP/VitaminE	Electrospinning	Antimicrobial packaging nanofiber membrane	[44]
Food packaging film	PHBV/CNC-me nanoomposite film	Solution casting method	O_2 and CO_2 barrier;	[64]

(continued)

Table 2 (continued)

Class of functional material	Nanostructure design	Synthesis method	Specific application	References
Food packaging film	PLA-Grafted Lignin Nanoparticle and PLA blend film	Solution casting method	Antioxidant activity and UV-Barrier properties	[13]
Food preservatives	Bacteriocin-capped Ag nanoparticles	Biosynthesis	Antibacterial activity (control the growth of food spoiling bacteria)	[58]
Edible food coating	Alginate/ascorbic acid-layered double hydroxide bio-nanohybrids	Mechanochemical grinding method	Reduce fruit metabolism, slow acid consumption; therefore, extended storage period	[39]
Food delivery	Poly-3-hydroxybutyrate-co-hydroxyhexanoate (PHB-HHx) encapsulated essential oil	Solvent evaporation technique	Slow release of essential oils, high inhibitory and bactericidal activity	[19]
Food delivery	Polydopamine coated curcumin nanoparticle	In situ polymerization	Improved target delivery (pH responsive release)	[48]

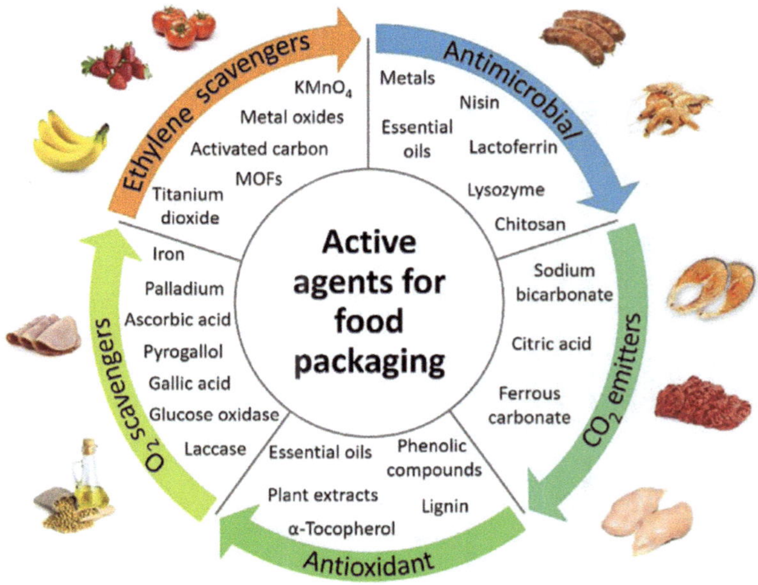

Fig. 2 Active agents for active food packaging [61]

when exposed to different chemicals—eg. HCl, pH 1 & queous ammonia solution ($NH_3 \cdot H_2O$, pH 10), high self-cleaning property, as well as reduced liquid food residue.

2.3 Preservation and Food Safety

Food is highly sensitive to pathogenic microorganisms those could cause contamination which in turn increase food waste and food spoilage thus increasing the risk of foodborne illness and death [11]. More commonly chemical preservatives such as sulfur dioxide, acids, benzoic acid and nitrates are utilized to reduce and/or prevent food contamination and increase its shelf life [47]. But these chemical preservatives are limited to preserving application because they lose their efficacy as microorganisms develop antimicrobial resistance to them. Nanotechnology has played a significant role in developing innovative nanoparticles, antimicrobial-loaded nanoparticles, antimicrobial nanoemulsions from essential oils, plant based preservatives/antimicrobials that are efficacious and environmentally friendly with new or improved functionality [41].

Bacteriocin is ribosomally synthesized antimicrobial peptide produced by lactic acid bacteria and known to inhibit growth of microorganisms and food pathogens. It is safe and appreciated food preservative because of its efficiency in controlling food pathogens without any side effects. However, it is limited to low production yield, exhibiting a narrow antimicrobial spectrum and having a high-dosage requirement [49, 58]. Silver nanoparticles capped bacteriocin is proved to overcome the drawbacks of this bacteriocin. Recently reported work confirmed that silver nanoparticles conjugated bacteriocin showed enhanced antibacterial activity upto 1.5 fold against all the tested bacterial strains (both Gram-negative and Gram-positive food spoiling bacteria) as compared to bacteriocin alone [58]. Another threat in food industry is accumulation of antibiotic residues in food staff and increasing drug resistance microbes when chemical preservatives are frequently used [14]. Nevertheless, use of nano-antimicrobials like chitosan nanoparticles and bioactive peptides are becoming a promising antibiotic in food processing due to their excellent broad-spectrum antimicrobial activity without increasing the mutation rate. Experimental output indicated that chitosan nanomaterial made of chitosan nanoparticles and peptides demonstrated strong anti-bactericidal against pathogenic drug resistant bacteria, Escherichia coli (E. coli). Additionally chitosan nanomaterial significantly reduced generation of reactive oxygen species by E. coli infection; thus reducing or preventing food spoilage [35].

Fresh food with unaltered characteristics and functionalities are highly demanded among consumers. To achieve such consumer needs, the food surface need to be covered by a thin edible layer (pronounced as nano-coating) for shelf-life extension and for preserving the characteristics and functionalities of food articles [40]. Food nanocoating is application of thin nanosized materials on the surface of food substrate. They could offer good barrier properties to the gases and moisture and also prevent the loss of natural volatile flavour compounds. It may also be utilized to maintain physicochemical properties of the food and antioxidant properties of the constituent or ingredients [50]. Nilwala Kottegoda developed alginate/ascorbic acid-layered double hydroxide bio-nanohybrids edible food coating and proved that it is efficient food coating material [39]. To demonstrate efficacy of developed bio-nanohybrids, the author coated strawberries by placing in four control coating solutions (water, alginate, glycerol, alginate-glycerol mixture, and ascorbic acid -glycerol mixture) and the novel bionanohybrid by adopting similar coating procedure for all coating materials at refrigerated temperature condition (Fig. 3). Interestingly, treatment with alginate/ascorbic acid-layered double hydroxide (AA-LDH) novel nanohybrid minimized pH increment, reduced fruit metabolism, and slowed down organic acid consumption, and therefore, extended storage period. In addition, this novel nanohybrid demonstrated lower weight loss (showing excellent barrier property), preserved ascorbic acid content (important in reducing the microbial growth in the case of aerobic microorganisms and molds) in the fruit and showed good efficacy in maintaining the psychochemical properties of strawberries in comparison to the controls during the storage time. This signifies the relevance of integrating nanotechnology in food industrial systems.

Fig. 3 Images of strawberries at refrigerated temperature: (1) water dipped, (2) alginate and glycerol mixture dipped, (3) alginate, glycerol, and ascorbic acid mixture dipped; (4) glycerol dipped; and (5) glycerol and alginate -AA-LDHs dipped. [Reprinted with permission from Madhusha et al. [39]. Copyright 2020 American Chemical Society]

2.4 *Food Delivery*

Some bioactive compounds are important in improving health and wellbeing of consumers. Therefore, it is important to incorporate them in food. Nanotechnology provides technical encapsulation matrix of such bioactive compounds to produce nanonutraceuticals [45]. Nanoencapsulation provides maximum stability, protection and longer shelf life, combat deterioration during processing and permit the release of encapsulated components during mastication and digestion for efficient absorption into the body system [1, 53]. Nano-emulsions are nano-sized droplet of such encapsulation matrix which is important in improving food bioavailability, ensuring food safety and nutrition, improving the quality of food and target release of sensitive bioactive functional compounds [1, 31].

Here, as it is repeatedly stated, controlling the microbial contamination in food chain is the main challenge of food industries, such as the issue of utilizing synthetic preservatives. i.e., the chance of developing bacterial resistance and the subsequent adverse impact on consumer health and the environment [51]. Fortunately, scientific advancement rewarded the use of plant based antimicrobials such as plant essential oils (EOs) that have intrinsic antimicrobial activity against food-borne pathogens; and can offer a greener and safer alternative to the utilization of chemical additives [19]. But their uses were under question because of their undesirable properties like volatility and instability. Their application, thus, becomes dependent on the use of efficient encapsulating agents ensuring their protection and release control. For instance, controlled release and antimicrobial test of polyhydroxyalkanoates, polyhydroxybutyrate (PHB) and poly-3-hydroxybutyrate-co-hydroxyhexanoate (PHB-HHx)-based nanoparticles encapsulated essential oil extracted from Mexican oregano showed slow delivery, assuring the prolonged release of EO over time in all the tested food samples, high inhibitory and bactericidal activity, confirming higher effectiveness of the polyhydroxyalkanoates encapsulated EO than the pure EO [19].

Bioavailability and target delivery is another sensitive issue raised regarding plant bioactive functional molecules [57]. For instance, curcumin is multi-purpose turmeric extract, which is commonly used as a food preservative, spice, coloring agents, and have extensive health benefits including anti-inflammatory, antimicrobial, and antitumor effects. However, limited absorption and bioavailability, low aqueous solubility at physiological pH, light sensitivity and rapid metabolism limits the use of curcumin [48, 57]. Encapsulation of Curcumin in nanocarrier-based delivery systems is the most promising emerging technology; especially for improved bioavailability and target delivery. Polydopamine (PDA) coated curcumin nanoparticle improved the drawbacks of curcumin alone. PDA encapsulated curcumin nanoparticle showed higher light stability, improved target delivery (pH responsive release; higher curcumin release from PDA-coated curcumin NPs at hither pH and decreased release at lower pH), high aqueous dispersibility and bioavailability and reduced chemical degradation of curcumin in alkaline conditions [48].

3 Opportunities

Emergence of foodborne disease, increasing food waste, unbalanced world population and annual crop production is forcing universities, governments, and industry to collaborate on advancements of nanotechnology which can further facilitate the transfer of knowledge from discovery to industry implementation, which in turn can allow sustained development and utilization of nano-enabled products in the consumer market [59]. In his respect, governments in both developed and developing world have established policies and initiatives to integrate nanotechnologies in their growth and innovation strategies [20]. Such opportunities calls for research and innovation interventions among the international scientific community, demanding the realization of safe and sustainable nanoproducts and nanosystems.

4 Challenges

Continuous and responsible innovation of technology used in food industries is important to pursue the present experience and to overcome the future challenges regarding improved food safety. Nanotechnology experts are on the way of illustrating the barrier for responsible innovation of nano-agrifoods. Insufficient long-term risk evaluations, lack of consistent product control, low reproducibility of studies, lack of regulatory guidance, lack of training and lack of product trust are some of the main barriers agreed by nanomaterial experts [21]. Lack of sufficient funding in the food sector, which can ultimately land the nano-based products in the market, is another main challenge to research and development of nanotechnology application in the food industry [59].

Product labeling is critically important to provide food information to increase consumer's confidence, for food traceability and as a precautionary measure to prevent and minimize the safety and health risks of food. Yet, nanofood labeling is not recognized by international food standardizing body, Codex Alimentarius, because there are controversies on the health effect of nanomaterials and their risk nature. For example, there is uncertainty whether nanomaterials currently used in the food industry are new materials or existing materials, magnitude of risks regarding safety and health, environmental damage of such tiny materials, and definition of nanomaterials itself to determine the scope of definition to include materials that pose a potential risk and risk-creating properties. Thus, it is challenging to determine which product could be considered as nanofood to implement mandatory labeling and take regulatory action [26].

Another main challenge of using nanomaterials in food is their health and environmental impact. For instance, it is frequently reported that TiO_2 nanomaterial is useful in manufacturing some consumer products (mostly as food and beverage colorant) and decomposition of organic compounds in water treatment. Unfortunately, recent research pointed out TiO_2 has adverse health and environmental impact; viz. it has ability to penetrate human skin, inhibition of growth of plants, causes the production of reactive oxygen species which are known to cause genetic damage in human and generation of free radicals of TiO_2 nanomaterials on human bronchial epithelial cells [55]. It is also confirmed that the structural and physicochemical properties of food-grade TiO_2 particles such as their interfacial composition, surface charge and aggregation state could be changed as solution condition is altered after food ingestion. This could promote toxicity to the consumer when it passes through human gastro intestinal tract [66]. These and other societal concerns need to be addressed for the wider application of nanomaterials and nanotechnology in the food sector.

5 Future Prospects

Application of nanotechnology in the food industry is highly increasing than any other sectors. Simultaneously, in addition to positive contribution to the sector its health and environmental threat is becoming an issue on every science stages. Thus, it is critically important to clarify every aspect of utilization of nanomaterials in food industries to answer the pending consumer's health and environmental issue questions. These pending questions calls for clear global action to further expand standards, certifications, and regulations for the upcoming realization of food nanoproducts in order to ensure their reliability and safety [59].

It is also important to putdown the common framework for successful continuous improvement; to address the numerous challenges and questions regarding application of nanotechnology in the food industry. Scholars suggested the idea of convergence technology which involves integration of nanotechnology, biotechnology, information science, and cognitive science to overcome the upcoming challenges [56]. This needs convergence thinking approach of problem solving that focuses on creation of new collaborations from academia, industry, government, foundations, national laboratories, non-governmental organizations, and a diverse set of stakeholders from producers to consumers.

References

1. Abubakar AS, Bashir MS (2014) Nanoparticles: a delivery system for bioactive food components. Int J Curr Res 6(08):2012–2015
2. Agunos RIF, Mendoza DVM, Rivera MAS (2020) Anthocyanin colorimetric strip for volatile amine determination. Int J Food Sci. https://doi.org/10.1155/2020/1672851
3. Alamer S, Chinnappan R, Zourob M (2017) Development of rapid immuno-based nanosensors for the detection of pathogenic bacteria in poultry processing plants. Procedia Technol. https://doi.org/10.1016/j.protcy.2017.04.012
4. Alizadeh Sani M, Tavassoli M, Salim SA, Azizi-lalabadi M, McClements DJ (2022) Development of green halochromic smart and active packaging materials: TiO_2 nanoparticle- and anthocyanin-loaded gelatin/κ-carrageenan films. Food Hydrocolloids 124. https://doi.org/10.1016/j.foodhyd.2021.107324
5. Arole VM, Munde SV (2014) Fabrication of nanomaterials by top-down and bottom-up approaches-an overview. JAAST: Mater Sci (Special Issue) 1(2):2–89
6. Bachheti A, Bachheti RK, Abate L, Husen A (2022) Current status of Aloe-based nanoparticle fabrication, characterization and their application in some cutting-edge areas. S Afr J Bot 147:1058–1069
7. Bachheti RK, Abate L, Bachheti A, Madhusudhan A, Husen A (2021) Algae-, fungi-, and yeast-mediated biological synthesis of nanoparticles and their various biomedical applications. In Handbook of greener synthesis of nanomaterials and compounds. Elsevier, Amsterdam, pp 701–734. https://doi.org/10.1016/B978-0-12-821938-6.00022-0
8. Bachheti RK, Fikadu A, Bachheti A, Husen A (2020) Biogenic fabrication of nanomaterials from flower-based chemical compounds, characterization and their various applications: a review. Saudi J Biol Sci 27(10):2551–2562
9. Bachheti RK, Godebo Y, Bachheti A, Yassin MO, Husen A (2020) Root-based fabrication of metal/metal-oxide nanomaterials and their various applications. In: Husen A, Jawaid M (eds) Nanomaterials for agriculture and forestry applications. Elsevier, pp 135–166 https://doi.org/10.1016/B978-0-12-817852-2.00006-8
10. Bachheti RK, Konwarh R, Gupta V, Husen A, Joshi A (2019) Green synthesis of iron oxide nanoparticles: cutting edge technology and multifaceted applications. In: Nanomaterials and plant potential. Springer, Cham, pp 239–259. https://doi.org/10.1007/978-3-030-05569-1_9
11. Bajpai VK, Kamle M, Shukla S, Mahato DK, Chandra P, Hwang SK, Kumar P, Huh YS, Han YK (2018) Prospects of using nanotechnology for food preservation, safety, and security. J Food Drug Anal 26(4):1201–1214. https://doi.org/10.1016/j.jfda.2018.06.011
12. Besha AT, Tsehaye MT, Tiruye GA, Gebreyohannes AY, Awoke A, Tufa RA (2020) Deployable membrane-based energy technologies: the Ethiopian prospect. Sustainability (Switzerland) 12(21):1–34. https://doi.org/10.3390/su12218792

13. Boarino A, Schreier A, Leterrier Y, Klok HA (2022) Uniformly dispersed poly(lactic acid)-grafted lignin nanoparticles enhance antioxidant activity and UV-barrier properties of poly(lactic acid) packaging films. ACS Appl Polym Mater 4(7):4808–4817. https://doi.org/10.1021/acsapm.2c00420
14. Capita R, Alonso-Calleja C (2013) Antibiotic-resistant bacteria: a challenge for the food industry. Crit Rev Food Sci Nutr 53(1). https://doi.org/10.1080/10408398.2010.519837
15. Cassano A, Conidi C, Castro-Muñoz R (2019) Current and future applications of nanofiltration in food processing. In: Separation of functional molecules in food by membrane technology. Elsevier, pp 305–348. https://doi.org/10.1016/B978-0-12-815056-6.00009-7
16. Chaudhry Q, Castle L (2011) Food applications of nanotechnologies: an overview of opportunities and challenges for developing countries. Trends Food Sci Technol 22(11):595–603. https://doi.org/10.1016/j.tifs.2011.01.001
17. Chausali N, Saxena J, Prasad R (2022) Recent trends in nanotechnology applications of bio-based packaging. J Agric Food Res. https://doi.org/10.1016/j.jafr.2021.100257
18. Cheng HP, Chuang HS (2019) Rapid and sensitive nano-immunosensors for botulinum. ACS Sensors 4(7). https://doi.org/10.1021/acssensors.9b00644
19. Corrado I, Di Girolamo R, Regalado-González C, Pezzella C (2022) Polyhydroxyalkanoates-based nanoparticles as essential oil carriers. Polymers 14(1). https://doi.org/10.3390/polym14010166
20. Cozzens S, Cortes R (2013) Nanotechnology and the millennium development goals: water, energy, and agri-food. https://doi.org/10.1007/s11051-013-2001-y
21. Cummings CL, Kuzma J, Kokotovich A, Glas D, Grieger K (2021) Barriers to responsible innovation of nanotechnology applications in food and agriculture: a study of US experts and developers. NanoImpact. https://doi.org/10.1016/j.impact.2021.100326
22. Cushen M, Kerry J, Morris M, Cruz-Romero M, Cummins E (2012) Nanotechnologies in the food industry—recent developments, risks and regulation. Trends Food Sci Technol 24(1):30–46. https://doi.org/10.1016/j.tifs.2011.10.006
23. FAO/WHO (2010) FAO/WHO Expert meeting on the application of nanotechnologies in the food and agriculture sectors: potential food safety implications. http://www.fao.org/docrep/012/i1434e/i1434e00.pdf
24. Fu ZJ, Wang ZY, Liu ML, Cai J, Yuan PA, Wang Q, Xing W, Sun SP (2021). Dual-layer membrane with hierarchical hydrophobicity and transport channels for nonpolar organic solvent nanofiltration. AIChE J 67(4). https://doi.org/10.1002/aic.17138
25. Guo H, Li X, Yang W, Yao Z, Mei Y, Peng LE, Yang Z, Shao S, Tang CY (2022) Nanofiltration for drinking water treatment: a review. Front Chem Sci Eng 16(5):681–698. https://doi.org/10.1007/s11705-021-2103-5
26. Hasmin NA, Zainol ZA, Ismail R, Mahmood A (2020) Regulatory challenges of nanofood labelling. J Polit Law 13(2):241. https://doi.org/10.5539/jpl.v13n2p241
27. Ho JY, Rabbi KF, Khodakarami S, Ma J, Boyina KS, Miljkovic N (2022) Opportunities in nano-engineered surface designs for enhanced condensation heat and mass transfer. J Heat Transf 144(5). https://doi.org/10.1115/1.4053454
28. Hossain A, Skalicky M, Brestic M, Mahari S, Kerry RG, Maitra S, Sarkar S, Saha S, Bhadra P, Popov M, Islam MT, Hejnak V, Vachova P, Gaber A, Islam T (2021) Application of nanomaterials to ensure quality and nutritional safety of food. In: Journal of nanomaterials, vol 2021. https://doi.org/10.1155/2021/9336082
29. Husen A, Rahman QI, Iqbal M, Yassin MO, Bachheti RK (2019) Plant-mediated fabrication of gold nanoparticles and their applications. In: Nanomaterials and plant potential. Springer, Cham, pp 71–110. https://doi.org/10.1007/978-3-030-05569-1_3
30. Husen A, Siddiqi KS (2023) Advances in smart nanomaterials and their applications. Elsevier Inc., 50 Hampshire St., 5th Floor, Cambridge, MA 02139, USA
31. Islam F, Saeed F, Afzaal M, Hussain M, Ikram A, Khalid MA (2022) Food grade nanoemulsions: promising delivery systems for functional ingredients. J Food Sci Technol. https://doi.org/10.1007/s13197-022-05387-3

32. Jagtiani E (2022) Advancements in nanotechnology for food science and industry. Food Front 3(1):56–82. https://doi.org/10.1002/fft2.104
33. Khan N, Tabasi ZA, Liu J, Zhang BH, Zhao Y, Husain T (2022) Recent advances in functional materials for wastewater treatment: from materials to technological innovations. J Mar Sci Eng 10(4). https://doi.org/10.3390/jmse10040534
34. Kim HS, Hong SG, Woo KM, Teijeiro Seijas V, Kim S, Lee J, Kim J (2018) Precipitation-based nanoscale enzyme reactor with improved loading, stability, and mass transfer for enzymatic CO_2 conversion and utilization. ACS Catal 8(7). https://doi.org/10.1021/acscatal.8b00606
35. Kuang M, Yu H, Qiao S, Huang T, Zhang J, Sun M, Shi X, Chen H (2021) A novel nano-antimicrobial polymer engineered with chitosan nanoparticles and bioactive peptides as promising food biopreservative effective against foodborne pathogen *E. coli* o157-caused epithelial barrier dysfunction and inflammatory responses. Int J Mol Sci 22(24). https://doi.org/10.3390/ijms222413580
36. Kuzminova A, Dmitrenko M, Zolotarev A, Korniak A, Poloneeva D, Selyutin A, Emeline A, Yushkin A, Foster A, Budd P, Ermakov S (2022) Novel mixed matrix membranes based on polymer of intrinsic microporosity PIM-1 modified with metal-organic frameworks for removal of heavy metal ions and food dyes by nanofiltration. Membranes 12(1). https://doi.org/10.3390/membranes12010014
37. Kumar A, Choudhary A, Kaur H, Mehta S, Husen A (2021) Metal-based nanoparticles, sensors and their multifaceted application in food packaging. J Nanobiotechnol 19(256):1–25. https://doi.org/10.1186/s12951-021-00996-0
38. Li C, Ma Y, Gu J, Zhi X, Li H, Peng G (2019) A green separation mode of synephrine from *Citrus aurantium* L. (Rutaceae) by nanofiltration technology. Food Sci Nutr 7(12):4014–4020. https://doi.org/10.1002/fsn3.1265
39. Madhusha C, Munaweera I, Karunaratne V, Kottegoda N (2020) Facile mechanochemical approach to synthesizing edible food preservation coatings based on alginate/ascorbic acid-layered double hydroxide bio-nanohybrids. J Agric Food Chem 68(33). https://doi.org/10.1021/acs.jafc.0c01879
40. Mahela U, Rana DK, Joshi U, Tariyal YS (2020) Nano edible coatings and their applications in food preservation. J Postharvest Technol 8(4):52–63. https://www.researchgate.net/publication/348931409
41. McClements DJ (2020) Nanotechnology approaches for improving the healthiness and sustainability of the modern food supply. ACS Omega 5(46):29623–29630. https://doi.org/10.1021/acsomega.0c04050
42. Mersha A, Fujikawa S (2019) Mechanical reinforcement of free-standing polymeric nanomembranes via aluminosilicate nanotube scaffolding. ACS Appl Polym Mater 1(2):112–117. https://doi.org/10.1021/acsapm.8b00104
43. Mersha A, Fujikawa S (2021) Self-supporting functional nanomembranes of metal oxide/polymer blends. In: Lecture notes of the institute for computer sciences, social-informatics and telecommunications engineering, vol 385. LNICST. https://doi.org/10.1007/978-3-030-80618-7_30
44. Munteanu BS, Aytac Z, Pricope GM, Uyar T, Vasile C (2014) Polylactic acid (PLA)/Silver-NP/VitaminE bionanocomposite electrospun nanofibers with antibacterial and antioxidant activity. J Nanoparticle Res 16(10). https://doi.org/10.1007/s11051-014-2643-4
45. Nile SH, Baskar V, Selvaraj D, Nile A, Xiao J, Kai G (2020) Nanotechnologies in food science: applications, recent trends, and future perspectives. Nano-Micro Lett 12(1). https://doi.org/10.1007/s40820-020-0383-9
46. Nuruddin M, Korani DM, Jo H, Chowdhury RA, Montes FJ, Howarter JA, Youngblood JP (2020) Gas and water vapor barrier performance of cellulose nanocrystal-citric acid-coated polypropylene for flexible packaging. ACS Appl Polym Mater 2(11). https://doi.org/10.1021/acsapm.0c00483
47. Oladapo AS, Akinyosoye FA, Abiodun OA (2014) The inhibitory effect of different chemical food preservatives on the growth of selected food borne pathogenic bacteria. Afr J Microbiol Res 8(14). https://doi.org/10.5897/ajmr2013.6370

48. Pan H, Shen X, Tao W, Chen S, Ye X (2020) Fabrication of polydopamine-based curcumin nanoparticles for chemical stability and pH-responsive delivery. J Agric Food Chem 68(9):2795–2802. https://doi.org/10.1021/acs.jafc.9b07697
49. Pang X, Song X, Chen M, Tian S, Lu Z, Sun J, Li X, Lu Y, Yuk HG (2022) Combating biofilms of foodborne pathogens with bacteriocins by lactic acid bacteria in the food industry. Compr Rev Food Sci Food Saf 21(2). https://doi.org/10.1111/1541-4337.12922
50. Poonia A, Mishra A (2022) Edible nanocoatings: potential food applications, challenges and safety regulations. Nutr Food Sci 52(3):497–514. https://doi.org/10.1108/NFS-07-2021-0222
51. Prakash B, Kujur A, Yadav A, Kumar A, Singh PP, Dubey NK (2018) Nanoencapsulation: an efficient technology to boost the antimicrobial potential of plant essential oils in food system. Food Control 89:1–11. https://doi.org/10.1016/j.foodcont.2018.01.018
52. Painuli S, Semwal P, Bacheti A, Bachheti RK, Husen A (2020) Nanomaterials from nonwood forest products and their applications. In: Husen A, Jawaid M (eds) Nanomaterials for agriculture and forestry applications. Elsevier, pp 15–40. https://doi.org/10.1016/B978-0-12-817852-2.00002-0
53. Quintanilla-Carvajal MX, Camacho-Díaz BH, Meraz-Torres LS, Chanona-Pérez JJ, Alamilla-Beltrán L, Jimenéz-Aparicio A, Gutiérrez-López GF (2010) Nanoencapsulation: a new trend in food engineering processing. Food Eng Rev 2(1). https://doi.org/10.1007/s12393-009-9012-6
54. Reshmy R, Philip E, Madhavan A, Sindhu R, Pugazhendhi A, Binod P, Sirohi R, Awasthi MK, Tarafdar A, Pandey A (2021) Advanced biomaterials for sustainable applications in the food industry: updates and challenges. Environ Pollut 283:117071. https://doi.org/10.1016/j.envpol.2021.117071
55. Sadiku B, Ogunshe AAO, Falode O, Soneye T (2022) Impacts of titanium dioxide nanomaterials on health , safety and environment
56. Scott NR, Chen H, Cui H (2018) Nanotechnology applications and implications of agrochemicals toward sustainable agriculture and food systems. J Agric Food Chem 66(26):6451–6456. https://doi.org/10.1021/acs.jafc.8b00964
57. Shanmugam H, Rengarajan C, Nataraj S, Sharma A (2022) Interactions of plant food bioactives-loaded nano delivery systems at the nano-bio interface and its pharmacokinetics: an overview. Food Front 3(2):256–275. https://doi.org/10.1002/fft2.130
58. Sidhu PK, Nehra K (2020) Bacteriocin-capped silver nanoparticles for enhanced antimicrobial efficacy against food pathogens. IET Nanobiotechnol 14(3). https://doi.org/10.1049/iet-nbt.2019.0323
59. Talebian S, Rodrigues T, das Neves J, Sarmento B, Langer R, Conde J (2021) Facts and figures on materials science and nanotechnology progress and investment. ACS Nano 15(10):15940–15952. https://doi.org/10.1021/acsnano.1c03992
60. United Nations (2010) United Nations, department of economic and social affairs, policy brief: population, food security, nutrition and sustainable development. Available from: https://www.un.org/development/desa/pd/sites/www.un.org.development.desa.pd/files/undesa_pd_2021_pol
61. Vilela C, Kurek M, Hayouka Z, Röcker B, Yildirim S, Antunes MDC, Nilsen-Nygaard J, Pettersen MK, Freire CSR (2018) A concise guide to active agents for active food packaging. Trends Food Sci Technol 80:212–222. https://doi.org/10.1016/j.tifs.2018.08.006
62. Wang F, Chang R, Ma R, Qiu H, Tian Y (2021) Eco-friendly and pH-responsive nano-starch-based superhydrophobic coatings for liquid-food residue reduction and freshness monitoring. ACS Sustain Chem Eng 9(30):10142–10153. https://doi.org/10.1021/acssuschemeng.1c02090
63. Yildirim S, Röcker B, Pettersen MK, Nilsen-Nygaard J, Ayhan Z, Rutkaite R, Radusin T, Suminska P, Marcos B, Coma V (2018) Active packaging applications for food. Compr Rev Food Sci Food Saf 17(1):165–199. https://doi.org/10.1111/1541-4337.12322
64. Yu H, Yan C, Yao J (2014) Fully biodegradable food packaging materials based on functionalized cellulose nanocrystals/poly(3-hydroxybutyrate-co-3-hydroxyvalerate) nanocomposites. RSC Adv 4(104). https://doi.org/10.1039/c4ra12691b

65. Zhang Q-Q, Zhu Y-J, Wu J, Dong L-Y (2019) Nanofiltration filter paper based on ultra-long hydroxyapatite nanowires and cellulose fibers/nanofibers. ACS Sustain Chem Eng 7(20):17198–17209. https://doi.org/10.1021/acssuschemeng.9b03793
66. Zhou H, Pandya JK, Tan Y, Liu J, Peng S, Muriel Mundo JL, He L, Xiao H, McClements DJ (2019) Role of mucin in behavior of food-grade TiO_2 nanoparticles under simulated oral conditions. J Agric Food Chem 67:5882–5890. https://doi.org/10.1021/acs.jafc.9b01732

MIX
Papier aus verantwortungsvollen Quellen
Paper from responsible sources
FSC® C105338

If you have any concerns about our products,
you can contact us on
ProductSafety@springernature.com

In case Publisher is established outside the EU,
the EU authorized representative is:
Springer Nature Customer Service Center GmbH
Europaplatz 3, 69115 Heidelberg, Germany

Printed by Libri Plureos GmbH
in Hamburg, Germany